AF379039

Poincaré Seminar 2003

Bose-Einstein Condensation – Entropy

Jean Dalibard
Bertrand Duplantier
Vincent Rivasseau
Editors

Birkhäuser Verlag
Basel · Boston · Berlin

Editors:

Jean Dalibard
Laboratoire Kastler Brossel
24, rue Lohomond
F-75005 Paris
e-mail: jean.dalibard@lkb.ens.fr

Bertrand Duplantier
Service de Physique Théorique
Orme des Merisiers
CEA - Saclay
F-91191 Gif-sur-Yvette Cedex
e-mail: bertrand.duplantier@spht.saclay.cea.fr

Vincent Rivasseau
Laboratoire de Physique Théorique
Université Paris XI
F-91405 Orsay Cedex
e-mail: Vincent.Rivasseau@th.u-psud.fr

2000 Mathematics Subject Classification 82Cxx, 82Dxx, 83Cxx

This book is also available as a softcover edition (ISBN 3-7643-7116-1).

A CIP catalogue record for this book is available from the Library of Congress,
Washington D.C., USA

Bibliographic information published by Die Deutsche Bibliothek
Die Deutsche Bibliothek lists this publication in the Deutsche Nationalbibliografie;
detailed bibliographic data is available in the Internet at <http://dnb.ddb.de>.

ISBN 3-7643-7106-4 Birkhäuser Verlag, Basel – Boston – Berlin

© 2004 Birkhäuser Verlag, P.O. Box 133, CH-4010 Basel, Switzerland
Part of Springer Science+Business Media
Printed on acid-free paper produced of chlorine-free pulp. TCF ∞
Printed in Germany
ISBN 3-7643-7106-4 (hardcover edition)
ISBN 3-7643-7116-1 (softcover edition)

9 8 7 6 5 4 3 2 1   www.birkhauser.ch

Contents

Foreword

This book is the second in a series of Proceedings for the *Séminaire Poincaré*, intended for a broad audience of physicists and mathematicians.

The goal of this Seminar is to provide up-to-date information about general topics of great interest in physics. Both theoretical and experimental aspects are covered, with some historical background. Inspired by the Bourbaki Seminar in mathematics in its organization, hence nicknamed "Bourbaphy", the Poincaré Seminar is held twice a year at the Institut Henri Poincaré in Paris, with written contributions prepared in advance. Particular care is devoted to the pedagogical nature of the presentation so as to fulfill the goal of being accessible to a broad audience of scientists.

This volume contains the third and fourth such seminars, both held in 2003. The third one is devoted to Bose-Einstein Condensation: it covers the physics of superfluid liquid helium as well as the recently discovered atomic Bose-Einstein condensates. Major experimental results are presented, together with relevant theoretical approaches and remaining open questions. The fourth one is devoted to Entropy, giving a comprehensive account of the history and various realizations of this concept, from thermodynamics to black holes, and includes theoretical and experimental discussions of the corresponding fluctuations for mesoscopic systems near equilibrium.

We hope the publication of this series will serve the community of physicists and mathematicians at professional or graduate student levels.

We thank the *Centre National de la Recherche Scientifique* (SPM), the *Commissariat à l'Énergie Atomique* (DSM), and the Daniel Iagolnitzer Foundation for sponsoring the Seminar. Special thanks are due to Chantal Delongeas for the preparation of the manuscript.

Jean Dalibard
Bertrand Duplantier
Vincent Rivasseau

Part I

Bose-Einstein Condensation

SÉMINAIRE POINCARÉ

Samedi 29 mars 2003

La condensation
de Bose-Einstein

C. Cohen-Tannoudji : Introduction (10h00-10h30)
S. Balibar : De la découverte de la superfluidité de l'hélium
à celle des condensats gazeux (10h30-11h30)
G. Shlyapnikov : Condensed Matter Approaches for Quantum Gases (12h00-13h00)
J. Dalibard et C. Salomon : Expériences sur les atomes ultra-froids (14h30-16h00)
P. Nozières : Condensation de Bose-Einstein et cohérence quantique (16h30-17h30)

Amphi Hermite, Institut Henri Poincaré - 11, rue Pierre et Marie Curie Paris 5e
www.lpthe.jussieu.fr/poincare   Avec le soutien du CNRS et du CEA

Poincaré Seminar 2003, 3 – 15
© Birkhäuser Verlag, Basel, 2004

Bose-Einstein Condensation: An Introduction

C. Cohen-Tannoudji

Abstract. The goal of this first lecture is to introduce the notion of Bose-Einstein condensation in the simple case of a perfect Bosonic gas, which means a set of Bosons trapped in an external potential and without mutual interactions. After briefly recalling some statistical physics results, we study the case of a perfect gas of Bosons, first trapped in a harmonic potential (section 2), then in a box (section 3). We underline the singularity discovered by Einstein [1], which appears when the density in phase space exceeds a critical value.

1 Statistical Mechanics Reminders

1.1 Grand-canonical Partition Function

In order to describe a set of indistinguishable quantum particles, the most convenient statistical ensemble is the grand-canonical ensemble [2, 3, 4]. It is obtained by assuming that the considered system can exchange energy and particles with a much bigger reservoir [5]. The presence of the reservoir fixes the mean number of particles N and the mean energy U. The equilibrium state is then determined by choosing the system's density operator $\hat{\rho}$ to maximize the missing information, or *statistical entropy*:

$$S(\hat{\rho}) = -k_B \mathrm{Tr}\left(\hat{\rho}\ln(\hat{\rho})\right) , \tag{1}$$

given the two constraints:

$$\langle \hat{N}\rangle = N \qquad \langle \hat{H}\rangle = U . \tag{2}$$

This maximization under constraints can be easily solved using Lagrange multipliers. One obtains:

$$\hat{\rho} = \frac{e^{-\alpha\hat{N}-\beta\hat{H}}}{Z_G} \qquad \text{avec} \qquad Z_G = \mathrm{Tr}\left(e^{-\alpha\hat{N}-\beta\hat{H}}\right) . \tag{3}$$

The function Z_G is called the *grand canonical partition function*. Lagrange multipliers α and β are associated to the constraints on $\langle\hat{N}\rangle$ and $\langle\hat{H}\rangle$. The β parameter is linked to the temperature T by the formula $\beta = (k_B T)^{-1}$. The α parameter is linked to the chemical potential μ (energy needed to add a particle) by $\alpha = -\beta\mu$, and to the fugacity $z = e^{-\alpha} = e^{\beta\mu}$. Determining the values of z and β which satisfy the constraints (2) can be done through:

$$N = z\frac{\partial}{\partial z}\ln Z_G(z,\beta,V)\,, \tag{4}$$

$$U = -\frac{\partial}{\partial \beta}\ln Z_G(z,\beta,V)\,. \tag{5}$$

It is sufficient to invert in order to get z and β as functions of N and U.

Once Z_G set, all the thermodynamic variables can be obtained through simple derivation. Thus pressure is determined from:

$$P = k_B T\frac{\partial}{\partial V}\ln Z_G(z,\beta,V)\,. \tag{6}$$

1.2   The Perfect Quantum Gas

Computing Z_G explicitly, which allows to derive explicit results from expressions such as (6), is very harsh in the general case of an interacting fluid. However, the perfect gas case can be exactly solved in a very simple way, as we shall see now.

For a system of N particles which do not interact, the complete Hamiltonian is the sum of one-body Hamiltonians:

$$\hat{H} = \hat{h}_1 + \hat{h}_2 + \ldots + \hat{h}_N\,. \tag{7}$$

Let us write $\{|\lambda\rangle\}$ for a basis of the one-body Hamiltonian $\hat{h}$ eigenvectors, and ϵ_λ for the energy associated to $|\lambda\rangle$:

$$\hat{h}\,|\lambda\rangle = \epsilon_\lambda\,|\lambda\rangle\,. \tag{8}$$

Let us now perform second quantization, and introduce the operators a_λ for destruction and $a_\lambda^\dagger$ for creation of a particle in the individual state λ. The complete Hamiltonian and the *number of particles* operator can be written as:

$$\hat{H} = \sum_\lambda \epsilon_\lambda\, a_\lambda^\dagger a_\lambda \qquad\qquad \hat{N} = \sum_\lambda a_\lambda^\dagger a_\lambda\,. \tag{9}$$

A basis of eigenstates in the Fock space is $\{|N_\lambda, N_{\lambda'}, N_{\lambda''}, \ldots\rangle\}$ where the occupation numbers N_λ of the individual quantum states (i) are equal to 0 or 1 in the case of fermions, (ii) are any positive or null integers in the case of bosons. For convenience, we write ℓ a given set $\{N_\lambda\}$: $|\ell\rangle \equiv |N_\lambda, N_{\lambda'}, N_{\lambda''}, \ldots\rangle$. We thus get

$$\hat{N}\,|\ell\rangle = N_\ell\,|\ell\rangle \qquad \text{with} \qquad N_\ell = \sum_\lambda N_\lambda\,, \tag{10}$$

$$\hat{H}\,|\ell\rangle = E_\ell\,|\ell\rangle \qquad \text{with} \qquad E_\ell = \sum_\lambda N_\lambda\,\epsilon_\lambda\,. \tag{11}$$

The grand-canonical partition function Z_G given by (3) is easily derived in the basis $|\ell\rangle$:

$$Z_G = \sum_\ell e^{-\alpha N_\ell - \beta E_\ell} = \sum_{N_\lambda, N_{\lambda'}, \ldots} e^{-(\alpha + \beta \epsilon_\lambda) N_\lambda} \times e^{-(\alpha + \beta \epsilon_{\lambda'}) N_{\lambda'}} \times e^{-(\alpha + \beta \epsilon_{\lambda''}) N_{\lambda''}} \times \ldots$$

$$= \prod_\lambda \zeta_\lambda \, , \tag{12}$$

where we wrote:

$$\zeta_\lambda = \sum_{N_\lambda} e^{-(\alpha + \beta \epsilon_\lambda) N_\lambda} \, . \tag{13}$$

This factorization of Z_G as a product of partition functions, each of them related to an individual quantum state λ, is the major advantage in using the grand-canonical formalism.

The Fermionic Case: Fermi-Dirac Statistics

For Fermions, the possible values of N_λ in the sum (13) are $N_\lambda = 0$ or $N_\lambda = 1$. We thus have:

$$\zeta_\lambda^{(F)} = 1 + e^{-\alpha - \beta \epsilon_\lambda} = 1 + z e^{-\beta \epsilon_\lambda} \, . \tag{14}$$

The partition function satisfies:

$$\ln Z_G = \sum_\lambda \ln \left(1 + z e^{-\beta \epsilon_\lambda} \right) \, . \tag{15}$$

By looking at (4), the expression of the total number of particles in the system can be obtained:

$$N = \sum_\lambda N_\lambda \qquad \text{with} \qquad N_\lambda = \frac{1}{e^{\beta(\epsilon_\lambda - \mu)} + 1} \, . \tag{16}$$

For a system at fixed temperature, the chemical potential can take any value, positive or negative. A big negative value corresponds to a mean number of particles very small, thus to a system well described by classical Boltzmann statistics:

$$\mu \longrightarrow -\infty \qquad : \qquad N_\lambda \simeq z e^{-\beta \epsilon_\lambda} \, . \tag{17}$$

On the contrary, a positive value which is big with respect to $k_B T$ corresponds to a very large number of particles, and thus to a highly degenerate Fermi gas. The occupation numbers N_λ are almost equal to 1 if $\epsilon_\lambda < \mu$, and to 0 otherwise.

The Bosonic Case: Bose-Einstein Statistics

For Bosons, the computation of (13) leads to the sum of a geometrical series, or:

$$\zeta_\lambda^{(B)} = \frac{1}{1 - e^{-\alpha - \beta \epsilon_\lambda}} = \frac{1}{1 - z e^{-\beta \epsilon_\lambda}} \, . \tag{18}$$

The number of particles is then given by:

$$N = \sum_\lambda N_\lambda \qquad \text{with} \qquad N_\lambda = \frac{1}{e^{\beta(\epsilon_\lambda - \mu)} - 1} \ . \tag{19}$$

In this case the chemical potential can take all values from $-\infty$ up to $\epsilon_{\min}$, which represents the energy of the fundamental level of $\hat{h}$'s. For a chemical potential beyond this value, the population of this fundamental level would become negative, which of course doesn't make any sense. As for the Fermi gas, the big negative values of μ correspond to a gas well described by classical physics (Boltzmann distribution):

$$\mu \longrightarrow -\infty \qquad : \qquad N_\lambda \simeq z e^{-\beta \epsilon_\lambda} \ . \tag{20}$$

2   Bose-Einstein Condensation in a Harmonic Trap

2.1   The saturation of the excited levels

Let us consider Bose-Einstein statistics given by (19). When μ goes to $\epsilon_{\min}$, at a fixed temperature, the number of particles N_0 in the fundamental level of $\hat{h}$ becomes infinite:

$$\mu \longrightarrow \epsilon_{\min} \qquad : \qquad N_0 \simeq \frac{k_B T}{\epsilon_{\min} - \mu} \tag{21}$$

If the gas is restricted to a finite box or trapped in a harmonic pit, the spectrum of $\hat{h}$'is discrete. The number of particles N' in the excited levels of $\hat{h}$ is bounded above:

$$N' = {\sum_\lambda}' \frac{1}{e^{\beta(\epsilon_\lambda - \mu)} - 1} < N'_{\max} = {\sum_\lambda}' \frac{1}{e^{\beta(\epsilon_\lambda - \epsilon_{\min})} - 1} \ , \tag{22}$$

where $\sum'$ represents the sum over all the eigenstates λ of $\hat{h}$ except the fundamental state.

In what follows, we will call *saturation number* the value of $N'_{\max}$. The existence of that number, which represents an upper bound for the number of particles which can be put in states other than the fundamental, may be considered as a signature of Bose-Einstein condensation: if, at fixed temperature, we put in the trap a number of particles N greater than $N'_{\max}$, we are sure that at least $N - N'_{\max}$ particles must belong to the fundamental state. This effect is sufficient to explain the phenomena observed in a harmonic trap.

The saturation of the excited levels of $\hat{h}$' should not be considered a phase transition. To introduce this notion, one should perform the thermodynamic limit of the system considered, and observe if the density corresponding to N' atoms placed on the excited levels also remains bounded. The answer to this question will highly depend on the system's dimensionality. Before, we are going to study the application of (22) to the case of a harmonic trap, for which this notion of thermodynamical limit is not necessary [6, 7, 8, 9].

2.2 Saturation in a Harmonic Isotropic Trap

For an isotropic trap of frequency $\nu = \omega / 2\pi$, the energy levels of $\hat{h}$' are characterized by the three quantum numbers $(n_x, n_y, n_z) \equiv \boldsymbol{n}$, which characterize the oscillator's state of vibration upon the three axes. The corresponding energy is:

$$\epsilon_{\boldsymbol{n}} = \hbar\omega \left(n_x + n_y + n_z + \frac{3}{2} \right) \qquad \epsilon_{\min} = \frac{3}{2}\hbar\omega \ . \tag{23}$$

Moreover, each energy level has a degeneracy g_n given by:

$$g_n = \frac{(n+1)(n+2)}{2} \qquad n = n_x + n_y + n_z \ . \tag{24}$$

The saturation number can be simply written:

$$N'_{\max} = \sum_{(n_x, n_y, n_z) \neq (0,0,0)} \frac{1}{e^{(n_x + n_y + n_z)\xi} - 1} = \sum_{n=1}^{\infty} \frac{g_n}{e^{n\xi} - 1} \qquad \text{with} \qquad \xi = \frac{\hbar\omega}{k_B T} \ . \tag{25}$$

In the limit where the vibration quantum $\hbar\omega$ is very small compared to the thermal energy $k_B T$, that is when $\xi \ll 1$, this discrete sum can be approximated by replacing it with an integral (see appendix). We obtain:

$$N'_{\max} \simeq 1.202 \left(\frac{k_B T}{\hbar\omega} \right)^3 \ . \tag{26}$$

The quality of this approximation can be evaluated on figure 1, which gives the variation of the discrete sum (25) and of the approximated result (26) as a function of $k_B T / \hbar\omega$. When $k_B T$ becomes larger than $25\,\hbar\omega$, the two results differ by less than 5%. In practice a typical harmonic trap frequency is of magnitude 100 Hz, or $h\nu/k_B \sim 5$ nK. For a gas cooled to 200 nK, the maximal number of atoms out of the fundamental state is then about 80 000.

2.3 The Equilibrium Distribution in a Harmonic Trap

Given the saturation number for a harmonic trap, we can describe the equilibrium state of a N bosons system in this trap, when its temperature changes [10]. We will limit ourselves, in the following discussion, to the case of a number of atoms large in front of 1. This case corresponds to the solid line of figure 2. The dotted line of this Figure 2 gives some indication on the necessary modifications for lower numbers of atoms ($N = 100$ respectively). To simplify notations, we shall shift the origin of energies by $3/2\,\hbar\omega$ to bring the energy of the fundamental level to 0.

The critical temperature for which $N'_{\max} = N$ can be deduced from (26). It is given by:

$$k_B T_c = 0.94\,\hbar\omega\,N^{1/3} \tag{27}$$

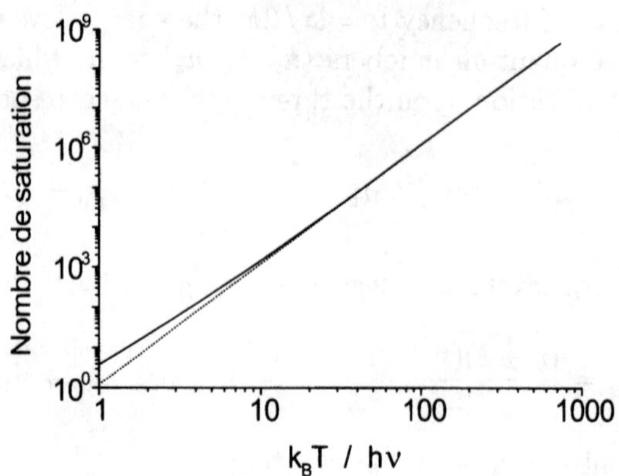

Figure 1: *Saturation number in a three-dimensional isotropic harmonic trap. The solid line gives the exact result (25) and the dotted line represents the approximated result (26).*

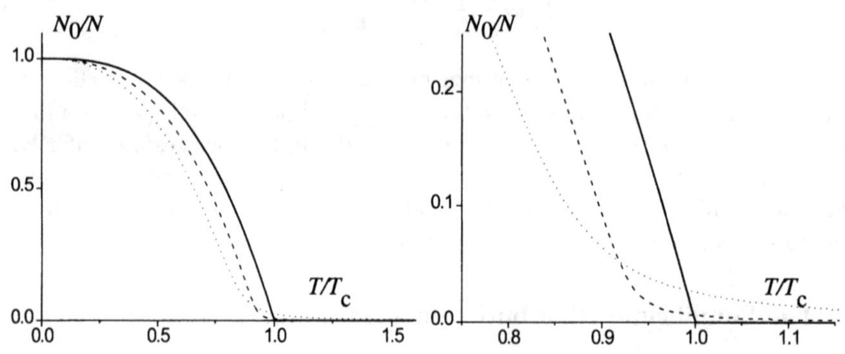

Figure 2: *Fraction of condensed atoms N_0/N as a function of the reduced temperature T/T_c. The solid curve corresponds to the limit of a large number of atoms (eq. 32). The dotted and dashed curves are the exact solutions of (19) for $N = 100$ and $N = 1000$ respectively. The right figure is an enlargement near the condensation transition.*

and is therefore large with respect to $\hbar\omega/k_B$: the approximate expressions computed in the previous paragraph are valid.

At high temperature, the saturation number $N'_{\max}$ (proportional to T^3) is much larger N. The gas is then only very weakly degenerate and one can apply Boltzmann's statistics (20). The fugacity is determined using $N = \sum_n N_n \simeq \sum_n g_n z e^{-\xi n}$, which gives after summation of a triple geometric series:

$$z = N \left(1 - e^{-\xi}\right)^3 \simeq N \, \xi^3 \simeq 1.202 \frac{N}{N'_{\max}} \ll 1 \, , \tag{28}$$

form which one deduces the occupation of each level:

$$N_{n_x,n_y,n_z} = N \, e^{-\xi(n_x+n_y+n_z)} \left(1 - e^{-\xi}\right)^3 \simeq N \, \xi^3 \, e^{-\xi(n_x+n_y+n_z)} \, . \tag{29}$$

In particular, the proportion of atoms in the fundamental state N_0/N is given by $\xi^3 = (\hbar\omega/k_B T)^3$ and is very small compared to 1. The distributions in position and velocity of the atoms are Gaussian, with respective variances $k_B T/(m\omega^2)$ and $k_B T/m$.

When temperature goes down and nears T_c, the fugacity z increases and nears 1. One can compute it by solving the transcendental equation:

$$N = \sum_{n=0}^{\infty} \frac{g_n}{z^{-1} e^{\xi n} - 1} \simeq N_0 + \xi^{-3} g_3(z) \qquad \text{avec} \qquad N_0 = \frac{z}{1-z} \tag{30}$$

where one performed an approximation similar to (50-53). Let us remark that it is essential to exclude the contribution of the fundamental level in this transformation of a discrete sum into an integral. If this is not done, the lower bound of the integral which replaces (51) is $-1/2$ and the integral can diverge when z is close enough to 1.

For $T = T_c$, the proportion of atoms in the fundamental state is still small compared to 1, but the population of excited levels is quasi saturated. Let us remark that very many levels have a significant occupation at this point. The vibration quantum number n_v after which the occupation rate becomes smaller than 1 is of order of $k_B T/\hbar\omega$, hence $n_v \sim N^{1/3} \gg 1$. One should therefore not confuse the condensation that occurs at that point with the more trivial phenomenon that is expected in the regime of extremely low temperature, i.e., $k_B T \ll \hbar\omega$, for which only the fundamental level has a significant population, no matter how many particles are present.

If the temperature comes below T_c, one observes a redistribution of particles from the excited levels towards the fundamental level, the fugacity remaining almost equal to 1. The population of excited levels decreases according to the previously determined saturation law:

$$N'(T) = 1.202 \left(\frac{k_B T}{\hbar\omega}\right)^3 = N \left(\frac{T}{T_c}\right)^3 \, , \tag{31}$$

and the population of the fundamental level is therefore:

$$N_0(T) = N \left(1 - \left(\frac{T}{T_c} \right)^3 \right) . \tag{32}$$

The result (30) should therefore be understood in the following way in the regime $\xi \ll 1$:

- either $T > T_c$, and the number of atoms in the fundamental state is negligible; one has then:

$$N \simeq \xi^{-3} g_3(z) ; \tag{33}$$

- or $T < T_c$ and the distribution of excited states is saturated:

$$N \simeq N_0 + \xi^{-3} g_3(1) . \tag{34}$$

Once the temperature comes below the critical temperature, the spatial distribution and the velocity distribution of the atoms each show two well-separated components. For instance the spatial distribution is the superposition of a narrow peak, of width $\Delta x_0 = (\hbar / m\omega)^{1/2}$ corresponding to the fundamental state of the harmonic trap, and a broader peak corresponding to the fraction of non condensed atoms, of width $\Delta x' = (k_B T / m\omega^2)^{1/2}$. The ratio of these two widths is:

$$\frac{\Delta x_0}{\Delta x'} = \left(\frac{\hbar \omega}{k_B T} \right)^{1/2} \qquad \text{soit, pour} \quad T = T_c : \qquad \frac{\Delta x_0}{\Delta x'} \simeq N^{-1/6} \ll 1 . \tag{35}$$

Similarly, in velocity space, one finds $\Delta v_0 = (\hbar\omega/m)^{1/2}$ and $\Delta v' = (k_B T/m)^{1/2}$, which leads to a ratio $\Delta v_0/\Delta v'$ equal to the ratio $\Delta x_0/\Delta x'$.

3 Bose-Einstein Condensation in a Box

We come now to the description of Bose-Einstein condensation for a gas confined in a parallelepipedic box, which is the situation initially considered by Einstein in 1924, to discover this phenomenon. This geometry corresponds to the experiments performed on macroscopic samples confined in real containers, like the study of helium's superfluidity in a cryostat (see S. Balibar's lecture in this book), or of an exciton gas in a semi-conductor, or of a bidimensional gas of hydrogen atoms maintained in levitation above a surface of liquid helium.

3.1 Energy Levels

Let us consider a parallelepipedic box, with sides L_x, L_y, L_z. One notes $V = L_x L_y L_z$ the volume of the box. We choose here periodic boundary conditions. The eigenstates λ of the one-body Hamiltonian are plane waves $|\lambda\rangle \equiv |k\rangle$:

$$\langle r|k \rangle = \frac{e^{ik \cdot r}}{\sqrt{V}} \qquad k_i = \frac{2\pi n_i}{L_i} \qquad i = x, y, z , \tag{36}$$

where the n_i's are positive or negative integers. One has:

$$\epsilon_k = \frac{2\pi^2\hbar^2}{m}\left(\frac{n_x^2}{L_x^2} + \frac{n_y^2}{L_y^2} + \frac{n_z^2}{L_z^2}\right) \tag{37}$$

et $\epsilon_{\min} = 0$.

The problem that we now want to solve is the following one: when one takes the thermodynamic limit for this system, by letting the size of the box tend to infinity and keeping the particle density, the temperature, and the chemical potential fixed, how do particles fill the various energy levels? More precisely, the interesting question is to find whether the saturation of the excited levels of $\hat{h}$, found in the previous paragraph for a system of finite size, will "survive" to this limit, although the gap between $\epsilon_{\min}$ and the first excited levels of $\hat{h}$ tends to 0 as the size of the box increases towards infinity.

3.2   The Quasi One Dimensional Bose Gas

Let us start with the simplest situation, which corresponds to a strong confinement along the axes x and y: the transverse dimensions L_x and L_y of the box are assumed small and fixed. More precisely, one assumes that the energy necessary to excite the motion of a particle along these directions $\hbar^2\pi^2/(2mL_i^2)$ (with $i = x, y$) is much larger than the thermic energy k_BT. The thermodynamic limit is taken by letting L_z tend to infinity, keeping N/L_z constant, and one wants to find the behavior of the maximal linear density of the particles which are not condensed in the fundamental state of the box $N'_{\max}/L_z$.

The explicit computation of $N'_{\max}$ from (22) for the energy levels (37) is performed in a very simple way if one neglects the population of the transversally excited levels ($n_x \neq 0$ or $n_y \neq 0$). One gets:

$$N'_{\max} \simeq 2 \sum_{n_z=1}^{+\infty} \frac{1}{e^{2\pi^2\hbar^2 n_z^2/(mk_BTL_z^2)} - 1} \,. \tag{38}$$

Let us introduce the thermic wave length:

$$\lambda = \frac{h}{\sqrt{2\pi mk_BT}} \,, \tag{39}$$

and define:

$$f(u) = \frac{1}{e^{\pi u^2\lambda^2/L_z^2} - 1} \tag{40}$$

Using the result (50) of the Appendix, which is valid if $\lambda \ll L_z$, one finds:

$$N'_{\max} \simeq 2 \int_{1/2}^{\infty} f(u)\, du = \frac{2}{\sqrt{\pi}} \frac{L_z}{\lambda} \int_{\epsilon}^{\infty} \frac{dv}{e^{v^2} - 1} \,, \tag{41}$$

where the lower bound of the integral is $\epsilon = \sqrt{\pi}\,\lambda/(2L_z)$. The linear density not condensed in the fundamental level is therefore at most:

$$\Lambda'_{\max} = \frac{N'_{\max}}{L_z} = \frac{2}{\sqrt{\pi}}\,\frac{1}{\lambda}\int_\epsilon^\infty \frac{dv}{e^{v^2}-1}\,. \tag{42}$$

Let now the size L_z of the box tend to infinity. The integral which appears in (42) diverges like $1/\epsilon$, hence more precisely:

$$\Lambda'_{\max} \simeq \frac{4}{\pi}\,\frac{L_z}{\lambda^2}\,. \tag{43}$$

There is therefroe no finite limit for the saturation linear density when one takes the thermodynamic limit $L_z \to \infty$. In other words, for a fixed linear density N/L_z and a fixed temperature, there is a size L_z above which atoms will essentially fill excited levels, the fraction of atoms in the fundamental level being negligible. In this one dimensional case, the saturation of excited levels did not survive the thermodynamic limit.

3.3   The Three Dimensional Bose Gas

Let us take now $L_x = L_y = L_z = L$. The number of non condensed atoms can be written as:

$$N'_{\max} = \sum_{(n_x,n_y,n_z)}^{'} \frac{1}{e^{\pi\lambda^2(n_x^2+n_y^2+n_z^2)/L^2}-1} \simeq \frac{4}{\sqrt{\pi}}\,\frac{L^3}{\lambda^3}\int_\epsilon^\infty \frac{v^2\,dv}{e^{v^2}-1} \tag{44}$$

The discrete sum is performed over all triplets which differ from zero triplet $(0,0,0)$ corresponding to the fundamental state. Similarly, the lower bound ϵ of the integral corresponds to exclude a sphere with radius of order λ/L, corresponding to the contribution of this fundamental level. In the three dimensional case here at hand, the integral that should be computed converges, even when its lower bound is replaced by 0, and its value therefore does not depend on ϵ in the limit $L \to \infty$. One finds:

$$n'_{\max}(T) = \frac{N'_{\max}}{L^3} = \frac{g_{3/2}(1)}{\lambda^3} \qquad \text{soit} \qquad n'_{\max}(T)\,\lambda^3 \simeq 2.612\,. \tag{45}$$

In this case, the saturation of excited levels "survived" passing to the thermodynamic limit. The study of this gas is done in detail in many textbooks on statistical physics and we shall only recall the major results, the road to follow in order to obtain them being similar to that followed in the case of the harmonic trap [2, 3, 4].

Consider a gas of fixed density n. In the high temperature region, characterized by $n\lambda^3 \ll 1$, the critical density $n'_{\max}(T)$, which varies as $T^{3/2}$, is much larger than n, and the gas is only very weakly degenerated. Physically, this temperature region corresponds to the case where the distance between particles, $n^{-1/3}$, is very large compared to their thermic wave length. Quantum effects are therefore in general hidden and the gas can be validly described by Boltzmann statistics. Its velocity distribution is well described by a Gaussian, of variance $k_B T/m$, and the fugacity is given by $z = n\lambda^3 \ll 1$.

When the temperature of this gas is lowered, the Gaussian approximation for the velocity distribution becomes worse and worse, and the fugacity z must be determined through the transcendental equation:

$$n = n_0 + \frac{g_{3/2}(z)}{\lambda^3} \qquad \text{avec} \qquad n_0 = \frac{1}{L^3} \frac{z}{1-z} \, . \tag{46}$$

Similarly to what we saw for (30), this equation must be understood in the following way:

 – When $T > T_c$, where the critical temperature T_c is such that $n = n'_{\max}(T_c)$, the density in the fundamental state n_0 is negligible and one has:

$$n = \frac{g_{3/2}(z)}{\lambda^3} \, . \tag{47}$$

 – When $T < T_c$, the excited levels are saturated, and one has:

$$n = n_0 + \frac{g_{3/2}(1)}{\lambda^3} \, . \tag{48}$$

The $N - N'$ remaining atoms accumulate in the fundamental state $\boldsymbol{p} = 0$, and the condensed density is:

$$n_0(T) = n - n'_{\max}(T) = n \left(1 - \left(\frac{T}{T_c} \right)^{3/2} \right) \, . \tag{49}$$

In contrast with the harmonic trap, this condensation happens only in the velocity space. The distribution of the atoms in position space remains uniform, as it should be taking into account the translation invariance of the system.

Appendix: Proof of Approximation (26)

When a function $f(x)$ varies slowly on an interval of length 1, one has the approximation:

$$f(1)+f(2)+f(3)+\ldots \simeq \int_{1/2}^{3/2} f(u)\,du + \int_{3/2}^{5/2} f(u)\,du + \int_{5/2}^{7/2} f(u)\,du + \ldots \tag{50}$$

For the sum (25), this hypothesis of slow variation corresponds to the case $\xi \ll 1$. One has then:

$$N'_{\max} = \frac{1}{2} \int_{1/2}^{\infty} \frac{(u+1)(u+2)\,du}{e^{u\xi} - 1} \, , \tag{51}$$

hence, writing $x = u\xi$:

$$N'_{\max} \simeq \frac{1}{2\xi^3} \int_{\xi/2}^{\infty} \frac{(x+\xi)(x+2\xi)\,dx}{e^x - 1} . \tag{52}$$

When one expands the numerator of the integrand as $x^2 + 3x\xi + 2\xi^2$, it is easy to show that, for $\xi \ll 1$, the essential contribution comes from x^2 since the function to sum takes significant values only for $x \sim 1$. Let us keep only this term, and let us send the lower bound of the integral $(\xi/2)$ to 0, which is possible since the function to sum is continuous at 0. One then finds:

$$N'_{\max} \simeq \frac{1}{2\xi^3} \int_0^{\infty} \frac{x^2\,dx}{e^x - 1} , \tag{53}$$

which is

$$N'_{\max} \simeq \left(\frac{k_B T}{\hbar\omega}\right)^3 g_3(1) . \tag{54}$$

We introduced here the two special functions:

$$g_\alpha(z) = \sum_{\ell=1}^{\infty} \frac{z^\ell}{\ell^\alpha} \qquad I_\alpha(z) = \int_0^{\infty} \frac{x^{\alpha-1}\,dx}{z^{-1}e^x - 1} , \tag{55}$$

linked through the relation:

$$I_\alpha(z) = \Gamma(\alpha)\, g_\alpha(z) \qquad \text{avec} \qquad \Gamma(\alpha) = \int_0^{+\infty} y^{\alpha-1}\, e^{-y}\, dy . \tag{56}$$

One has $\Gamma(\alpha + 1) = \alpha\, \Gamma(\alpha)$ with in particular:

$$\Gamma(1) = \Gamma(2) = 1 \quad , \quad \Gamma(3/2) = \frac{\sqrt{\pi}}{2} \quad , \quad \Gamma(3) = 2 ,$$

and

$$g_{3/2}(1) \simeq 2.612 \quad , \quad g_2(1) = \frac{\pi^2}{6} \quad , \quad g_3(1) \simeq 1.202 .$$

To obtain a better precision on the value of the critical temperature, one can try to evaluate the contributions of the terms $3x\xi$ and $2\xi^2$ which appear in the numerator of (52). For the term in $3x\xi$ one can still send the lower bound of the integral to 0 and one obtains a correction in ξ^{-2} to $N'_{\max}$. For the term in $2\xi^2$, the function to sum diverges like $1/x$ in 0, and one must keep the lower bound equal to $\xi/2$. The leading contribution for this last term is $\xi^{-1} \ln \xi$. Remark however that these expansions improved with respect to (54) are not very interesting in practice. If one wants to determine the saturation number with a very good precision in the case where $k_B T$ is not very large compared to $\hbar\omega$, it is faster and more secure to return to the series (25).

References

[1] A. Einstein, *Sitzber. Klg. Preuss. Akad. Wiss.* **261** (1924); A. Einstein, *Sitzber. Klg. Preuss. Akad. Wiss.* **3** (1925).

[2] K. Huang, *Statistical Mechanics* (Wiley, New-York, 1963).

[3] B. Diu, C. Guthmann, D. Lederer, B. Roulet, *Physique statistique* (Hermann, Paris, 1989).

[4] M. Le Bellac and F. Mortessagne, *Thermodynamique statistique* (Dunod, Paris, 2001).

[5] Even if this reservoir does not exist in practice, it does not matter. Indeed, one can show that the predictions of the various differents ensembles (micro-canonical, canonical, or grand-canonical) coincide, when restricted to the computation of mean values for thermodynamic quantities. In contrast, differences may appear when computing fluctuations, and one should then return precisely to the concrete physical situation under consideration in order to make relevant predictions.

[6] V. Bagnato, D. Pritchard, D. Kleppner, *Phys. Rev. A* **35**, 4354 (1987).

[7] V. Bagnato and D. Kleppner, *Phys. Rev. A* **44**, 7439 (1991).

[8] C. Cohen-Tannoudji, course at Collège de France 1997–98, http://www.ens.fr/cct.

[9] DEA de Physique Quantique, option *atomes froids*, lecture notes: http://www.lkb.ens.fr/~dalibard/Notes_de_cours/DEA_atomes_froids.pdf.

[10] W. Ketterle and N. van Druten, *Phys. Rev. A* **54**, 656 (1996).

C. Cohen-Tannoudji,
Laboratoire Kastler Brossel[1]

and

Collège de France,
24, rue Lhomond
F-75005 Paris, France
email: cct@lkb.ens.fr

---

[1] Unité de Recherche de l'Ecole normale supérieure et de l'Université Pierre et Marie Curie, associée au CNRS.

Poincaré Seminar 2003, 17 – 29

Looking Back at Superfluid Helium

Sébastien Balibar

Abstract. A few years after the discovery of Bose Einstein condensation in several gases, it is interesting to look back at some properties of superfluid helium. After a short historical review, I comment shortly on boiling and evaporation, then on the role of rotons and vortices in the existence of a critical velocity in superfluid helium. I finally discuss the existence of a condensate in a liquid with strong interactions, and the pressure variation of its superfluid transition temperature.

The discovery of superfluidity

In January 1938, J.F. Allen and A.D. Misener [1] published the experimental evidence that the hydrodynamics of liquid helium was not classical below 2.2 K. In the same issue of Nature, Kapitza introduced the word "superfluid" to qualify this anomalous behavior [2]. Soon after this discovery, F. London suggested that superfluid helium forms a macroscopic liquid matter wave, as a consequence of "Bose-Einstein condensation" (BEC). 65 years later, it is generally accepted that London was right, but it happened to be very difficult to prove. On the contrary, when BEC was discovered in cold alkali vapors (in 1995), the experimental evidence was immediately very clear. The superfluidity of these cold gases was soon demonstrated as well. It is thus interesting to look back at some properties of superfluid helium now. As we shall see, not everything is completely understood in the properties of superfluid helium. In this short review, I have selected only a few properties of superfluid helium. For a complete description, one can look at the historical book written by J. Wilks [3], or the one by P. Nozières and D. Pines on Bose liquids [4]. Here, I only wish to recall some aspects of the early history of this subject in order to show that fundamental questions came up very soon in this field. Then, I wish to briefly discuss three particular topics:

1) boiling and evaporation because this is what makes superfluid helium look different from normal helium in a dewar.
2) the mechanisms which determine the "critical velocity".
3) the existence of a condensate in superfluid helium and the pressure variation of the superfluid transition temperature.

In the years 1928–32 in Leiden, W.H. Keesom and his co-workers had already realized that liquid helium exhibited surprising properties below about 2.2 K [5]. They had found a peak in the temperature variation of the specific heat at this temperature. The shape of this peak resembling the Greek letter lambda, Keesom called the superfluid transition temperature the lambda point or "lambda temperature T_λ". He also called "helium I" the liquid above T_λ and "helium II" the liquid below T_λ. In his laboratory in Leiden, Keesom had also discovered that helium II was able to flow through very tiny pores as soon as in 1930 [6]. In 1932 in Toronto, J.C. McLennan, H.D. Smith and J.O. Wilhelm then discovered that liquid helium ceased boiling when cooled down through T_λ [7]. This was soon attributed to an exceptionally high thermal conductivity, as evidenced by successive experiments by W.H. Keesom again [8] and by J.F. Allen in Cambridge [9]. J.F. Allen had been hired with R. Peierls to replace Kapitza there, because Kapitza had been forced by Stalin in 1934 to stay in Moscow and to quit his research position in Cambridge[10]. Since the high thermal conductivity was attributed to some kind of convection or turbulence, researchers looked at the flow of liquid helium. In 1935 in Toronto, J.O. Wilhelm, A.D. Misener and A.R. Clark had measured the viscosity of liquid helium with a torsion pendulum and found that it decreased sharply below T_λ [11]. When A.D Misener moved to Cambridge in order to prepare his doctorate with J.F. Allen, they started together a systematic study of the flow of liquid helium through capillaries. Classical fluids obey the Poiseuille law: the flow rate is proportional to the pressure difference across the capillary and to the fourth power of the capillary radius. Instead of this, Allen and Misener found that, below T_λ, the flow rate was not only high, it was independent of the pressure and independent of the capillary radius which they had changed by a factor 50! Clearly, this liquid was not classical. In a sense, this was also what Kapitza claimed in his article [2], since he introduced the word "superfluid" in connection with the already known "superconductivity". However, Kapitza did not justify his remarkable intuition that superfluidity and superconductivity were related phenomena, and his 1938 article contained no quantitative measurements. He performed remarkable experimental measurements during the following years.

In February 1938, J.F. Allen and H. Jones published another astonishing discovery. They had found a remarkable thermomechanical effect [12]. When heating superfluid helium on one side of a porous medium or a thin capillary, the pressure increased sufficiently to produce a little fountain at the end of the tube which contained the liquid. The "fountain effect" was another spectacular phenomenon which was impossible to understand within classical thermodynamics. This is really what triggered Fritz London's thinking [13]. London new that, in this liquid, there were large quantum fluctuations responsible for the large molar volume, and he proposed that the superfluid transition at T_λ was a consequence of Bose-Einstein condensation (BEC). 65 years later, the connection between superfluidity and BEC is still a matter of debate and study. Before discussing this (in Section III), I wish to say a few words about boiling and evaporation (section I), and , in Section II, I briefly review the problem of the critical velocity, in connection with Landau's "two fluid model" [14], the existence of "rotons" and quantized vortices. One of my motivations in section II is to allow a comparison with observations of a critical velocity and vortices in gaseous superfluids.

1   Boiling, evaporation and cavitation

The easiest way to see superfluidity is to look at liquid helium in a dewar, while pumping on it so as to cool it below $T_\lambda = 2.2$ K (see Fig. 1). When crossing T_λ, the liquid stops boiling. This is because the thermal conductivity of liquid helium has suddenly increased, so that the temperature is very homogeneous. The walls are no longer warmer than the surface. They no longer provide efficient nucleation sites for bubbles. As a result, superfluid helium evaporates from its free surface, instead of showing bubble nucleation on hot defects. Superfluid helium looks quiet, and it is tempting, although not rigorous, to take this as an illustration of quantum order getting established. At least this shows a striking reality: a liquid made of very simple atoms can exist in two different states! It happens that I worked on the evaporation of superfluid helium [15] so that i would like to add a few comments on the evaporation of superfluid helium, an interesting property which is rarely mentioned. We have verified the prediction by P.W. Anderson that, at low enough temperature, this evaporation is a quantum process which is analogous to the photoelectric effect [16]. A photon incident on a metal surface can eject one electron elastically: the electron kinetic energy is the difference between the photon energy and the binding energy of the electron to the metal. Similarly, we have shown that atoms can be ejected from superfluid helium with a kinetic energy which is close to 1.5 K, the difference between the energy of incident rotons (whose minimum is 8.65 K) and the binding energy of atoms to the liquid, which is nothing but the latent heat of evaporation per atom (7.15 K). This phenomenon was later called "quantum evaporation" by A.F.G. Wyatt who made a long quantitative study of it. A few open issues remain in this subject, such as the exact measurement and calculation of evaporation probabilities by phonons or rotons as a function of energy, momentum and incidence angle, or the nature of possible inelastic processes as well.

Furthermore, one should not believe that boiling, more generally cavitation or bubble nucleation, is impossible in superfluid helium. On the contrary, this is another interesting subject on which we also worked [34]. A strong local heat input can generate bubbles in superfluid helium, of course. Moreover, one can apply a large stress, that is a large negative pressure to liquid helium at low temperature and also look for the nucleation of bubbles. This is a way to test the internal cohesion of this liquid. We have observed this phenomenon near -9.5 bar, the stability limit of liquid helium under stress, also called the "liquid-gas spinodal limit". It allowed us also to obtain some information on superfluidity at reduced density as we shall see briefly in Section III.

2   The critical velocity: rotons and vortices

The vanishing of boiling is spectacular but the existence of flow without dissipation is of course the more fundamental property which signals superfluidity. In may 1938, one month only after London had published his suggestion of a Bose-

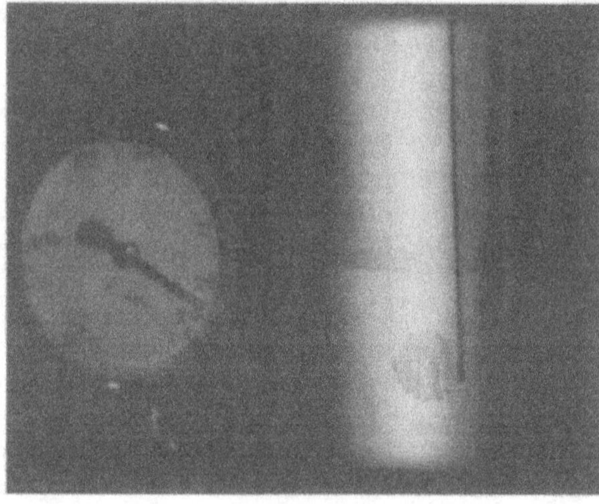

Figure 1: As shown by these two images from a film by J.F. Allen and J. Armitage, superfluid helium stops boiling below T_λ. This is due to its large thermal conductivity. The top image is taken at 2.4 K as indicated by the needle of the thermometer on the left. The bottom image is taken just below the lambda transition.

Einstein condensation, L. Tisza introduced a "two fluid model" which qualitatively explained the observations. He considered that "the atoms belonging to the lowest energy state do not take part in the dissipation. Thus the viscosity of the system is entirely due to the atoms in the excited states". He understood that this liquid could be considered as a mixture of two components: a superfluid component which carries no entropy and has zero viscosity, plus a normal component, made of atoms in excited states, which is viscous and conducts heat. The ratio of one to the other component was determined by the temperature: at zero temperature there is only the superfluid and at the lambda transition there is only the normal component. The important point was to understand that these two components could move independently. This very pure liquid had two independent velocity fields, something London could not easily admit. From these considerations Tisza predicted that when superfluid helium flowed through a capillary, the normal component, being viscous, was blocked while the superfluid component could flow without dissipation. As a consequence, viscosity measurements could not give the same result when done in an open geometry, as in the torsion pendulum used in Toronto, or in a restricted geometry like a thin capillary or a porous medium. This resolved the apparent discrepancy between different experiments at that time. Moreover, he made an even more surprising prediction, namely that "a temperature gradient should arise during the flow of helium II through a thin capillary". Tisza was right! Helium cools down when flowing so. Furthermore, he explained the fountain effect as the reverse of the previous one: the superfluid component moves in the direction of the hot side where the pressure increases. His 1938 article was only one column long. He developed his "two fluid model" in two subsequent articles [18] where he made another fundamental prediction: thermal waves should exist where the superfluid and normal components move out of phase.

Nearly all the predictions by Tisza were proved correct by subsequent experiments. However, his model appeared slightly incorrect because the superfluid and normal components could not be defined in relation to a Bose gas. The two fluid model was subsequently developed by Landau in his 1941 article, with no reference, this time, to BEC. He considered energy states of the liquid instead of individual atoms. He attributed a mass density ρ_s to the part of the fluid which occupies the ground state. The rest of the density $\rho_n = \rho - \rho_s$ corresponded to the excited states. Furthermore, he had a remarkable intuition about the nature of these excited states. He assumed that single particle states were replaced by collective modes of two different kinds: phonons, i.e., sound quanta, and quantized vortices which he called "rotons". Phonons had basically a linear dispersion relation $\omega = ck$ and rotons had a quadratic one with a gap ($\epsilon = \Delta + p^2/2\mu$), so that no excitation had a phase velocity ϵ/p less than a certain minimum value v_c. From this assumption, Landau calculated the thermodynamics of helium II, the magnitude of thermal waves called "second sound", etc. The major breakthrough in this article is the introduction of elementary excitations as quantized collective modes which led him to the crucial prediction of a critical velocity v_c. He explained that, due to the necessary conservation of energy and momentum, the superfluid component in motion at a velocity v could not slow down by exciting collective modes if v was less than v_c. In the absence of rotons, v_c would have been the sound velocity c. Landau modified the dispersion relation of rotons in his 1947 article [19] in order to obtain a better agreement with thermodynamic properties. The phonon branch evolved continuously into a roton branch $\epsilon = \Delta + (p - p_0)^2/2\mu$ around the finite momentum p_0. Given this phonon-roton spectrum, Landau's critical velocity was predicted to be about 60 m/s at low pressure.

M. Cohen and R.P. Feynman [20] then proposed to verify Landau's dispersion relation in neutron scattering experiments. This was done with great accuracy by several groups, in particular by Henshaw and Woods [21]. Of course, it was a strong support to Landau's theory. In the mean time, superfluidity was proved not to exist in liquid helium 3 at similar temperatures, and this was a strong support to London and Tisza. As for the existence of Landau's critical velocity, things appeared difficult. In fact, Landau had proposed this mechanism but he had not ruled out the possibility that other mechanisms could exist at lower velocity. The existence of quantized vortex lines was proposed in 1949 by Onsager [22] who predicted that the circulation of their velocity should be quantized as multiples of h/m. Independently in 1955, Feynman [23] explained that such vortices should lead to a much lower critical velocity of order $v_c = (\hbar/md)\ln(d/a)$ where d is the size of the container (a vessel or a capillary diameter) and a the atomic size of the vortex core. Feynman was right. The most beautiful evidence for the vortex mechanism was obtained by Avenel and Varoquaux [24] who could see quantized dissipation events in a flow through a small aperture (see Fig. 2). Moreover, their analysis of vortex nucleation shows a remarkable crossover from thermally activated nucleation above a crossover temperature of order 0.2 K to nucleation by quantum tunneling in the low temperature limit. A similar crossover was found in studies of superconducting Josephson junctions, also in our studies of cavitation in superfluid helium 4.

As for Landau's critical velocity, it was also observed, but in cases where the system size was very small only. P. McClintock et al. measured the mobility of electrons moving in superfluid helium under special conditions. These electrons repel the electronic clouds of neighbouring atoms so as to form little empty cavities around them, whose typical radius is 10 Å. At low pressure and small electric field, these electron bubbles trap vortex rings as was checked from by their energy vs momentum curve. At high pressure (25 bar, near the melting pressure) and high electric field (from 20 to 2000 V/cm), the electron bubbles could in fact be accelerated to velocities slightly larger then Landau's prediction. This was understood by Bowley and Sheard as a consequence of rotons being emitted in pairs [25]. This whole subject would need a much longer review, but the comparison with gaseous superfluids is interesting. Superfluid gases have no rotons in their excitations. Landau's critical velocity is thus associated with phonons. It is observed if the moving object is small (an atom), but if it is larger (a laser beam for example), then dissipation occurs at lower velocity. Just as in superfluid helium the nucleation of vortices is invoked to understand the low value of the critical velocity in this case.

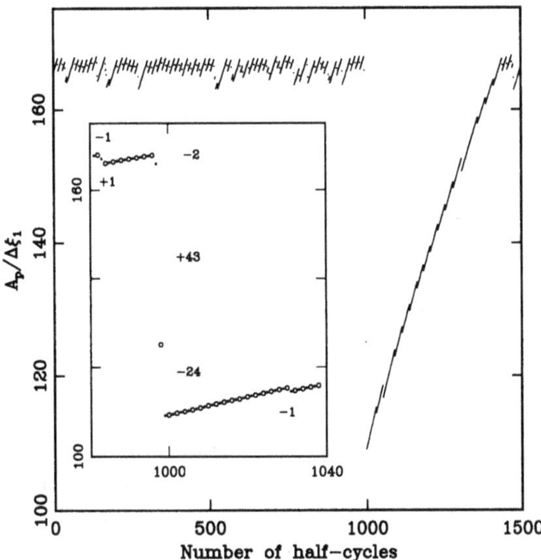

Figure 2: The quantized dissipation in the experiment by O. Avenel and E. Varoquaux. The vertical coordinate is the amplitude of oscillation of a superfluid flow through a small aperture. It is proportional to an average velocity in the hole. The horizontal coordinate is proportional to time. The velocity changes by finite jumps which are quantized. Single jumps correspond to one quantized vortex line crossing the hole. Multiple jumps are also observed.

The comparison between quantized vortices in superfluid helium and in superfluid gases has been beautifully extended to the observation of similar vortex arrays in both cases (see J. Dalibard and C. Salomon, this conference).

3   The critical temperature and the condensate fraction in liquid helium

London had noticed that T_λ was close to the transition temperature of an ideal Bose gas with the same density.

$$T_{BEC} = \left(\frac{2\pi\hbar^2}{1.897 m k_B} \right) n^{2/3} \tag{1}$$

Inserting a number density $n = 2.18 \times 10^{22} \mathrm{cm}^{-3}$ in Eq. 1 leads to $T_{BEC} = 3.1$ K while $T_\lambda = 2.2$ K. London had also noticed that the molar volume of liquid helium was large, due to the large kinetic energy corresponding to the quantum fluctuations of atoms in the cage formed by their neighbours. He had further noticed that the anomaly of the specific heat of liquid helium near T_λ was not very different from the cusp one should expect in the case of the BEC of an ideal gas. He thus claimed that , for the understanding of the superfluid transition in helium, "it seems difficult not to imagine a connection with the condensation of the Bose-Einstein statistics". However, in the next sentence, he immediately added that "a model which is so far away from reality that it simplifies liquid helium to an ideal gas" cannot lead to the right value of its specific heat.

In fact, the difficulty is much deeper than a question of specific heat only. As claimed by A.J. Leggett [26], it is not rigorously proven that a condensate, i.e., a macroscopic occupation of a ground state, should necessarily exist in a Bose liquid. It is not obvious either that there is a continuous path from the ideal Bose gas to the strongly interacting Bose liquid via the weakly interacting Bose gas where the existence of a condensation has been demonstrated both theoretically and, now, experimentally. The total absence of reference to BEC in Landau's work is obviously intentional. Landau deliberately ignored London [27]. He must have considered that there was no continuity from the gas to the liquid. It is now generally accepted that London was right, but, as we shall see, this issue is quite subtle.

Let us keep a few important steps only in the long history of this controversy. In 1947, Bogoliubov [28] showed that, in a Bose gas with weak repulsive interactions, below the BEC transition, the low energy excitations are collective modes with a non-zero velocity. This is what allows such a weakly interacting Bose gas to be superfluid. Bogoliubov justified Landau's assumption that there are no single particle excitations at low energy, so that a superfluid current cannot dissipate its kinetic energy if its velocity is less than a critical value, here the velocity of this collective mode, which is identical to the ordinary sound velocity.

In my opinion, the next crucial step is the historical paper by O. Penrose and L. Onsager [29] who generalized BEC to an interacting system. The single particle states are no longer eigenstates. They considered the one-body density matrix $\rho_1(r)$ and its eigenvalues in the limit where the distance r tends to infinity. They correspond to the occupation of the various eigenstates of the whole liquid. If the occupation of the ground state is macroscopic, meaning that the "condensate fraction" n_0 is of order unity (when r tends to infinity), then BEC has taken place. Since the momentum distribution is the Fourier transform of this one-body density matrix, a finite n_0 means a delta function at $p = 0$ in the momentum distribution: there is a ground state with zero momentum which is occupied by a macroscopic fraction of the total number of particles. As explained by P. Sokol [30], the physical interpretation of $\rho(r)$ is the overlap of a wave function of the system when a particle is removed at the origin and replaced a distance r away. The finite value of n_0 indicates a long range coherence which is often called "ODLRO" for "Off diagonal long range order". The corresponding eigenfunction is the ground state wave function which is the order parameter for the transition. It is remarkable to see that, in the same article, Penrose and Onsager used Feynmann's wave function for the condensate, approximated liquid helium as a gas of hard spheres, and ended with $n_0 \approx 8$ % at low pressure, in agreement with more recent results. The ground state wave function has an amplitude and a phase ϕ. The superfluid velocity is

$$v_s = \frac{\hbar}{m}\nabla\phi \tag{2}$$

More theoretical efforts have been done in order to calculate n_0. In his review [30], Sokol cites several numerical methods from "Path integral Monte Carlo" (PIMC) to "Green's function Monte Carlo" (GFMC) which are more or less consistent with the rough estimate first made by Penrose and Onsager. His GFMC predicts 9.2 % at low temperature and pressure. As for measuring the condensate fraction in liquid helium it has been another challenge. Contrary to what happens with trapped Bose gases, where the condensate is directly observed and easily measured, there is no experimental evidence for the existence of a condensate in liquid helium. The quantitative analysis of "Deep Inelastic Neutron scattering" experiments is very delicate. Reliable experimental values for n_0 were obtained in the last decade only. They depend not only on the assumption that a condensate exists, but also on the particular shape of the distribution function for excited states with non-zero momentum. Given this "caveat", the presently accepted experimental value for n_0 in liquid helium at zero pressure and at low temperature is 10 ± 1.5 % , in agreement with theory.

It is now interesting to consider the pressure (or density) dependence of the condensate fraction. As shown by Sokol [30] again, theory and experiment agree about this also: n_0 decreases from about 10 % at $P = 0$ bar where the density of liquid helium is $\rho = 0.145$ g.cm^{-3} to about 5 % only at the melting pressure $P_m = 25.3$ bar where $\rho = 0.172$ g.cm^{-3} (Remember that liquid helium does not solidify at low temperature except if a pressure greater than 25.3 bar is applied). One understands the decrease of n_0 as a result of the exchange between atoms becoming more difficult as the density increases in this range. A probably equivalent explanation is given by Bauer et al. [31] who write that "exchange is inhibited by the increase in effective mass of the particles". This variation is consistent with the transition temperature T_λ being also a decreasing function of pressure. One now realizes that this behaviour is opposite to the prediction from Eq. 1. Does it mean that interactions always decrease the critical temperature for condensation? Certainly not, as shown by Gruter, Ceperley and Laloe [33]. Gruter et al. calculated T_{BEC} in a hard sphere fluid as a function of the parameter na^3 which describes the interaction strength (n is the number density and a is either a scattering length in the dilute case or a hard core in the dense limit of liquid helium). They found that the variation of T_{BEC} is non-monotonic, a quite remarkable result. At low density, some interactions help the system to be homogeneous and to establish a long range coherence. On the contrary, large interactions make exchange difficult at large density. On Fig. 3, one sees that they were able to put the transition temperature of liquid helium on the same curve as their prediction for the critical temperature of BEC in dilute Bose gases. This curve would have surprised Landau!

I would like to make two further remarks on the pressure variation of T_λ in liquid helium. Suppose now that, by some clever mean, one could study liquid

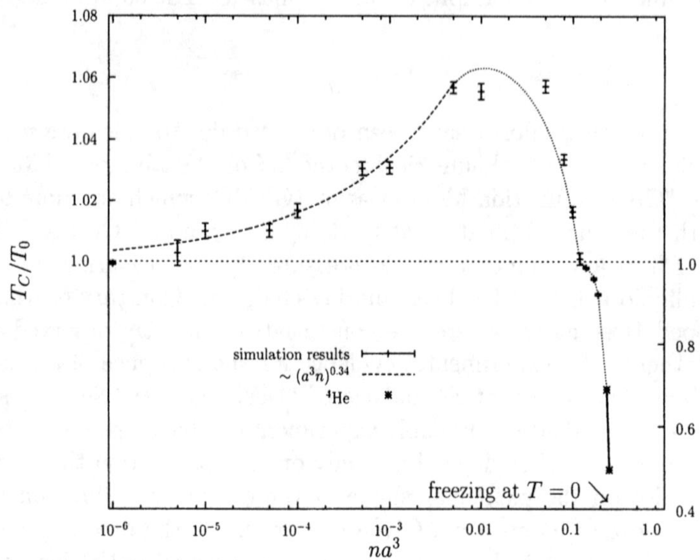

Figure 3: The calculation by Gruter et al. of the BEC transition temperature in a hard sphere model as a function of the strength of the interactions. Note that the temperature scale is not the same for the low interaction part and for the large interaction part, so that an artificial cusp shows up where T_c crosses the value T_0 of the ideal Bose gas. The star-like points on the right correspond to liquid helium. They lie on the same curve (dotted line) as for dilute gases.

helium at a density lower than its equilibrium value. Would T_λ come closer to the ideal gas value T_{BEC}? We have shown that this is possible by using acoustic pulses of high intensity and short duration [34]. In the absence of walls and impurities, it is possible to drive liquid helium to a metastable state at a negative pressure which approaches the "spinodal limit" at -9.5 bar. Under such conditions, the density of liquid helium can be lowered to 0.10 g.cm^3, about 30 % less than in equilibrium at the saturated vapor pressure. Apenko [32] and Bauer [31] have calculated T_λ in this metastable region of the phase diagram (see Fig. 4). They both predicted that T_λ should reach a maximum value of about 2.2 K for a density of 0.12 g.cm^3 corresponding to -8 bar. In summary, in liquid helium at negative pressure, T_λ ceases to be a decreasing function of pressure and comes closer to the ideal gas behaviour. This is consistent with our observation of a cusp in the cavitation pressure at 2.2 K [34]. We naturally attributed this cusp to the crossing of the lambda transition at negative pressure, and we found it where it is predicted by recent theories, not on a linear extrapolation of the behavior at positive pressure. It is the sign that rotons are no longer dominating the thermodynamics of liquid helium near its spinodal limit.

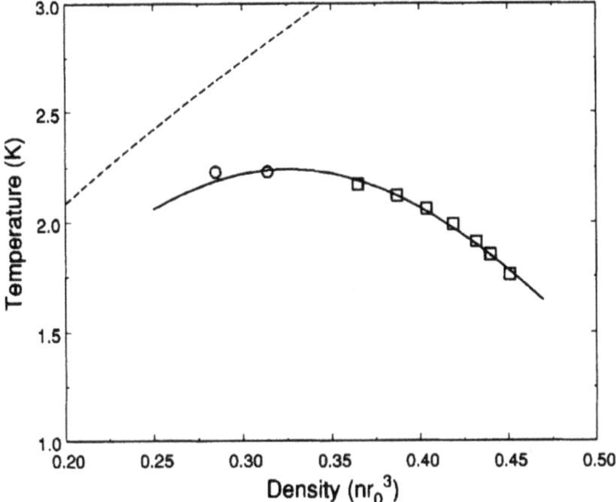

Figure 4: The density variation of the lambda transition according to Apenko (solid line and symbols). The lambda temperature approaches the BEC temperature of the ideal Bose gas (dotted line) in the metastable region of liquid helium at low density. This behavior is consistent with the results obtained by Bauer et al.

We are also using the same acoustic technique to study overpressurized liquid helium. Our main motivation is to look for a possible limit of instability for the metastable liquid at high pressure with respect to the formation of the crystalline phase which is the stable one. In the course of this search, we have already reached about 120 bar without seeing nucleation of helium crystals. This shows that it is possible to study liquid helium at much higher densities than previously thought. In his review, Sokol [30] considers this region as "unaccessible" but he was not aware of our experiments. It is thus not absurd to consider what happens to superfluidity in liquid helium at high density. A crude extrapolation of Sokol's calculation of n_0 would predict that it becomes vanishingly small near a density of about 0.19 g.cm^{-3}. Given the known equation of state [34], it would correspond to a pressure of order 55 bar, i.e., 30 bar above the equilibrium melting pressure. We have driven liquid helium up to a much higher density. It is hard to imagine that, if n_0 is very small, T_λ is not very small as well. How exactly does the lambda line extrapolate in this metastable region? Is it possible that a helium glass exists? It is certainly not easy to study such speculative properties but we might obtain some information on it from our experiments. It would also be interesting to calculate the properties of a quantum hard sphere model in the very high density limit where the system jams, not only in the very dilute case where the system approaches the ideal gas.

With these few comments and ideas, I hope that I have provided some information for a comparison between superfluid liquids and superfluid gases. My purpose was also to show that a few questions deserve further study.

References

[1] J.F. Allen, *Nature* **141**, 75 (1938).

[2] P. Kapitza, *Nature* **141**, 74 (1938).

[3] J. Wilks, *The Properties of Liquid and Solid Helium*, Clarendon Press, Oxford (1967). More recent developments of the field can be found in J. Wilks and D.S. Betts, *An introduction to liquid helium*, Clarendon Press, Oxford (1987).

[4] P. Nozières and D. Pines, *The Theory of Quantum Liquids, Vol. II, Superfluid Bose Liquids*, Perseus Books, Cambridge, massachussetts (1999).

[5] For a review on this work, see W.H. Keesom, *Helium*, Elsevier, Amsterdam (1942).

[6] W.H. Keesom and J.N. van der Ende, *Proc. Roy. Ac. Amsterdam* **33**, 24 (1930).

[7] J.C. Mclennan, H.D. Smith and J.O. Wilhelm, *Phil. Mag.* **14**, 161 (1932).

[8] W.H. Keesom and A.P. Keesom, *Physica* **3**, 359 (1936).

[9] J.F. Allen, R. Peierls and M.Z. Uddin, *Nature* **140**, 62 (1937).

[10] For more details on this historical period, see, for exemple, S. Balibar, *Qui a découvert la superfluidité?*, *Bulletin de la SFP* **128**, 15 (Paris, 2001), and L. P. Pitaevskii, *J. Low Temp. Phys.* **87**, 127 (1992).

[11] J.O. Wilhelm, A.D. Misener and A.R. Clark, *Proc. Roy. Soc. A* **151**, 342 (1935).

[12] J.F. Allen and H. Jones, *Nature* **141**, 243 (1938).

[13] F. London, *Nature* **141**, 643 (1938).

[14] L.D. Landau, *J. Phys. USSR* **5**, 71 (1941) and **11**, 91 (1947).

[15] S. Balibar, J. Buechner, B. Castaing, C. Laroche and A. Libchaber, *Phys. Rev.* **B18**, 3096 (1978).

[16] P.W. Anderson, *Phys. Lett.* **A29**, 563 (1969).

[17] L. Tisza, *Nature* **141**, 913 (1938).

[18] L. Tisza, *J. Physique et le Radium* **1**, 164 (1940) and **1**, 350 (1940).

[19] L.D. Landau, *J. Phys. USSR* **11**, 91 (1947).

[20] M. Cohen and R.P. Feynman, *Phys. Rev.* **107**, 13 (1957).

[21] D.G. Henshaw and A.D.B. Woods, *Phys. Rev.* **121**, 1266 (1961).

[22] Noticed by F. London in his book *Superfluids*, Wiley (1954).

[23] R.P. Feynman in *Progress in Low temperature Physics*, Vol. 1, ed. by C.J. Gorter, North Holland (1955).

[24] O. Avenel and E. Varoquaux, *Phys. Rev. Lett.* **55**, 2704 (1985).

[25] See D.R. Allum, R.M. Bowley and P.V.E. McClintock, *Phys. Rev. Lett.* **36**, 1313 (1976) and references therein.

[26] A.J. Leggett, *Superfluidity, Rev. Mod. Phys.* **71**, S318 (1999).

[27] Given his apparent contempt for London's ideas, it is somewhat astonishing that Landau received the prestigious "London memorial award" in 1960, soon after London's death!

[28] N.N. Bogoliubov, *J. Phys. USSR* **11**, 23 (1947).

[29] O. Penrose and L. Onsager, *Phys. Rev.* **104**, 576 (1956).

[30] P. Sokol, *Bose-Einstein Condensation in liquid helium*, in *Bose-Einstein Condensation*, ed. by A Griffin, D.W. Snoke and S. Stringari, Cambridge University Press, p.51 (1995).

[31] G.H. Bauer, D.M. Ceperley and N. Godenfeld, *Phys. Rev.* **B61**, 9055 (2000).

[32] S.M. Apenko, *Phys. Rev.* **B60**, 3052 (1999).

[33] P. Gruter, D. Ceperley and F. Laloe, *Phys. Rev. Lett.* **79**, 3549 (1997).

[34] For a review and other references, see S. Balibar, *J. Low Temp. Phys.* **129**, 363 (2002).

Sébastien Balibar
Laboratoire de Physique Statistique
Ecole Normale Supérieure
associé au CNRS et aux Universités Paris 6 et 7,
24, rue Lhomond
F-75231 Paris Cedex 05
email: balibar@lps.ens.fr

Poincaré Seminar 2003, 31 – 52
© Birkhäuser Verlag, Basel, 2004

Condensed Matter Approaches to Quantum Gases

Gora V. Shlyapnikov

1 Introduction

The discovery of Bose-Einstein condensation (BEC) in dilute atomic gases of Rb
[1], Na [2], and Li [3] in magnetic traps has stimulated an enormous revival of
the interest in macroscopic quantum behavior of dilute gases at low temperature.
Up to this discovery the main emphasis had been on the development of efficient
evaporative [4] and optical cooling [5] methods to reach the critical temperature
$T_c \lesssim 1\ \mu K$ and density $n \sim 10^{14}\ cm^{-3}$ for the observation of BEC. Experiments
with trapped Bose-condensed gases have revealed profound condensed matter behavior of these extremely dilute systems. The goal of this lecture is to describe
the key features of this behavior and discuss theoretical approaches that are being
used in the field of quantum gases.

The condensed matter behavior of quantum Bose gases originates from the
dominant role of interparticle interactions once a single quantum state becomes
macroscopically occupied [6]. So was the difference in free expansion between condensate and thermal gas clouds after switching off the trap, important supporting
evidence for the presence of a Bose-condensed state. Differences from the non-
degenerate behavior were strongly pronounced in studies of eigenfrequencies and
temperature-dependent damping of the lowest excitations. The most profound
features of the macroscopic quantum nature of dilute Bose-condensed gases were
found in the MIT experiment on interference of two independently prepared condensates [7], and in the JILA experiment [8] on a strong reduction of 3-body
recombination due to a change of local correlation properties in the presence of
a condensate. In a later stage, superfluid character of Bose-condensed gases was
demonstrated in experiments on creating vortex [9] and soliton [10] structures, in
the studies of scissors excitation modes [11], and in the measurement of the critical
velocity for superfluidity [12].

Theoretical and numerical studies of trapped Bose-condensed gases were first
focused on the ground-state properties and elementary excitations of a static gas
or on coherently evolving condensates in the mean-field approach. Studies beyond
the mean-field succeeded in describing temperature-dependent damping rates and
frequency shifts of low-energy excitations of a trapped condensate [6, 13, 14]. These
studies were followed by and done in parallel with investigations of the zero- and
finite-temperature dynamics of vortices [9] and solitons [10, 15]. Theoretical developments in the field of trapped Bose-condensed gases have become successful

due to a wide use of condensed matter approaches, such as the Gross-Pitaevskii equation for a trapped condensate, Bogoliubov equations for the excitations, finite-temperature perturbation theory, etc. The presence of the trapping potential and a finite size of the system required a serious reformulation of these approaches. Investigations of a sharp cross-over to the BEC regime [16] revived an interest in the general question of how the transition temperature depends on the interaction between particles and stimulated theoretical and Monte Carlo studies in this direction. On the basis of both experiment and theory, equilibrium properties and dynamics of trapped Bose-condensed gases are now rather well understood.

In the last years a lot of attention has been focused on phase coherence phenomena. These studies are expected to provide new fundamental insights into the nature of macroscopic quantum states and are important for future applications, such as the creation atom lasers – devices for generation of coherent matter waves. Recent theoretical studies [17] have revealed that in elongated 3D traps the finite-temperature equilibrium state can be a *quasicondensate* characterized by suppressed density fluctuations and axially fluctuating phase. The existence of these phase-fluctuating BEC states has been found in Hannover [18] and Orsay [19] experiments. In the present stage, the phenomenon of quasicondensation is one of the important issues in the studies of quantum gases.

2   Scaling approach

Time-dependent variations of the trapping potential lead to the evolution of a trapped condensate. This evolution is quite different from that of a classical gas under the same conditions. The simplest example is a free expansion of the gas after abruptly switching off the trap as in the first JILA [1] and MIT [2] experiments. In this lecture we discuss the scaling approach [20, 21, 22] for describing the evolution of a condensate with a fixed number of particles in a harmonic potential

$$V(\mathbf{r}) = \sum_i m\omega_i^2 r_i^2/2 \tag{1}$$

under time-dependent variations of the frequencies $\omega_i(t)$. An initially static condensate is assumed to be in equilibrium in an external potential $V(\mathbf{r})$ with constant frequencies $\omega_{0i} = \omega_i(0)$.

We assume that the mean interparticle separation greatly exceeds the radius of interaction between them and $n|a|^3 \ll 1$, where a is the scattering length. Therefore one can use a contact potential of pair interaction characterized by a single parameter, the scattering length a. Then the Hamiltonian of the system takes the form

$$\hat{H} = \int d\mathbf{r}\, \hat{\psi}^\dagger \{-(\hbar^2/2m)\Delta + V(\mathbf{r}) + (g/2)\hat{\psi}^\dagger\hat{\psi}\}\hat{\psi}, \tag{2}$$

and the Schrödinger equation for the Heisenberg field operator of atoms, $\hat{\psi}(\mathbf{r}, t)$, reads

$$i\hbar(\partial\hat{\psi}/\partial t) = -(\hbar^2/2m)\Delta\hat{\psi} + V(\mathbf{r})\hat{\psi} + g\hat{\psi}^\dagger\hat{\psi}\hat{\psi}, \tag{3}$$

where the last term in the right-hand side of Eq. (3) describes the interaction between particles, and the coupling constant is $g = 4\pi\hbar^2 a/m$. The field operator $\hat{\psi}$ can be represented as a sum of the non-condensed part $\hat{\psi}'$ and the condensate wave function ψ_0 which is a c-number (see, e.g., [23]):

$$\hat{\psi} = \psi_0 + \hat{\psi}'. \tag{4}$$

Averaging both sides of Eq. (3) and omitting contributions originating from the non-condensed part $\hat{\psi}'$, we obtain the familiar mean-field Gross-Pitaevskii equation:

$$i\hbar(\partial\psi_0/\partial t) = -(\hbar^2/2m)\Delta\psi_0 + V(\mathbf{r})\psi_0 + g|\psi_0|^2\psi_0. \tag{5}$$

The condensate wave function is normalized by the condition

$$\int d\mathbf{r}\,|\psi_0|^2 = N_0, \tag{6}$$

where N_0 is the number of particles in the condensate.

In equilibrium the time dependence of the condensate wave function is reduced to $\psi_0 \propto \exp(-i\mu t)$, where μ is the chemical potential. Then Eq. (5) takes a stationary form describing the initial static condensate:

$$-(\hbar^2/2m)\Delta\psi_0 + V(\mathbf{r})\psi_0 + g|\psi_0|^2\psi_0 - \mu\psi_0 = 0. \tag{7}$$

From this point on we consider a repulsive interaction between particles ($a > 0$). The shape of ψ_0 is determined by a balance between the interparticle repulsion and the confining potential. In the so-called Thomas-Fermi regime the mean-field interaction greatly exceeds the spacing between the trap levels, and the kinetic energy term in Eq. (7) is not important. One then has the well-known algebraic solution for the condensate wave function [24, 25]:

$$\psi_0 = \sqrt{(\mu - V(\mathbf{r})/g}\exp(-i\mu t). \tag{8}$$

This solution is valid in the spatial region where the argument of the square root is positive, and $\psi_0 = 0$ otherwise. The chemical potential is given by $\mu = n_0 g$, with n_0 being the maximum condensate density. In a harmonic confining potential the condensate density $|\psi_0|^2$ has a shape of an inverted parabola, with the size in the i-th direction $R_{0i} = (2\mu/m\omega_{0i}^2)^{1/2}$.

For analyzing the evolution of the condensate wave function under time-dependent variations of the frequencies ω_i at $t \geq 0$, we return to Eq. (5) and introduce scaling parameters $b_i(t)$. Turning to rescaled coordinates $\rho_i = r_i/b_i(t)$ we search for the solution of Eq. (5) in the form

$$\psi_0(\mathbf{r}, t) = \mathcal{V}^{-1/2}(t)\chi_0(\rho_i, \tau(t))\exp(i\Phi(\mathbf{r}, t), \tag{9}$$

where the dimensionless volume is $\mathcal{V}(t)\prod_i b_i(t)$, and the rescaled time $\tau(t) = \int^t dt'/\mathcal{V}(t')$. Substituting Eq. (9) into Eq. (5) we require the cancellation of $\nabla_\rho \chi_0$ terms, which gives the phase

$$\Phi(\mathbf{r},t) = (m/2\hbar)\sum_i r_i^2[\dot{b}(t)/b(t)]. \tag{10}$$

Then, for the scaling parameters governed by equations

$$\ddot{b}_i + \omega_i^2(t)b_i = \omega_{i0}^2/b_i\mathcal{V}(t), \tag{11}$$

with initial conditions $b_i(0) = 1$, $\dot{b}_i(t) = 0$, we arrive at the equation of motion

$$i\hbar(\partial\chi_0/\partial t) = K[\chi_0] + (m/2)\sum_i \omega_{0i}^2\rho_i^2\chi_0 + g|\chi_0|^2\chi_0, \tag{12}$$

where the kinetic energy term is given by

$$K[\chi_0] = -\frac{\hbar^2}{2m}\sum_i \frac{\mathcal{V}(t)}{b_i^2(t)}\frac{\partial^2\chi_0}{\partial\rho_i^2}. \tag{13}$$

In the Thomas-Fermi regime the ratio of the kinetic energy term to the nonlinear interaction term in Eq. (12) is initially very small and scales as $\eta(t) = \sum_i[\hbar\omega_{0i}/\mu b_i(t)]^2\mathcal{V}(t)$. The condition $\eta(t) \ll 1$ is satisfied on a long (or even infinite) time scale. Then the kinetic energy term can be omitted and in rescaled variables ρ,τ the equation of motion is reduced to an equation for the initial static Thomas-Fermi condensate. The latter equation is nothing else than Eq. (7) in which the kinetic energy term is neglected. The solution is given by Eq. (8), with $V(\mathbf{r})$ (1). Thus, Eqs. (9) and (12) give a universal scaling solution for $\psi_0(\mathbf{r},t)$ under arbitrary variations of the frequencies and anisotropy of the external potential:

$$\chi_0(\rho_i,\tau(t)) = \frac{1}{g}\left(\mu - \frac{m}{2}\sum_i \frac{\omega_{0i}^2 r_i^2}{b_i^2(t)}\right)^{1/2}\exp\left[-i\mu\tau(t)\right]. \tag{14}$$

The condensate preserves its shape, and at time t the ratio of the condensate size in the i-th direction, $R_i(t)$, to the initial size R_{0i} is given by the value of the scaling parameter $b_i(t)$.

In symmetrical two-dimensional traps, or in infinitely long cylindrical traps, the scaling solution (9) is exact [20, 22]. In these cases the scaling parameters $b_x^2(t) = b_y^2(t) = \mathcal{V}(t)$, and the kinetic energy term in equation of motion (12) becomes $K[\chi_0] = -(\hbar^2/2m)\Delta_\rho\chi_0$. Accordingly, in rescaled variables this equation is the same as Eq. (7) for the initial static condensate, irrespective of the shape of the condensate wave function. Thus, any initial shape governed by Eq. (7) is preserved and the condensate size is rescaled as $b(t)$.

Interestingly, the evolution dynamics of the quantum coherent state (condensate) is governed by *classical* equations of motion (11). These equations follow from the *classical* Hamiltonian of "scaling dynamics" [20]

$$H_{sd} = \frac{1}{2}\sum_i (p_i^2 + \omega_i^2(t)q_i^2) + \frac{\bar{\omega}_0^2}{\prod_i q_i},\tag{15}$$

where $\bar{\omega}_0$ is the geometrical mean of the initial trap frequencies, $q_i = (\bar{\omega}_0/\omega_{0i})b_i$, $p_i = \dot{q}_i$, and $\prod_i q_i = \mathcal{V}$. The Hamiltonian H_{sd} describes harmonic oscillators coupled to each other through the non-linear term of volume scaling $\bar{\omega}_0^2/\mathcal{V}$. This Hamiltonian and scaling equations (11) are independent of the interaction between particles, although we are considering the Thomas-Fermi regime where the interaction is very important. This is a consequence of harmonicity of the trapping potential, which at the same time is a major reason for the existence of the scaling approach. One thus sees that the evolution of Thomas-Fermi condensates in the scaling approach is governed only by the time dependence of the trap frequencies $\omega_i(t)$.

Solutions of Eqs. (11) determine the evolution of phase $\Phi(\mathbf{r},t)$ and the condensate density. For example, resonance frequencies of small shape oscillations of the condensate are the eigenfrequencies of small oscillations around the minimum value of the Hamiltonian (15) with $\omega_i(t) = \omega_{0i}$. This Hamiltonian is minimized at $q_i = \bar{\omega}_0/\omega_{0i}$, $p_i = 0$. In the vicinity of this point we arrive at the quadratic form which in the case of cylindrical symmetry gives the frequency of a quadrupole oscillation $\Omega_0 = \sqrt{2}\omega_{0r}$ (orbital angular momentum $M = 2$) and two frequencies of coupled monopole oscillations ($M = 0$):

$$\Omega_{\pm} = \omega_{0r}[9\beta^4 + 3\beta^2 \pm \sqrt{9\beta^4 - 16\beta62 + 16})/2]^{1/2},\tag{16}$$

where $\beta = \omega_{0z}/\omega_{0r}$ is the ratio of the axial to radial trap frequency. Resonance frequencies Ω_0 and Ω_- have been measured for Rb condensate in the JILA experiment [26] for $\beta = \sqrt{8}$ ($\Omega_- \approx 1.8\omega_{0r}$), and the frequencies $\Omega_{\pm}$ in the experiment at MIT [27] for $\beta = 0.08$ ($\Omega_+ \approx 2\omega_{0r}$, $\Omega_- \approx 1.58\omega_{0z}$).

The scaling equations (11) determine the character of the expansion of the condensate after the trap is abruptly switched off ($\omega_i = 0$ for $t \geq 0$). At times t greatly exceeding the lowest oscillation period the expansion becomes free in all directions and $b_i = $ const. A characteristic expansion velocity is governed by the velocity of sound in the initial condensate. However, due to anisotropy in the initial density gradient the velocity depends on the direction of expansion. In cylindrical traps the asymmetry of free expansion is characterized by the ratio of the axial to radial size, R_z/R_r. In the limiting case of $\beta \ll 1$, studied in the MIT experiment [28], the solution of scaling equations (11) gives $R_z/R_\rho = \pi\beta/2$ for $t \to \infty$. The expansion predominantly occurs in the radial direction, and the initially cigar-shaped condensate becomes pancake-shaped.

Collisionless thermal (non-condensed) gases expand symmetrically with thermal velocities $v_T \sim \sqrt{T/m}$ for any initial anisotropy of the trap. Therefore, ob-

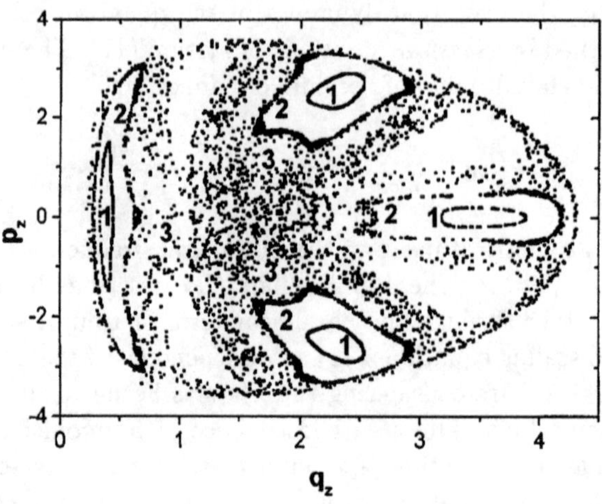

Figure 1: Poincaré map for an abrupt change of the frequencies $\omega_{0i} \rightarrow \omega_{1i}$. The phase-space trajectories for the initial sets $\omega_{0i} = \{5.4, 5.4, 4.6\}$, $\omega_{0i} = \{6.6, 6.6, 4.2\}$, and $\omega_{0i} = \{6.5, 6.5, 4.2\}$ are labeled as 1, 2, and 3, respectively.

servation of asymmetry in the expansion of dilute clouds in the first JILA [1] and MIT [2] experiments was the key evidence for BEC in the initial trapped cloud.

In spherical traps the solution of Eqs. (11) for a fast and strong change of the trap frequencies shows large undamped oscillations of the condensate density and phase. In anisotropic traps, the coupling between different degrees of freedom through the non-linear term of volume scaling in the Hamiltonian H_{sd} can lead to stochastization of motion of the scaling parameters b_i [20]. Accordingly, the evolution of the condensate becomes stochastic. We will give an example where without changing the cylindrical symmetry the frequencies were abruptly changed from ω_{0i} to $\omega_{1i} = \{1.7, 1.7.1\}$. Fig. 1 shows the Poincare map for three phase-space trajectories corresponding to three different sets of initial frequencies. The mapping points for the initial set $\omega_{0i} = \{5.4, 5.4, 6\}$ describe almost regular quasiperiodic motion in the vicinity of a second-order non-linear resonance. Points for $\omega_{0i} = \{6.6, 6.6, 4.2\}$ show that a large part of the phase space is occupied by stochastic motion. For the initial set $\omega_{0i} = \{6.5, 6.5, 4.2\}$ one finds an unstable trajectory intermediate between quasiperiodic motion and chaos.

Stochastic evolution of a condensate has been found in numerical calculations beyond the scaling approach [29, 31]. Various aspects of stochastization in the dynamics of trapped condensates have been discussed in literature (see, e.g., [30, 32]). In the cases 1 and 2 in Fig. 1 the time dependence of the axial and radial sizes of the condensate is very irregular and can even "imitate" the relaxation behavior. It is important to emphasize that chaotic evolution of the condensate density and, especially, of the phase $\Phi(\mathbf{r}, t)$ makes the system vulnerable to the appearance of real relaxation and irreversibility under a small external influence.

3 Elementary excitations

Elementary excitations of a Bose-Einstein condensate, i.e., small oscillations around the equilibrium value of the condensate wave function ψ_0, represent a primary issue for understanding the macroscopic quantum behavior of the system. In particular, the character of the excitations determines the response of the system to external perturbations and is responsible for quantum depletion of the condensate. The presence of the trapping potential introduces a finite size of the system and provides a discrete structure of the excitation spectrum. In this section we discuss the Bogoliubov-de Gennes mean-field approach for finding the spectrum and wave functions of excitations of a trapped condensate.

We consider an equilibrium Bose-condensed gas in an external potential $V(\mathbf{r})$ (1) with constant frequencies ω_i. We then use the separation (4) of the field operator into the condensed and non-condensed parts and turn to the grand-canonical Hamiltonian $\hat{H}_\mu = \hat{H} - \mu\hat{\psi}^\dagger\hat{\psi}$, where $\hat{H}$ is given by Eq. (2). Assuming that the condensate density greatly exceeds the density of non-condensed particles we omit terms proportional to $\hat{\psi}^3$ and $\hat{\psi}^4$ in the grand-canonical Hamiltonian and write it as

$$\hat{H}_\mu = H_0 + \int d\mathbf{r}\Big\{\hat{\psi}'^\dagger\Big[-(\hbar^2/2m)\Delta + V(\mathbf{r}) - \mu +$$
$$2g|\psi_0|^2\Big]\hat{\psi}'\Big\} + (g/2)\Big[\psi_0^2\hat{\psi}'^\dagger\hat{\psi}'^\dagger + \psi_0^{*2}\psi'\psi'\Big], \tag{17}$$

where

$$H_0 = \int d\mathbf{r}\psi_0^*\left[-(\hbar^2/2m)\Delta + V(\mathbf{r}) - \mu + (g/2)|\psi_0|^2\right]\psi_0.$$

Owing to Eq. (7) the part of the Hamiltonian, which is linear in $\hat{\psi}'$, is equal to zero. The bilinear Hamiltonian $\hat{H}_\mu$ can be reduced to a diagonal form

$$\hat{H}_\mu = H_0 + \sum_\nu \varepsilon_\nu \hat{a}_\nu^\dagger \hat{a}_\nu \tag{18}$$

by using the Bogoliubov transformation generalized to the spatially inhomogeneous case [33]:

$$\psi'(\mathbf{r}, t) = \sum_\nu \left[u_\nu\hat{a}_\nu(\mathbf{r})e^{-i\varepsilon_\nu t} - v_\nu^*(\mathbf{r})\hat{a}_\nu^\dagger e^{i\varepsilon_\nu)t}\right]e^{-i\mu t}.$$

Here $\hat{a}_\nu$ and $\hat{a}_\nu^\dagger$ are (Schrödinger) annihilation and creation operators of an excitation characterized by a set of quantum numbers ν. The Hamiltonian $\hat{H}_\mu$ takes the form (18) if the functions u_ν, v_ν satisfy the Bogoliubov-de Gennes equations

$$\left(-\frac{\hbar^2}{2m}\Delta + V(\mathbf{r}) - \mu\right)u_\nu + g|\psi_0|^2(2u_\nu - v_\nu) = \varepsilon_\nu u_\nu, \tag{19}$$

$$\left(-\frac{\hbar^2}{2m}\Delta + V(\mathbf{r}) - \mu\right)v_\nu + g|\psi_0|^2(2v_\nu - u_\nu) = -\varepsilon_\nu v_\nu. \tag{20}$$

The condensate wave function in Eqs. (19), (20) is taken to be real, and the functions u_ν, v_ν are normalized by the condition

$$\int d\mathbf{r}\,(u_\nu u_{\nu\prime}^* - v_\nu v_{\nu\prime}^*) = \delta_{\nu\nu\prime}.$$

Taking into account that ψ_0 obeys the Gross-Pitaevskii equation (7) we reduce Eqs. (19) and (20) to equations for the functions $f_{\nu\pm} = u_\nu \pm v_\nu$:

$$\frac{\hbar^2}{2m}\left(-\Delta + \frac{\Delta\psi_0}{\psi_0}\right) f_{\nu+} = \varepsilon_\nu f_{\nu-}, \tag{21}$$

$$\frac{\hbar^2}{2m}\left(-\Delta + \frac{\Delta\psi_0}{\psi_0}\right) f_{\nu-} + 2g|\psi_0|^2 f_{\nu-} = \varepsilon_\nu f_{\nu+}. \tag{22}$$

In the case of Thomas-Fermi condensates one has a small parameter

$$\zeta = \hbar\bar{\omega}/\mu \ll 1, \tag{23}$$

and the Bogoliubov-de Gennes equations for low-energy excitations ($\varepsilon_\nu \ll \mu$) are significantly simplified. The condition $\varepsilon_\nu \ll \mu$ corresponds to the hydrodynamic limit for the excitations. The functions $f_{\nu\pm}$ describe the phase and density fluctuations related to the excitation mode ν:

$$\hat{\phi}_\nu = (i/2\psi_0)f_{\nu+}\hat{a}_\nu \exp\left(-i\varepsilon_\nu t\right) + h.c., \tag{24}$$

$$\hat{\delta n} = \psi_0\, f_{\nu-}\hat{a}_\nu \exp\left(-i\varepsilon_\nu t\right) + h.c. \tag{25}$$

One then sees that Eqs. (21) and (22) are nothing else than the continuity and Euler equations. Inequality (23) allows one to omit the quantum pressure term, that is the first term in left-hand side of Eq. (22). This term scales as ζ^2/μ and is small compared to the second term in this equation, except near the border of the condensate spatial region. Then, turning to reduced coordinates $y_i = r_i/R_i$ we obtain $f_{\nu\pm} = [(2\mu(1 - y^2)/\varepsilon_\nu)^{\pm 1/2}W_\nu$, where $y^2 = \sum_i y_i^2$ and the function W_ν satisfies the equation

$$\sum_i \omega_i^2\left[(1 - y^2)\frac{d^2}{dy_i^2} - 2y_i\frac{d}{dy_i}\right] W_\nu + 2\varepsilon_\nu^2 W_\nu = 0. \tag{26}$$

This approach has been first developed by Stringari [34] directly from the consideration of the density and phase fluctuations. Eq. (26) shows that the spectrum of low-energy excitations of Thomas-Fermi condensates is independent of the interaction between particles and is governed by the trap frequencies, which is a consequence of harmonicity in the trapping potential. In any other trapping field the dependence on the interaction will be pronounced.

In spherical traps one has a complete separation of variables and the excitations are characterized by the orbital angular momentum l, its projection on the quantization axis m_l, and by the radial quantum number j which is a positive integer. We then have $W_\nu = w(y)Y_{lm_l}(\theta, \varphi)$ and Eq. (26) becomes a hypergeometrical differential equation for the function w:

$$x(1-x)\frac{d^2w}{dx^2} + \left[l + \frac{3}{2} - \left(l + \frac{5}{2}\right)x\right]\frac{dw}{dx} + \left(\frac{\varepsilon^2}{2} - \frac{l}{2}\right)w = 0,$$

where $x = y^2$. The solution of this equation, convergent at $x = 0$, is the hypergeometrical function that converges at the border of the condensate spatial region ($x = 1$) only when reduced to a polynomial. This immediately gives the excitation spectrum [34]

$$\varepsilon_{jl} = \hbar\omega(2j^2 + 2jl + 3j + l)^{1/2}, \tag{27}$$

and expresses the function w through classical Jacobi polynomials:

$$w_{jl} = [(4j + 2l + 3)/R^3]^{1/2}y^l P_j^{(l+1/2),0}(1 - 2y^2).$$

Due to the interaction between particles the low-energy excitations have collective character and the spectrum (27) is quite different from that for a collisionless thermal gas.

In cylindrical traps excitations are characterized by the projection of the orbital angular momentum on the cylinder axis, m, and their wave functions can be written as $W_\nu = y_\rho^{|m|}B_{jm}\exp im\varphi$, where B_{jm} is expressed in terms of polynomials of power j of the reduced radial (y_ρ) and axial (y_z) coordinates [35, 36]. The eigenstates are characterized by the axial parity and by the power of the polynomial. For a given m and odd j one has $(j+1)/2$ excitation modes, and for an even j the number of modes is equal to $(j+2)/2$. Actually, in elliptical coordinates one finds a complete separation of variables [36], which brings in a third quantum number for the eigenstates. A scheme for finding energies and wave functions of low-energy excitations in cylindrical traps has been described in Refs. [35, 36]. In particular, for quadrupole and monopole modes we arrive at the same eigenfrequencies as derived in the previous section from the scaling approach.

In non-symmetrical traps the functions W_ν are expressed in terms of polynomials of y_i and the eigenstates are characterized by a power of the polynomial [35, 37]. In this case one also finds a complete separation of variables [37].

A complete separation of variables in spherical traps is present at any excitation energy. This allows a straightforward numerical and, in certain limits, analytical solution of Eqs. (21), (22) at an arbitrary ε_ν [38, 39]. In cylindrical traps the situation is quite different as the problem is completely separable only for $\varepsilon_\nu \ll \mu$, or in the opposite limit $\varepsilon_\nu \gg \mu$ where the interaction between particles is not important. At intermediate energies $\varepsilon \sim \mu$ the spectrum becomes very irregular. For this case the study of classical dynamics of Bogoliubov-de Gennes quasiparticles shows stochastization of their motion. This allows the use of the statistical Wigner-Dyson approach [40, 41] for finding the distribution of energy levels at a given value of the projection of the angular momentum on the cylinder axis [13].

4   Critical temperature

Studies of a sharp cross-over to the BEC regime in trapped gases encounter a general problem related to the dependence of the critical temperature T_c on the interaction between particles. Actually, in trapped gases one has two reasons for this dependence. The first one is related to many-body effects beyond the mean-field theory, which are also present in the uniform case. Another reason is that the repulsive interaction between particles in a trap expands the gas cloud, with a consequent decrease of the density and critical temperature [6]. In the dilute limit, where $na^3 \ll 1$, both effects are expected to be small. Nevertheless, first measurements of T_c at JILA [16] indicate a negative shift of T_c by about 6% from the ideal gas value T_c^0.

We first discuss the interaction-induced shift of T_c in the uniform case. This problem has a long prehistory and most studies predict an increase of the critical temperature [42, 43, 44, 45, 46, 47, 48, 49, 50]. In the dilute limit the critical temperature rises linearly with the scattering length a. The relative shift of T_c is given by the relation

$$\frac{\delta T_c}{T_c} = C(na^3)^{1/2},\tag{28}$$

where $C > 0$ is a dimensionless constant. However, there is a large discrepancy in the values of the constant C presented in literature. This is not surprising as one is dealing with the region of critical fluctuations, where perturbation theory breaks down and the physics that determines T_c is non-perturbative.

The problem of finding δT_c for a weakly interacting gas is related to solving static three-dimensional $|\psi|^4$ field theory [46]. The key point here is universality of the long-wave behavior of this theory in the fluctuation region at the transition point [48]. All such theories lead to a generic long-wave Hamiltonian (see, e.g., [46])

$$\mathcal{H} = \int d\mathbf{r}\,\{(\hbar^2/2m)|\nabla\psi|^2 + (g/2)|\psi|^4\}.\tag{29}$$

From this universality one finds that the shift of the critical density, which is not sensitive to the ultraviolet cutoff of the theory, follows from the equation

$$\delta n_c(T) = -Dm^3T^2g/\hbar^6,\tag{30}$$

where D is a universal constant. Then, the shift of the critical temperature, which is sensitive to short-wave physics, can be obtained for a particular system from the relation

$$\frac{\delta T_c}{\delta n_c} = -\frac{dT_c^0(n)}{dn}.\tag{31}$$

For an ideal gas the critical temperature $T_c = 3.31\hbar^2n^{2/3}/m$ and we immediately arrive at Eq. (28). The positive sign of the constant C can be established from the change in the energy of low momentum particles near T_c [44].

The first numerical calculation of the shift δT_c was an *ab initio* simulation using a path-integral Monte Carlo method [43]. The results of this approach are consistent with Eq. (28). An alternative Monte Carlo approach [45] was based on an assumption that Eq. (28) can be obtained in a sophisticated perturbative way. However, these two calculations arrived at very different values of C. A reliable value of this constant, namely $C \approx 1.3$, has been obtained in recent lattice Monte Carlo studies of the $|\psi|^4$ model [48, 49].

Theoretical derivations of the shift δT_c are in a reasonably good agreement with the results of Refs. [48, 49]. On the basis of the $|\psi^4|$ model, self-consistent calculation of the quasiparticle spectrum at low momenta at the transition, give $C \approx 2.9$. Calculations using general renormalization group arguments [47] lead to $C \approx 2.3$ (see also [51]). The recent contributions [52, 53] find a nonanalytical correction $\propto a^2 \ln a$ to the previously calculated [46, 47] shift of the critical temperature (see also [42]). This correction does not introduce a new length scale beyond $n^{-1/3}$. It is negative and, because of its logarithmic character, leads to a strong dependence on a even in the very dilute limit. The presence of this nonanalytical correction, to a certain extent explains the discrepancy between theoretical derivations [46, 47] and Monte Carlo calculations [48, 49].

In a trapped gas the cross-over temperature T_c is well defined for a very large number of particles N. As already mentioned, in this case one has another contribution to the shift of T_c, originating from the dependence of the density profile of the gas on the interaction between particles. In particular, the interparticle repulsion decreases the central density and leads to a negative shift [6]

$$\frac{\delta T_c}{T_c^0} = -1.3\frac{a}{l_0}N^{1/6}, \tag{32}$$

where l_0 is the harmonic oscillator length of the trap. Note that the quantity $(a/l_0)N^{1/6}$ is of the order of $n_{\mathrm{max}}^{1/3}$, with n_{max} being the central density of the cloud. Therefore, the shift (32) has the same scaling as the non-perturbative shift (28) discussed above for the uniform case. More detailed discussions of the shift of the critical temperature in a trapped gas one finds in Refs. [54, 55].

5 Phase coherence

Phase coherence properties of Bose-condensed gases attract a great deal of interest as they provide deeper understanding of the nature of BEC states. The first phase coherence experiments relied on the interference of two independently prepared condensates [7] and on the measurement of the phase coherence length and/or single-particle correlations [56, 57, 58]. These experiments showed that trapped condensates are phase coherent, in accordance with a common understanding of BEC in 3D gases. In equilibrium, the fluctuations of density and phase are important only in a narrow temperature range near T_c and are suppressed outside this region.

In this section we show that the phase coherence properties of three-dimensional (3D) Thomas-Fermi condensates depend on the geometry of the system [17]. In particular, strong elongation of the gas in one direction brings in the interesting physics of one-dimensional (1D) systems. In very elongated 3D condensates, the axial phase fluctuations manifest themselves even at temperatures far below T_c. Then, as the density fluctuations are suppressed, the equilibrium state will be a *condensate with fluctuating phase* (quasicondensate) similar to that in 1D trapped gases [59]. Decreasing T below a sufficiently low temperature, the 3D quasicondensate gradually transforms into a true condensate.

We consider a 3D Bose gas in an elongated cylindrical harmonic trap and analyze the behavior of the single-particle correlation function. The natural assumption of the existence of a true condensate at $T = 0$ automatically comes out of these calculations. In the Thomas-Fermi regime, where the repulsive interparticle interaction greatly exceeds the radial (ω_ρ) and axial (ω_z) trap frequencies, the density profile of the zero-temperature condensate has the well-known shape $n_0(\rho, z) = n_{0m}(1 - \rho^2/R^2 - z^2/L^2)$, where $n_{0m} = \mu/g$ is the maximum condensate density. Under the condition $\omega_\rho \gg \omega_z$, the radial size of the condensate, $R = (2\mu/m\omega_\rho^2)^{1/2}$, is much smaller than the axial size $L = (2\mu/m\omega_z^2)^{1/2}$.

Fluctuations of the density of the condensate are dominated by the excitations with energies of the order of μ. The wavelength of these excitations is much smaller than the radial size of the condensate. Hence, the density fluctuations have the ordinary 3D character and are small. Therefore, one can write the total field operator of atoms as

$$\hat{\psi}(\mathbf{r}) = \sqrt{n_0(\mathbf{r})}\exp(i\hat{\phi}(\mathbf{r})), \tag{33}$$

where the operator of the phase is

$$\hat{\phi}(\mathbf{r}) = \sum_\nu \hat{\phi}_\nu(\mathbf{r}), \tag{34}$$

and the operator $\phi_\nu(\mathbf{r})$ is given by Eq. (24). The single-particle correlation function is then expressed through the mean square fluctuations of the phase (see, e.g. [60]):

$$\langle\hat{\psi}^\dagger(\mathbf{r})\hat{\psi}(\mathbf{r}')\rangle = \sqrt{n_0(\mathbf{r})n_0(\mathbf{r}')}\exp\{-\langle[\delta\hat{\phi}(\mathbf{r},\mathbf{r}')]^2\rangle/2\}, \tag{35}$$

with $\delta\hat{\phi}(\mathbf{r}, \mathbf{r}') = \hat{\phi}(\mathbf{r}) - \hat{\phi}(\mathbf{r}')$.

The excitations of elongated condensates can be divided into two groups: "low energy" axial excitations with energies $\epsilon_\nu < \hbar\omega_\rho$, and "high energy" excitations with $\epsilon_\nu > \hbar\omega_\rho$. The latter have 3D character as their wavelengths are smaller than the radial size R. Therefore, as in ordinary 3D condensates, these excitations can only provide small phase fluctuations. The "low-energy" axial excitations have wavelengths larger than R and exhibit a pronounced 1D behavior. These excitations give the most important contribution to the long-wave axial fluctuations of the phase.

The solution of the Bogolyubov-de Gennes equations (21), (22) for the low-energy axial modes gives the spectrum $\varepsilon_j = \hbar\omega_z\sqrt{j(j+3)/4}$ [61], where j is a positive integer. The wavefunctions f_j^+ of these modes have the form

$$f_j^+(\mathbf{r}) = \sqrt{\frac{(j+2)(2j+3)gn_0(\mathbf{r})}{4\pi(j+1)R^2L\varepsilon_j}}\,P_j^{(1,1)}\left(\frac{z}{L}\right),\tag{36}$$

where $P_j^{(1,1)}$ are Jacobi polynomials. Note that the contribution of the low-energy axial excitations to the phase operator (34) is independent of the radial coordinate ρ.

Relying on Eqs. (24), (34) and (36), we now calculate the mean square axial fluctuations of the phase at distances $|z - z'| \ll R$. As in 1D trapped gases [59], the vacuum fluctuations are small for any realistic axial size L. The thermal fluctuations are determined by the equation

$$\langle[\delta\hat{\phi}(z,z')]^2\rangle_T = \sum_{j=1}^{\infty}\frac{\pi\mu(j+2)(2j+3)}{15(j+1)\varepsilon_jN_0}\times$$
$$\left(P_j^{(1,1)}\left(\frac{z}{L}\right) - P_j^{(1,1)}\left(\frac{z'}{L}\right)\right)^2 N_j,\tag{37}$$

with $N_0 = (8\pi/15)n_{0m}R^2L$ being the number of Bose-condensed particles, and N_j the equilibrium occupation numbers for the excitations. Strictly speaking, to zero order in perturbation theory one should make the summation in Eq. (37) only over excitations with energies $\varepsilon_j < \mu$. This is, however, not a problem as the main contribution to the sum over j comes from several lowest excitation modes. At temperatures $T \gg \hbar\omega_z$ we may put $N_j = T/\varepsilon_j$, and in the central part of the cloud ($|z|, |z'| \ll L$) a straightforward calculation yields

$$\langle[\delta\hat{\phi}(z,z')]^2\rangle_T = \delta_L^2|z - z'|/L,\tag{38}$$

where the quantity δ_L^2 represents the phase fluctuations on a distance scale $|z-z'| \sim L$ and is given by

$$\delta_L^2(T) = 32\mu T/15N_0(\hbar\omega_z)^2.\tag{39}$$

Note that at any z and z' the ratio of the phase correlator (37) to δ_L^2 is a universal function of z/L and z'/L:

$$\langle[\delta\hat{\phi}(z,z')]^2\rangle_T = \delta_L^2(T)f(z/L, z'/L).\tag{40}$$

In Fig. 2 we present the function $f(z/L) \equiv f(z/L, -z/L)$ calculated numerically from Eq. (37).

The phase fluctuations decrease with temperature. As the chemical potential is $\mu = (15N_0g/\pi)^{2/5}(m\bar{\omega}^2/8)^{3/5}$ ($\bar{\omega} = \omega_\rho^{2/3}\omega_z^{1/3}$), Eq. (39) can be rewritten in the form

$$\delta_L^2 = (T/T_c)(N/N_0)^{3/5}\delta_c^2.\tag{41}$$

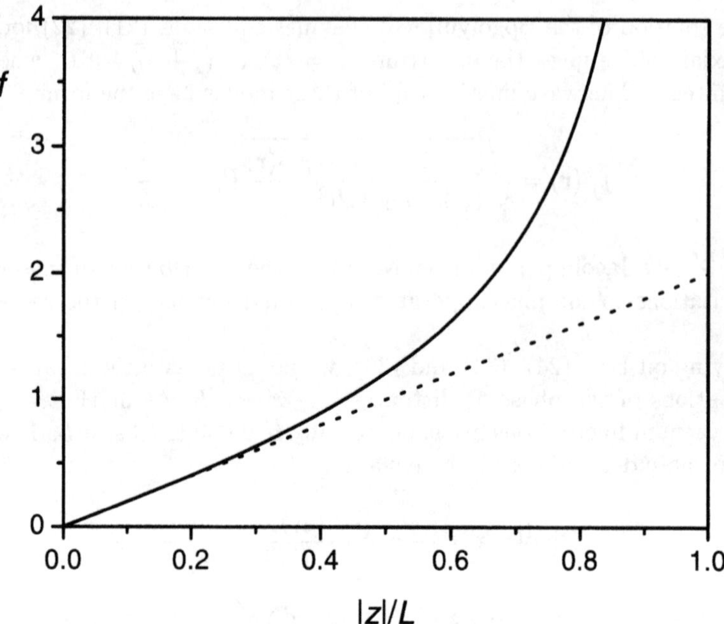

Figure 2: The function $f(z/L)$. The solid curve shows the numerical result, and the dotted line is $f(z) = 2|z|/L$ following from Eq. (38).

The presence of the 3D BEC transition in elongated traps requires the inequality $T_c \gg \hbar\omega_\rho$ and, hence, limits the aspect ratio to $\omega_\rho/\omega_z \ll N$. The parameter δ_c^2 is given by

$$\delta_c^2 = \frac{32\mu(N_0 = N)}{15 N^{2/3}\hbar\bar{\omega}} \left(\frac{\omega_\rho}{\omega_z}\right)^{4/3} \propto \frac{a^{2/5}m^{1/5}\omega_\rho^{22/15}}{N^{4/15}\omega_z^{19/15}}. \tag{42}$$

Except for a narrow interval of temperatures just below T_c, the fraction of non-condensed atoms is small and Eq. (41) reduces to $\delta_L^2 = (T/T_c)\delta_c^2$. Thus, the phase fluctuations can be important at large values of the parameter δ_c^2, whereas for $\delta_c^2 \ll 1$ they are small on any distance scale and one has a true Bose-Einstein condensate.

The single-particle correlation function is determined by Eq. (35) only if the condensate density n_0 is much larger than the density of non-condensed atoms, n'. Otherwise, this equation should be completed by terms describing correlations in the thermal cloud. For T close to T_c and $N_0 \ll N$, assuming $n' \ll n_0$ the density fluctuations are still suppressed, and Eq. (41) gives $\delta_L^2 = (N/N_0)^{3/5}$.

We will focus our attention on the case where $N_0 \approx N$ and the presence of the axial phase fluctuations is governed by the parameter δ_c^2. For $\delta_c^2 \gg 1$, the nature of the Bose-condensed state depends on temperature. In this case we can introduce a characteristic temperature

$$T_\phi = 15(\hbar\omega_z)^2 N/32\mu \tag{43}$$

at which the quantity $\delta_L^2 \approx 1$ (for $N_0 \approx N$). In the temperature interval $T_\phi < T < T_c$, the phase fluctuates on a distance scale smaller than L. Thus, as the density fluctuations are suppressed, the Bose-condensed state is a condensate with fluctuating phase or quasicondensate. The expression for the radius of phase fluctuations (phase coherence length) follows from Eq. (38) and is given by

$$l_\phi \approx L(T_\phi/T). \tag{44}$$

The phase coherence length l_ϕ greatly exceeds the correlation length $l_c = \hbar/\sqrt{m\mu}$. Eqs. (44) and (43) give the ratio $l_\phi/l_c \approx (T_c/T)(T_c/\hbar\omega_\rho)^2 \gg 1$. Therefore, the quasicondensate has the same density profile and local correlation properties as the true condensate. However, the phase coherence properties of quasicondensates are drastically different.

The decrease of temperature to well below T_ϕ makes the phase fluctuations small ($\delta_L^2 \ll 1$) and continuously transforms the quasicondensate into a true condensate. There is no sharp cross-over.

Most important is the dependence of δ_c^2 on the aspect ratio of the cloud ω_ρ/ω_z, whereas the dependence on the number of atoms and on the scattering length is comparatively weak. Fig. 3 shows $T_c/T_\phi = \delta_c^2$, μ/T_ϕ, and the temperature T_ϕ as functions of ω_ρ/ω_z for rubidium condensates at $N = 10^5$ and $\omega_\rho = 500$ Hz. From these results we see that 3D quasicondensates can be obtained in elongated geometries with $\omega_\rho/\omega_z \gtrsim 50$.

The phase fluctuations are very sensitive to temperature. From Fig. 3 we see that one can have $T_\phi/T_c < 0.1$, and the phase fluctuations are still significant at $T < \mu$, where only a tiny indiscernible thermal cloud is present.

This suggests a principle for thermometry of 3D Bose-condensed gases with indiscernible thermal clouds. If the sample is not an elongated quasicondensate by itself, it is first transformed to this state by adiabatically increasing the aspect ratio ω_ρ/ω_z. This does not change the ratio T/T_c as long as the condensate remains in the 3D Thomas-Fermi regime. Second, the phase coherence length l_ϕ or the single-particle correlation function are measured. These quantities depend on temperature if the latter is of the order of T_ϕ or larger. One thus can measure the ratio T/T_c for the initial cloud, which is as small as the ratio T_ϕ/T_c for the elongated cloud.

One can measure the phase fluctuations and distinguish between quasicondensates and true BEC's in various types of experiments. In a gedanken "juggling" experiment described in [59] one can directly measure the single-particle correlation function. The latter is obtained by repeatedly ejecting small clouds of atoms from the parts z and z' of the sample and averaging the pattern of interference between them in the detection region over a large set of measurements. As follows from Eqs. (39) and (40), for $z' = -z$ the correlation function depends on temperature as $\exp\{-\delta_L^2(T)f(z/L)/2\}$, where $f(z/L)$ is given in Fig. 2.

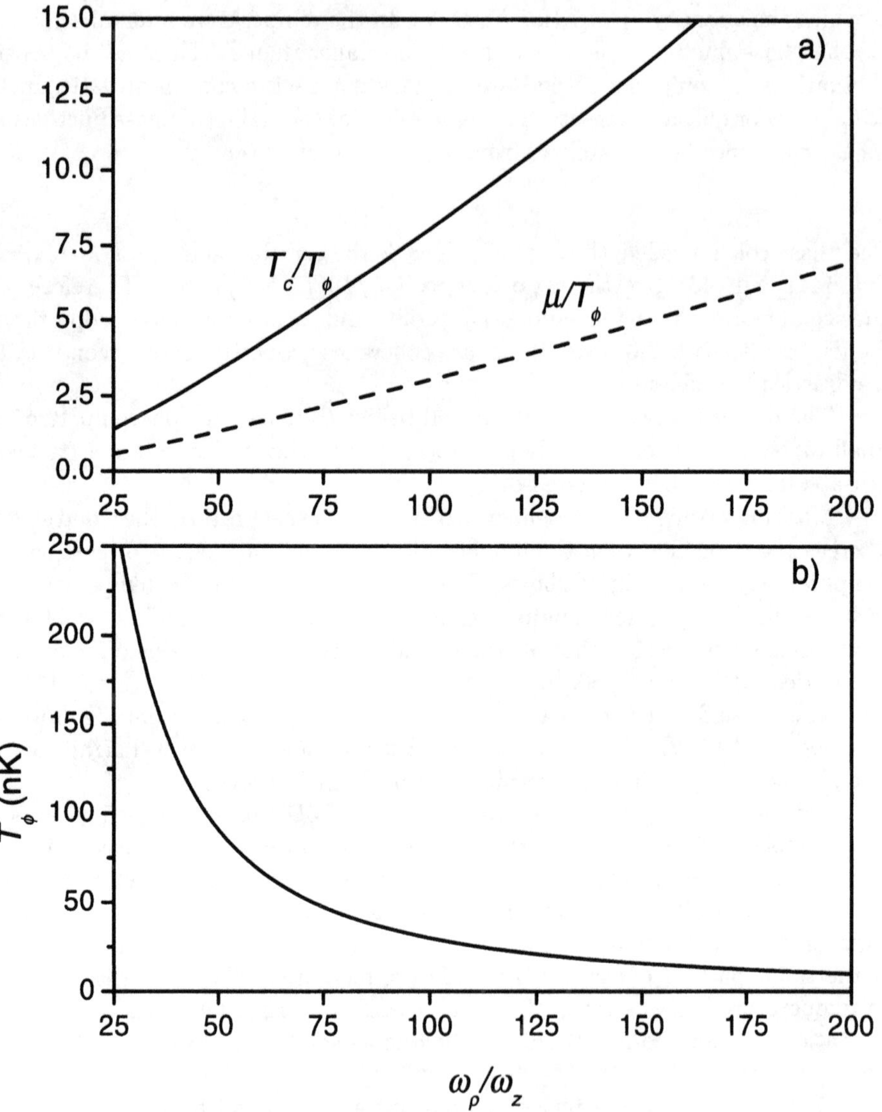

Figure 3: The ratios $T_c/T_\phi = \delta_c^2$ and μ/T_ϕ in (a) and the temperature T_ϕ in (b), versus the aspect ratio ω_ρ/ω_z for trapped Rb condensates with $N = 10^5$ and $\omega_\rho = 500$ Hz.

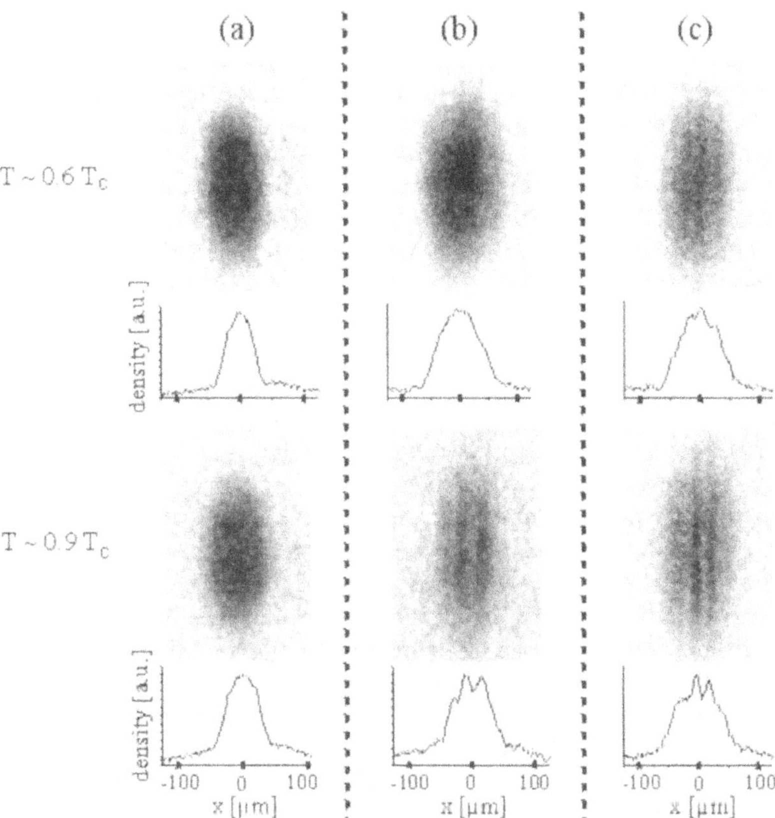

Figure 4: Absorption images and corresponding density profiles of BECs after 25 ms time-of-flight in the Hannover experiment for aspect ratios [$\omega_\rho/\omega_z = 10$ (a), 26 (b), 51 (c)].

Pronounced phase fluctuations have been first observed in Hannover experiments with very elongated cylindrical 3D condensates of up to 10^5 rubidium atoms [18]. The expanding cloud released from the trap was imaged after 25 ms of time of flight and the images showed clear modulations of the density (stripes) in the axial direction (see Fig. 4). The physical reason for the appearance of stripes is the following. In a trap the density distribution does not feel the presence of the phase fluctuations, since the mean-field interparticle interaction prevents the transformation of local velocity fields provided by the phase fluctuations into modulations of the density. After switching off the trap, the cloud rapidly expands in the radial direction, whereas the axial phase fluctuations remain unaffected. As the mean-field interaction drops to almost zero, the axial velocity fields are then converted into the density distribution.

The mean square modulations of the density in the expanding cloud provide a measure of the phase fluctuations in the initial trapped condensate. A direct relation between these quantities has been established from analytical and numerical solutions of the Gross-Pitaevskii equation for the expanding cloud, with explicitly included initial fluctuations of the phase [18]. The obtained phase coherence length was inversely proportional to T, in agreement with theory, and for most measurements it was smaller than the axial size L of the trapped Thomas-Fermi cloud. This implies that the measurements were performed in the regime of quasicondensation.

The properties of quasicondensates and the phase coherence length were measured directly in Bragg spectroscopy experiments with elongated rubidium BECs at Orsay [19]. In this type of experiment one measures the momentum distribution of particles in the trapped gas. The use of axially counter-propagating laser beams to absorb a photon from one beam and emit it into the other one, results in axial momentum transfer to the atoms which have momenta at Doppler shifted resonance with the beams. These atoms form a small cloud which will axially separate from the rest of the sample provided the mean free path greatly exceeds the axial size L. The latter condition is assured at Orsay by applying the Bragg excitation after abruptly switching off the radial confinement of the trap. To build the momentum distribution one measures the fraction of diffracted atoms versus the detuning between the counterpropagating beams.

The Orsay experiment [19] finds a Lorentzian momentum distribution characteristic of quasicondensates with axially fluctuating phase [62], whereas a true condensate has a Gaussian distribution. The width of the Lorentzian momentum distribution is related to the phase coherence length at the trap center as $\Delta p_\phi \approx 0.67 \hbar/l_\phi$. Therefore, the width of the Bragg spectrum is $\Delta \nu_M \approx 0.67 \Delta \nu_\phi$, where $\Delta \nu_\phi = \hbar k_L/\pi m l_\phi$, and k_L is the photon momentum. According to the theoretical analysis, the quantity $\Delta \nu_\phi$ should be proportional to the temperature. Fig. 5 shows the measured spectral width Δ_M versus Δ_ϕ calculated by using l_ϕ following from theoretical approaches of Refs. [17, 62]. The coherence length deduced from the measurements of $\Delta \nu_M$ was in perfect agreement with theory. It was ranging from $L/6$ to $L/36$, which shows a deep penetration into the quasi-condensate regime. The suppression of the density fluctuations was established through the measurement of the axial size of the cloud.

We believe that the studies of phase coherence in elongated condensates will reveal many new interesting phenomena. The measurement of phase correlators will allow one to study the evolution of phase coherence in the course of the formation of a condensate out of a non-equilibrium thermal cloud. This problem has a rich physics. For example, recent experiments on the formation kinetics of trapped condensates [63] indicate the appearance of non-equilibrium quasicondensates slowly evolving towards the equilibrium state.

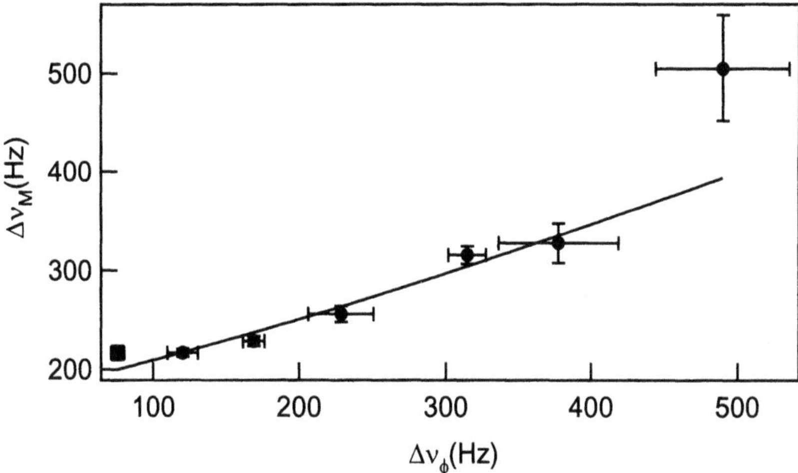

Figure 5: Half-width $\Delta\nu_M$ of the Bragg spectrum versus the parameter $\Delta_\phi \propto 1/l_\phi$ (see text) in the Orsay experiment. The solid line is a fit assuming a Voigt profile for the spectrum.

6 Concluding remarks

We see that the use of condensed matter approaches for dilute quantum gases provides us with remarkable physics closely related to ongoing experiments. In the near future, even more exciting developments are expected, in particular in new directions of cold atom physics. We mention a novel system of strongly correlated atoms in an optical lattice, where the Mott-insulator-Superfluid transition has been recently observed for bosons [64]. Studies of this system have strong ties to both condensed matter physics and quantum computing. Another hot topic is related to ultracold trapped Fermi gases which recently have been cooled to quantum degeneracy [65]. A search for superfluidity in this system requires advanced theoretical approaches for describing strong-coupling regimes. The progress in atom chip technologies and optical techniques provides unique possibilities for creating 1D atomic systems, and exploration of their physics requires the development of exactly solvable models of theoretical physics.

References

[1] M.H. Anderson *et al.*, *Science* **269**, 198 (1995).

[2] K.B. Davis *et al.*, *Phys. Rev. Lett.* **75**, 3969 (1995).

[3] C.C. Bradley *et al.*, *Phys. Rev. Lett.* **75**, 1687 (1995).

[4] See for review W. Ketterle and N.J. van Druten, in: Advances in Atomic, Molecular and Optical Physics, B. Bederson and H. Walther (Ed.), Vol.37, p.181 (1996); J.T.M. Walraven, in: Quantum Dynamics of Simple Systems, G.-L. Oppo (Ed.), vol. 44, p. 315 (1994).

[5] See for review C. Cohen-Tannoudji in: Atomic Physics XIV, C.E. Wieman, D.J. Wineland and S.J. Smith (Ed.), AIP, New York, p.193 (1995); W.D. Phillips, *ibid*, p.211 (1995).

[6] F. Dalfovo *et al*, *Rev. Mod. Phys.* **71**, 463 (1999).

[7] M.R. Andrews *et al*, *Science* **275**, 637 (1997).

[8] E.A. Burt *et al*, *Phys. Rev. Lett.* **79**, 337 (1997).

[9] A.L. Fetter and A.A. Svidzinsky, *J. Phys. – Condensed Matter* **13**, R135 (2001).

[10] Yu.S Kivshar and G.P. Aqrawal, *Optical solitons: From fibers to photonic crystals*, Chapter 14 (Elsevier Science, USA, 2003).

[11] O.M. Marago *et al*, Phys. Rev. Lett. **85**, 692 (2000).

[12] C. Raman *et al*, *Phys. Rev. Lett.* **83**, 2502 (1999).

[13] P.O. Fedichev, G.V. Shlyapnikov, and J.T.M. Walraven, *Phys. Rev. Lett.* **80**, 2269 (1998); P.O. Fedichev and G.V. Shlyapnikov, *Phys. Rev. A* **58**, 3146 (1998) and references therein.

[14] B. Jackson and E. Zaremba, *Phys. Rev. Lett.* **88**, 033606 (2002); *ibid* **89**, 150402 (2002) and references therein.

[15] A.E. Muryshev *et al*, *Phys. Rev. Lett.* **89**, 110401 (2002).

[16] J.R. Ensher *et al*, *Phys. Rev. Lett.* **75**, 4984 (1996).

[17] D.S. Petrov, G.V. Shlyapnikov, and J.T.M. Walraven, *Phys. Rev. Lett.* **87**, 050404 (2001).

[18] S. Dettmer *et al*, *Phys. Rev. Lett.* **87**, 160406 (2001).

[19] S. Richard *et al*, cond-mat/0303137.

[20] Yu. Kagan, E.L. Surkov, and G.V. Shlyapnikov, *Phys. Rev. A* **54**, R1753 (1996); *ibid* **55**, R18 (1997).

[21] Y. Castin and R. Dum, *Phys. Rev. Lett.* **77**, 5315 (1996).

[22] L.P. Pitaevskii and A. Rosch, *Phys. Rev. A* **55**, R853 (1996).

[23] E.M. Lifshitz and L.P. Pitaevskii, *Statistical Physics, Part 2* (Pergamon Press, Oxford, 1980).

[24] V.V. Goldman, I.F. Silvera, and A. Leggett, *Phys. Rev. B* **24**, 2870 (1981).

[25] D.A. Huse and E.D. Siggia, *J. Low Temp. Phys.* **46**, 137 (1982).

[26] D.S. Jin *et al*, *Phys. Rev. Lett.* **77**, 420 (1996).

[27] M.-O. Mewes *et al*, *Phys. Rev. Lett.* **77**, 988 (1996).

[28] M.-O. Mewes *et al*, *Phys. Rev. Lett.* **77**, 416 (1996).

[29] A. Smerzi and S. Fantoni, *Phys. Rev. Lett.* **78**, 3589 (1997).

[30] F. Dalfovo *et al*, *Phys. Lett. A* **227**, 259 (1997).

[31] A. Sinatra *et al*, *Phys. Rev. Lett.* **82**, 251 (1999).

[32] M. Fliesser and R. Graham, *Physica D* **131**, 141 (1999) and references therein.

[33] P.G. de Gennes *Superconductivity of Metals and Alloys* (Benjamin, New York, 1966).

[34] S. Stringari, *Phys. Rev. Lett.* **77**, 2360 (1996).

[35] P. Ohberg *et al*, *Phys. Rev. A* **56**, R3346 (1997).

[36] M. Fliesser *et al*, *Phys. Rev. A* **56**, R2533 (1997); ibid **56**, 4879 (1997).

[37] A. Csordas and R. Graham, *Phys. Rev. A* **59**, 1477 (1999).

[38] A. Csordas, R. Graham, and P. Szepfalusy, *Phys. Rev. A* **56**, 5179 (1997); *ibid* **57**, 4669 (1998).

[39] M. Guilleumas and L.P. Pitaevskii, *Phys. Rev. A* **61**, 013602 (1999).

[40] E.P. Wigner, *Math. Ann.* **53**, 36 (1951); *ibid* **62**, 548 (1955).

[41] F.J. Dyson, *J. Math. Phys.* **3**, 140 (1955).

[42] H.T.C. Stoof, *Phys. Rev. A* **45**, 8398 (1992); M. Bijlsma and H.T.C. Stoof, *Phys. Rev. A* **54**, 5085 (1996).

[43] P. Gruter, D. Ceperley, and F. Laloe, *Phys. Rev. Lett.* **79**, 3549 (1997).

[44] M. Holzmann, P. Gruter, and F. Laloe, *Eur. Phys. J. B* **10**, 739 (1999).

[45] M. Holzmann and W. Krauth, *Phys. Rev. Lett.* **83**, 2687 (1999).

[40] G. Baym *et al*, *Phys. Rev. Lett.* **83**, 1703 (1999).

[47] G. Baym, J.-P. Blaizot, and J. Zinn-Justin, *Europhys. Lett.* **49**, 150 (2000).

[48] V.A. Kashurnikov, N.V. Prokof'ev, and B.V. Svistunov, *Phys. Rev. Lett.* **87**, 120402 (2001).

[49] P. Arnold and G. Moore, *Phys. Rev. Lett.* **87**, 120401 (2001).

[50] See for review: G. Baym *et al*, *Eur. Phys. J. B* **24**, 104 (2001).

[51] P. Arnold and B. Tomasik, *Phys. Rev. A* **62**, 063604 (2000).

[52] M. Holzmann *et al*, *Phys. Rev. Lett.* **87**, 120403 (2001).

[53] P. Arnold, G. Moore, and B. Tomasik, *Phys. Rev. A* **65**, 013606 (2002).

[54] M. Houbiers, H.T.C. Stoof, and E.A. Cornell, *Phys. Rev. A* **56**, 2041 (1997).

[55] P. Arnold and B. Tomasik, *Phys. Rev. A* **64**, 053609 (2001).

[56] J. Stenger *et al.*, *Phys. Rev. Lett.* **80**, 4569 (1999).

[57] E.W. Hagley *et al*, *Science* **283**, 1706 (1999).

[58] I. Bloch *et al*, *Nature (London)* **403**, 166 (2000).

[59] D.S. Petrov, G.V. Shlyapnikov, and J.T.M. Walraven, *Phys. Rev. Lett.* **85**, 3745 (2000).

[60] V.N. Popov, *Functional Integrals in Quantum Field Theory and Statistical Physics*, (D. Reidel Pub., Dordrecht, 1983).

[61] S. Stringari, *Phys. Rev. A* **58**, 2385 (1998).

[62] F. Gerbier *et al*, cond-mat/0211094.

[63] I. Shvarchuk *et al*, *Phys. Rev. Lett.* **89**, 270404 (2002).

[64] M. Greiner *et al*, *Nature (London)* **415**, 39 (2002).

[65] B. De Marco and D.S. Jin, *Science* **285**, 1703 (1999); F. Schreck *et al*, *Phys. Rev. Lett.* **87**, 080403 (2001); A.G. Truscott *et al*, *Science* **291**, 2570 (2001); S.R. Granade *et al*, *Phys. Rev. Lett.* **88**, 120405 (2002); Z. Hadzibabic *et al*, *Phys. Rev. Lett.* **88**, 160401 (2002); G. Roatti *et al*, *Phys. Rev. Lett.* **89**, 150403 (2002); K.M. O'Hara *et al*, *Science* **298**, 2179 (2002).

G.V. Shlyapnikov
FOM Institute for Atomic and Molecular Physics,
Kruislaan 407
NL-1098 SJ Amsterdam
The Netherlands
email: shlyap@phys.uva.nl

Poincaré Seminar 2003, 53 – 83
© Birkhäuser Verlag, Basel, 2004

Experiments with Cold Atoms

Jean Dalibard and Christophe Salomon

Abstract. In this paper we present the methods for producing and detecting gaseous Bose-Einstein condensates. We then describe some characteristic experiments dealing with these quantum macroscopic systems. We also address the cooling of fermionic gases and of boson-fermion mixtures. Finally we briefly discuss the metrological applications of ultra-cold atoms.

1 Introduction

In 1924 Einstein, inspired by the work of the young Bengali physicist Satyendra Nath Bose [1], proved a remarkable result for an ideal gas of identical particles with spatial density n and temperature T [2]. For bosonic particles (integer spin), a condensation is expected in the ground state of the box confining the particles when the phase space density $n\lambda^3$ of the fluid is larger than the critical value:

$$n\lambda^3 = \zeta(3/2) \simeq 2.612 \qquad \text{with} \qquad \lambda = \frac{h}{\sqrt{2\pi m k_B T}} \, . \tag{1}$$

The parameter λ is the thermal wavelength of the particles with mass m, k_B and h being the Boltzmann and Planck constants.

This wavelength gives an estimate of the average size of the wave packet which is associated to each particle. Einstein's criterion (1) indicates that the condensation occurs when the various wave packets start to overlap (see the contribution of C. Cohen-Tannoudji to this book).

"It is a nice theory, but does it contain any truth?" These are the words that Einstein used to describe his result in a letter to his friend Ehrenfest, before stopping his research in this field [3]. Einstein's prediction remained quite controversial until 1938. At this date Kapitza [4], Allen and Misener [5] discovered the superfluidity of liquid helium (cf. S. Balibar's contribution to this book). London then noticed that the temperature of the superfluid transition, $T_s = 2.2$ K, is remarkably close to the condensation temperature of an ideal Bose gas with the same spatial density, $T_c = 3.2$ K. London then postulated that the two phenomena were linked.

London's idea is the starting point of all modern theories of liquid helium. However the connection between Bose-Einstein condensation and superfluidity is not obvious. Superfluidity originates from the interaction between particles, while Einstein considered an ideal gas. More quantitatively, neutron scattering experiments lead to a condensed fraction which does not exceed 10% at low pressure. By comparison, one predicts that all particles should be condensed for an ideal gas at zero temperature.

The research for systems closer to Einstein's ideal model has been very active over the last 20 years. The development of laser trapping and cooling of atoms, as well as magnetic trapping, have allowed to reach Einstein's criterion. In 1995, in Boulder, the group of E. Cornell and C. Wieman succeeded in producing a rubidium condensate [7].

Since this date, many other atoms have been added to the list. By increasing mass, one finds hydrogen [8], helium in a metastable level [9, 10], lithium [11], sodium [12], potassium [13], cesium [14] and ytterbium [15]. Excitons in semiconductors constitute another example of a dilute bosonic system in which quantum statistics can lead to spectacular effects [16, 17, 18]. In this article we describe the procedure for preparing a Bose-Einstein condensate and we present a few experiments which are characteristic of these quantum macroscopic systems. We also address the problem of ultra-cold fermions and mixtures of degenerate gases. Finally we give a short overview of the applications of these cold atomic gases in the domain of metrology and high precision sensors. We do not pretend to give here an exhaustive review of this field of research. We have selected a few examples which we consider as representative of the richness of this domain. For a more detailed presentation the interested reader will find useful information in the review articles and the books recently published in this field [19, 20, 21, 22, 23, 24].

2 Preparation and characterization of a condensate

2.1 Trapping of atoms

The basic idea behind this new class of experiments consists in getting closer to Einstein's criterion, i.e., working with a dilute gas, instead of a liquid as it is the case for helium. The price to pay can be read immediately in (1): when the spatial density of the system decreases, the transition temperature also decreases. These new atomic condensates are formed for densities around 10^{13} atoms/cm^3 (instead of 10^{21} atoms/cm^3 for liquid helium) and the transition temperature is on the order of one microkelvin.

All experiments share a common point because of this constraint on the temperature. The confinement of the gas cannot be ensured using material walls, since all atoms would immediately stick to these walls and stay there. The gas is generally trapped using a non homogenous magnetic field. It levitates at the center of an ultra-vacuum chamber (pressure below 10^{-9} Pa). Each atom has a magnetic moment μ, which couples to the local magnetic field. The magnetic energy $E = -\mu \cdot B$ is a potential energy for the center-of-mass motion and the resulting magnetic force ensures the desired trapping.

Consider for example an atom in the vicinity of the point O where the amplitude $|B(r)|$ of the magnetic field is minimum (figure 1). If the direction of the magnetic moment is opposed to the local magnetic field, the magnetic energy reads

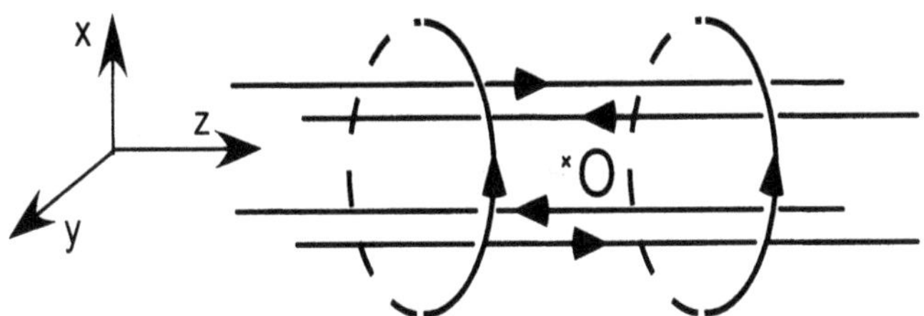

Figure 1: *An example of a magnetic trap (Ioffe Pritchard trap). This trap is formed by 4 current carrying wires which ensure the transverse trapping of the atoms. The longitudinal confinement is obtained using the two circular coils. The distance between these two coils must be larger than their radius, so that the modulus of the magnetic field is minimum in O. In usual experiments the magnetic field in O is of the order of 10^{-4} T.*

$E(r) = -\boldsymbol{\mu} \cdot \boldsymbol{B}(r) = |\boldsymbol{\mu}| \, |\boldsymbol{B}(r)|$. This potential energy is itself minimum in O, and the atom oscillates around this point. In usual working conditions, the oscillation frequency is in the range 100 to 1000 Hz, and the assumption that $\boldsymbol{\mu}$ and $\boldsymbol{B}$ stay opposite is well verified.

How can one store atoms in such a trap? Most experiments start with a laser cooling phase, in which one takes advantage of the exchange of momentum between light and atoms to decrease the temperature of the atomic vapor. Starting from room temperature, one reaches in this way sub-millikelvin temperatures [25, 26, 27]. This is done in a magneto-optic trap (MOT) or in an optical molasses. Laser cooled atoms do not yet form a condensate because their spatial density is too low. Indeed inelastic, light-assisted collisions limit the spatial density in a MOT to $n \sim 10^{11}$ cm^{-3}, and the phase space density is only $n\lambda^3 \sim 10^{-6}$, while it should be larger than 1 for the gas to enter in the degenerate regime.

2.2 Evaporative cooling

The phase space density does not change much during the transfer of atoms from the MOT to the magnetic trap. The necessary additional cooling is obtained using evaporation [28]. One truncates the potential well confining the atoms to a value slightly larger than the mean kinetic energy of the gas. Therefore the fastest atoms can escape and the remaining atoms thermalize to a temperature lower than the initial one. One can show that the central spatial density in the trap increases in the process. Therefore, if the evaporation is performed during a sufficient amount of time, one can hope to reach the condensation threshold.

One must sacrifice a lot of atoms to reach this goal. Since the parameter $n\lambda^3$ is only 10^{-6} after the loading of the magnetic trap, one must gain several orders of magnitude on the temperature and on the spatial density of the gas. In practice the evaporation is performed keeping constant the elastic collision rate between atoms. In this way the thermalization process remains efficient during the whole evaporative cooling sequence. The collision rate is proportional to the spatial density and to the thermal atomic velocity, and it scales as n/λ. One must then increase n et λ by a factor 30 in order to gain the six orders of magnitude on the degeneracy parameter $n\lambda^3$. In a harmonic trap this is obtained by dividing by 1000 the number of atoms during the evaporation. One starts with 10^9 atoms in the magnetic trap, to end with only 10^6. A Bose-Einstein condensate is then formed at the center of the trap.

2.3   Observing a condensate

The observation of gaseous condensates can be done by shining them with a short resonant light pulse, and by measuring the absorption or the dephasing of this pulse. One has access in this way to the spatial distribution of the atoms in the magnetic trap. One can also switch off the magnetic trap, leave the gas expand for some time, and then send the light flash. From the extension of the cloud after this ballistic expansion, one deduces the velocity distribution after the extinction of the trap.

Figure 2 gives an illustration of this principle. Figures 2a and 2b are *in situ* photographs, which show the distribution of the atoms inside the magnetic trap. The atom distribution is cigar-shaped, which results from the trap anisotropy. The presence of a condensate is revealed in a non-ambiguous manner by the time-of-flight pictures (2c,d). For the picture 2c, taken for a temperature above the critical temperature, one gets a quasi isotropic distribution, as expected from equipartition of energy for a gas described by classical statistical physics. On the opposite, the picture 2d, taken for a temperature below the critical temperature, shows a strongly anisotropic velocity distribution. The direction which was the most strongly confined in the magnetic trap is the one for which the velocity distribution is the widest. This anisotropy results from the conversion of the interaction energy into transverse kinetic energy when the confining potential is turned off [29, 30].

This type of pictures allows one to obtain quantitative information on the condensate, such as the atom number and the residual temperature associated with the uncondensed atoms. One has then checked with a very good precision (a few %) that the transition temperature was indeed given by (1). The small deviation with respect to the prediction for the ideal gas is due to atomic interactions (see the contribution of G. Shlyapnikov to this volume). It is also possible to produce quasi-pure condensates, in which the non condensed fraction is below 15% (below this value, it becomes very difficult to estimate it).

2.4   Coherence of condensates

To show that the state of the gas after evaporation indeed corresponds to a condensate, one must check the quantum coherence of this atomic assembly. In order to study the transverse coherence of a light beam, the standard procedure consists

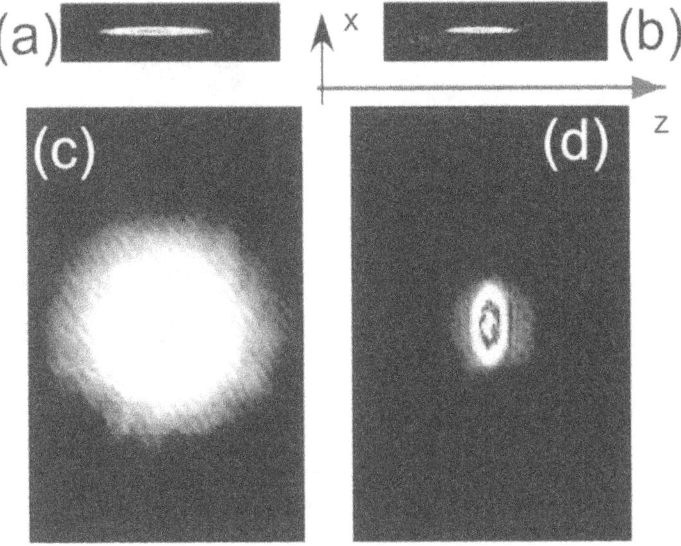

Figure 2: *Rubidium atoms cooled by evaporation in a magnetic trap with frequencies $\nu_x = \nu_y = 150$ Hz, $\nu_z = 11$ Hz. (a,b): in situ images of atoms in the magnetic trap, with $T = 2\,T_c$ et $T = 0.8\,T_c$. (c,d): Images obtained after a 30 ms ballistic expansion (c and d). The Bose-Einstein condensate corresponds to the elliptical structure at the center of figure (d) (photos by P. Desbiolles, D. Guéry-Odelin et J. Söding, ENS).*

in sending this beam onto a screen pierced with two holes. One then measures the contrast of the interferences between the waves diffracted by these holes. This scheme has been transposed to Bose-Einstein condensates by the Munich group. The equivalent of the hole is a radio-frequency wave which extracts the atoms from a given point of the magnetic trap [31] (see also [32, 33, 34]). When a single rf wave is applied, one gets a atomic beam which falls under the influence of gravity. By applying two waves, one obtains the desired situation, with two beams emerging from two different points of the condensate (see fig. 3a) [35].

The results of this experiment fully confirm the idea that the atoms are accumulated in the same quantum state and that the gas is coherent. When the temperature T is well below the critical temperature, one observes an interference with a large contrast in the region where the two atomic beams overlap (fig. 3b). This is true even for a large difference between the two radio-frequencies, corresponding to a distance between the two point sources of the order of the condensate size. On the opposite when $T > T_c$, no detectable interference is visible in the zone where the two beams overlap (fig. 3c), unless the distance between the two sources is below 200 nm, i.e., the coherence length of the gas in these experimental conditions.

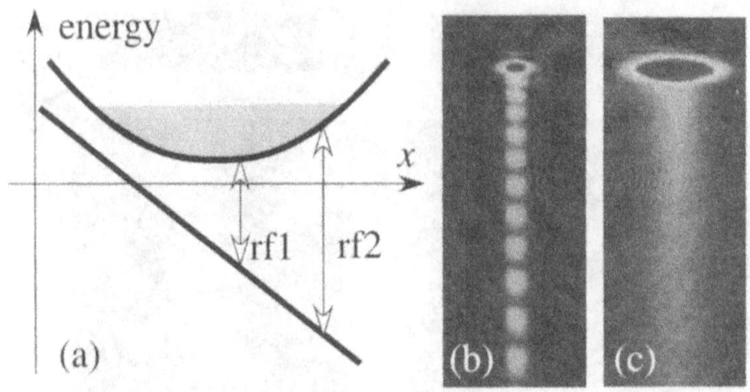

Figure 3: *(a) Extraction of two atomic beams ("atom lasers") from a cloud of rubidium atoms. Two radio-frequency waves, rf1 and rf2, flip the magnetic moment of the trapped atoms. The atoms fall under gravity (directed along x). (b) The atom cloud is a quasi-pure condensate and a strong interference contrast is observed, which reveals the phase coherence of the sample. (c) For a cloud above the critical temperature, no phase coherence is measured if the distance between the extraction points exceeds 200 nm (photograph from Immanuel Bloch, Munich)*

2.5   The atoms that can condense and the experimental setups

It is essential to note the atom clouds prepared with the techniques described above are metastable gases. The thermodynamic equilibrium of sodium or rubidium atoms at a temperature of the order of one microkelvin corresponds to a solid phase, with a negligible vapor pressure. In the experiments described here, we can produce a gaseous phase thanks to laser and evaporative cooling, without ever putting the atoms in contact with a material thermostat. The fact that the gas is metastable (and not stable in the strict sense) manifests itself in the existence of losses: in a 3-body collision, two atoms may form a molecule and the third atom carries away the released energy. Because of such losses the lifetime of a condensate is in practice restricted to a few tens of seconds.

Thanks to this metastability, many atomic species can be brought to condensation. Until now most studies have focused on alkali atoms. Indeed it is relatively easy to manipulate these atoms with lasers. Their resonance line is in the visible or near infrared range, for which reliable and relatively cheap laser sources exist. Rare gas atoms placed in a metastable electronic state are also good candidates (this metastability of the internal atomic state should not be confused with the metastability of the macroscopic state of the gas described above). Metastable helium has thus been condensed in Orsay (Institut d'Optique) [9] and in Paris (LKB) [10]. Similar studies are being pursued for neon. Other types of atoms, magnesium, strontium, chromium, are also actively studied. At last atomic hydrogen (a his-

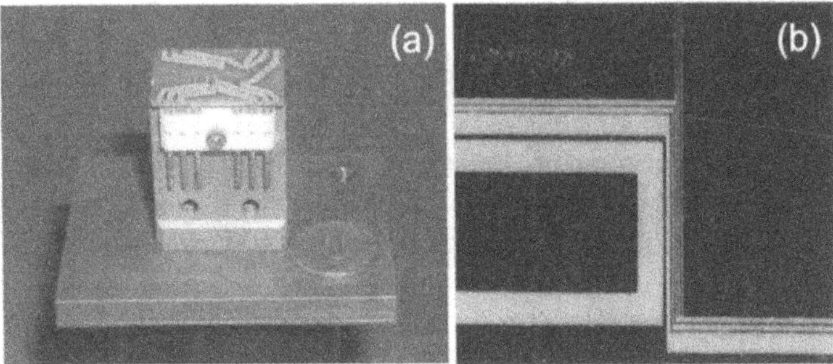

Figure 4: *Examples of "atom chips" on which all the elements necessary for magnetic trapping of atoms have been graved. (a) Setup used in Munich for the observation of one of the first condensates on a chip [36] (photo Jakob Reichel, Munich). (b) Chip used in the laboratoire Charles Fabry, developed in collaboration with the laboratoire de Photonique et de Nanostructures (Marcoussis). The gold wires deposited on the silicon substrate have a width of the order of 10 μm (photo Isabelle Bouchoule, Dominique Mailly, Chris Westbrook).*

torical atom, since this type of research started with it, more than 30 years ago) could be brought to condensation [8]. In this last case, laser light is used only for the diagnoses; the pre-cooling (before evaporation) is performed using standard cryogenic techniques.

Most of the experimental setups are relatively complex. They involve several vacuum chambers. In the first one, atoms are captured from residual vapor pressure or from a slow atomic beam. Atoms are then transferred in a second chamber containing the magnetic trap, where the evaporative cooling takes place. In some setups the condensate is transferred in a third chamber, in which its properties are studied. A second generation of experiments, much more compact, recently appeared. In these experiments, one puts on a chip most of the elements necessary for the trapping and the cooling of the atoms (see figure 4). This can lead to easily transportable devices, which may also be more reliable than the present ones since the alignment of the various components is ensured through the appropriate graving of the chip. [36, 37, 38, 39, 40, 41].

Another line of research which is actively studied concerns the fabrication of a continuous condensate. In a recent experiment the MIT group showed that one can inject a new condensate in a laser trap before the condensate which had been previously deposited has disappeared [42]. At ENS we are trying to transpose to the spatial domain what is usually done in the temporal domain: an intense and dense atomic beam slowly propagates along a magnetic guide. It is cooled by evaporation as it progresses [43]. One can then hope to reach after a reasonable distance (a few meters) a coherent atomic beam [44].

3 "Taylor made" condensates

The research on gaseous Bose-Einstein condensates has allowed for the exploration of a class of phenomena that were not accessible to the community working on liquid helium. The crucial point is that one can finely tune the interactions between atoms as well as the confining potential. The goal of this section is to give some examples of the possibilities opened by these original means of control.

3.1 Controlling interactions

In gaseous Bose-Einstein condensates, interactions are most often described using a mean-field approach. The state of the condensate is obtained by looking for the ground state of the Gross-Pitaevskii equation:

$$-\frac{\hbar^2}{2m}\Delta\psi + V(\boldsymbol{r})\psi + Ng|\psi|^2\psi = \mu\psi$$

where V is the potential confining the N atoms and g the coefficient describing the importance of the mean field. This mean field term can be either positive or negative, depending on the atomic species that is considered, and this point plays an essential role for the behavior of the condensate. Atomic interactions at very low energy are usually characterized by the scattering length a ($g = 4\pi\hbar^2 a/m$, see the contribution of G. Shlyapnikov to this book). A repulsive (resp. attractive) mean field corresponds to a positive (resp. negative) scattering length.

For sodium or rubidium 87 for example, the scattering length is positive for a low magnetic field. One can then place an arbitrary large number of atoms in the condensate, the equilibrium size increasing with the population. On the contrary, a negative scattering length, as it is the case for lithium 7 in a low magnetic field, limits the number of atoms in the condensate. If the atom number exceeds a critical value N_c, the condensate collapses. For an isotropic harmonic trap with pulsation ω, one finds (see, e.g., [21]):

$$N_c \sim \frac{a_0}{|a|} \qquad \text{with} \qquad a_0 = \sqrt{\frac{\hbar}{m\omega}}\,.$$

This number is of the order of 1000 for a standard trap in the case of lithium 7 ($a_0 \sim 10^{-6}$ m, $|a| \sim 10^{-9}$ m) [11].

The sign of the scattering length strongly depends on the atomic parameters. For example, it varies with the C_6 coefficient describing the long range Van der Waals interaction between atoms (see figure 5 and [45]). When one increases continuously C_6, one introduces new bound states in the two-body problem. Each time a new bound state is close to appear, the scattering length tends to $-\infty$ (Levinson theorem). When C_6 is adjusted juste above the critical value for the new bound state to appear, a is large and positive. It then slowly varies with C_6 until the next bound state is about to appear.

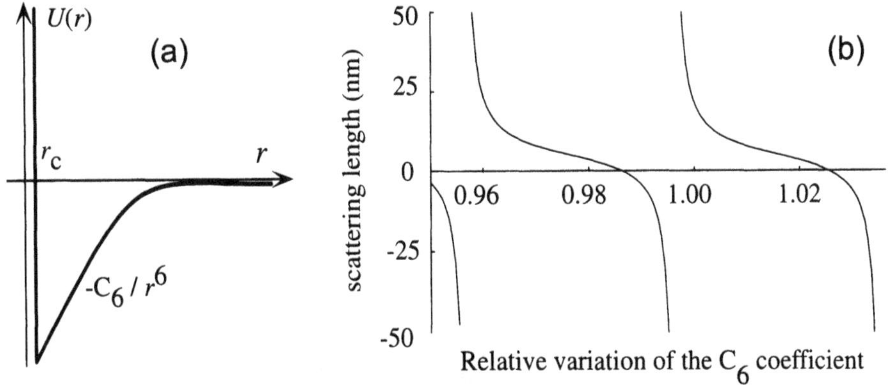

Figure 5: *(a) Van der Waals potential C_6/r^6 between two neutral atoms; the potential is truncated in r_c to take into account the repulsion between electronic clouds. (b) Variations of the scattering length a (in nanometers) for a small change of the C_6 coefficient (the data correspond to the cesium case.)*

The sensitivity of the scattering length to the atomic parameters allows to prepare a condensate with a given mean field term, with a positive sign to ensure the stability of the system. One can then change suddenly the sign of the mean field, by changing for example the magnetic field at the condensate location [46]. The future of the condensate strongly depends on the geometry of the experiment. If the condensate can compress in the 3 directions of space, one observes an implosion of the cloud and the formation of secondary products (molecules, aggregates). This phenomenon called "Bose Nova" has been studied in detail in Boulder [47].

In a quasi 1D geometry, the Gross-Pitaevskii with an attractive mean field possesses stable solutions which are solitons. This has been observed experimentally in our group at ENS and at Rice University [48, 49]. Figure 6 presents the ENS results. A lithium condensate (^{7}Li) is prepared using the "traditional" method described above, by choosing a magnetic field leading to a positive value of a, of the order of $+2.1$ nm. The magnetic potential is then slowly switched off and the condensate is released in an optical guide, formed by a focused laser beam propagating along the z axis. One then achieves a quasi-1D situation: the oscillation frequency $\omega_\perp$ in the xy plane is large and one can consider with a good approximation that the motion in this plane is frozen. In other words the energy quantum $\hbar\omega_\perp$ is large compared to the energy scales of the problem, temperature or mean field energy. If the mean field is negligible ($a \sim 0$, figure 6a), one then observes that the atom cloud expands as it propagate. This is the well-known *spreading of the wave packet.* On the contrary for a sufficiently large and attractive mean field ($a = -0.21$ nm, figure 6b), the propagation of the atom packet occurs without detectable deformation on a distance larger than a millimeter.

Note that grey or black solitons have also been observed in Bose-Einstein condensates [50, 51]. These solitons are produced with a repulsive mean field ($g >$

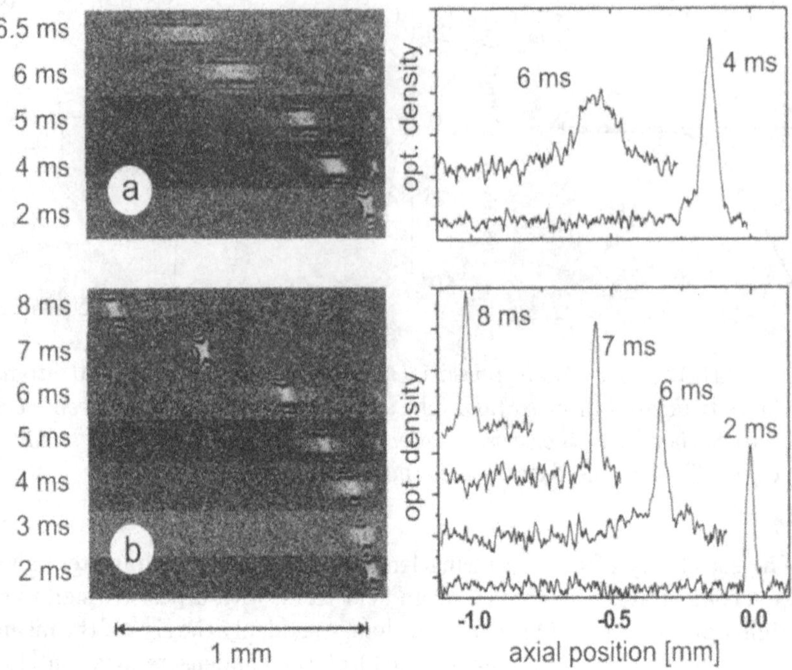

Figure 6: *Propagation of a packet of condensed atoms in a light guide formed by a focused laser beam. (a): When the mean field interaction is negligible, the packet spreads as it propagates. (b): If the mean field is large and attractive, the atom cloud propagates without deformation over more than 1 mm: this is a soliton. In both cases the center-of-mass motion is uniformly accelerated towards the left by the residual gradient of the magnetic field [48].*

0) and they are characterized by a hole in the density profile of the condensate, with a phase jump at the center of the soliton. They propagate along the condensate axis with a speed lower than the sound velocity and they disappear when they reach the edge of the condensate.

3.2 Control of the confining potential

3.2.1 Condensate in a rotating potential: quantized vortices

The study of rotating fluids is directly connected to the research on their super-fluid properties [52, 53, 54]. Consider a superfluid placed in a cylindrical bucket, rotating around its vertical axis with the frequency Ω. If Ω is smaller than Ω_c, the fluid will stay at rest. This is a direct manifestation of superfluidity: the motion of the rough walls of the bucket is too slow to set the superfluid in motion. When Ω is above Ω_c, the superfluid starts to move. As shown by Onsager [55] and Feynman [56], the resulting velocity field is strongly constrained due to the quantum nature of the fluid. From the macroscopic wave function of the condensate $\psi(\boldsymbol{r}) = \sqrt{\rho(\boldsymbol{r})} \, \exp i\theta(\boldsymbol{r})$, one deduces, in a point where the density $\rho(\boldsymbol{r})$ is not zero, the velocity field:

$$\boldsymbol{v}(\boldsymbol{r}) = \frac{\hbar}{m}\boldsymbol{\nabla}\theta(\boldsymbol{r}) \ . \tag{2}$$

It follows that the circulation of the velocity along any closed contour is quantized

$$\oint \boldsymbol{v} \cdot \mathrm{d}\boldsymbol{r} = n\frac{h}{m} \qquad \text{with } n \text{ integer.} \tag{3}$$

The motion of the fluid occurs *via* lines of singularity, or vortex lines, along which the density is zero, and around which the curl of the velocity is non zero. These vortices are universal structures associated with a rotating or turbulent velocity field. They appear in liquid helium physics (see the contribution of S. Balibar to this book), as well as in other macroscopic quantum systems: neutrons stars, superconductors.

Several groups have studied the properties of rotating condensates [57, 58, 59, 60, 61]. One starts with a condensate trapped in a cylindrically symmetric harmonic potential:

$$U(\boldsymbol{r}) - \frac{1}{2}m\omega_\perp^2(x^2 + y^2) + \frac{1}{2}m\omega_z^2 z^2 \ . \tag{4}$$

The condensate is then stirred using the additional potential in the xy plane:

$$\delta U(\boldsymbol{r}) = \frac{\epsilon}{2}m\omega_\perp^2(X^2 - Y^2) \ , \tag{5}$$

where ϵ is of the order of a few %. The eigenaxes of this potential rotate with the angular frequency Ω in the lab frame:

$$X = x\cos(\Omega t) + y\sin(\Omega t) \qquad Y = -x\sin(\Omega t) + y\cos(\Omega t) \ . \tag{6}$$

The rotation frequency Ω is chosen in the range $(0, \omega_\perp)$. When $\Omega = \omega_\perp$, the centrifugal force is equal to the confinement force and the trap become unstable.

In the experiment carried out in our laboratory, the stirring of the condensate uses a laser beam and the vortices are observed after the ballistic expansion of the condensate [57]. Typical results are shown in figure 7. When the frequency is chosen below the threshold frequency Ω_c (with $\Omega_c \simeq \omega_\perp/\sqrt{2}$ for $\epsilon \ll 1$), there is no detectable change in the condensate shape. Just above Ω_c, a hole appears in the condensate. For notably larger frequencies, a vortex lattice is formed. For a given ratio $\Omega/\omega_\perp$, the number of vortices depends on the condensate size. In our setup, we could see up to 50 vortices with 3×10^5 atoms. At MIT and then in

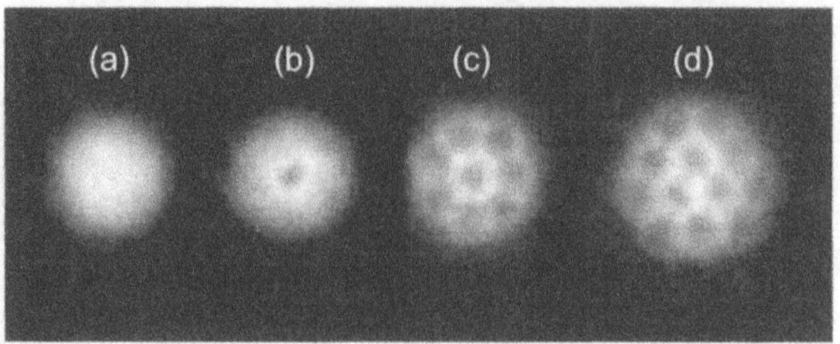

Figure 7: *Density profile observed after time-of-flight for a rotating condensate (a) The rotation frequency Ω is below the critical frequency Ω_c and nothing happens. (b) Just above Ω_c, the first vortex appears. (c,d) Ω is notably larger than Ω_c and a regular vortex lattice appears (Abrikosov triangular lattice) [57].*

Boulder, with condensates containing up to 10^7 atoms, lattices with 200 vortices have been observed [58, 62].

Many studies have been pursued during these last years concerning these vortices. Their nucleation and their decay have been characterized and the angular momentum of the rotating condensate has been measured (for a review, see [54]). The shape of the vortex line is another non trivial problem which has recently received some theoretical and experimental answers [63, 64, 65, 66]. Also the vibration modes of a single vortex and of a vortex lattice have been studied experimentally. Let us note that the nucleation of quantized vortices is not the only signature of superfluidity in these gases. Other evidences for superfluidity are presented in [69, 70, 71, 72, 73]. Let us mention also that vortices can be nucleated in a condensate without any stirrer, using the direct printing on the atoms of the desired pattern $e^{i\theta}$ [74, 75].

3.2.2   Condensate in a periodic potential: the Mott transition

Another type of easily achievable potential consists in a periodic spatial modulation. One irradiates the condensate with a laser standing wave and one thus achieves a potential with an adjustable period and amplitude.

Many studies have been pursued on the behavior of atoms in these optical lattices [77, 78]. A spectacular experiment concerns the Mott transition, e.g., the transition between a superfluid and an insulating state [79, 80]. This experiment has been performed in Munich [81] and its starting point consists in placing the condensate in a 3D periodic potential with cubic symmetry:

$$U(\boldsymbol{r}) = U_0 \left(\sin^2 kx + \sin^2 ky + \sin^2 kz \right) \ .$$

For a small amplitude U_0 of the potential, the tunnel coupling between adjacent sites is important. The ground state of the N atom system is then given by a

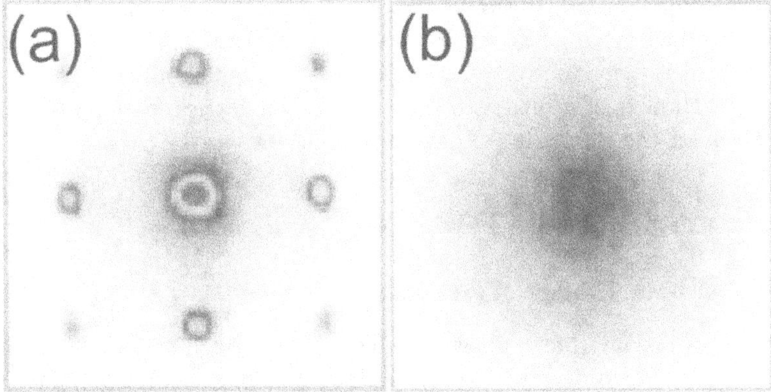

Figure 8: *Atom distribution in a 3D cubic lattice, measured after time-of-flight (a) Superfluid phase, obtained for a small value U_0 of the confining periodic potential. The various peaks reveal the coherence of the wave function over several sites of the lattice. (b) Insulating phase, obtained for a value of U_0 twice as large as the previous one. The atoms are then localized in the various sites of the lattice.*

macroscopic wave function:

$$|\Psi\rangle \propto \left(\sum_{i=1}^{M} a_i^\dagger\right)^N |0\rangle \,. \tag{7}$$

The operator $a_i^\dagger$ creates an atom on the i-th site of the lattice and the sum runs over the M sites. In this superfluid state, each atom is delocalized over all the sites of the lattice. This can be checked experimentally by measuring the momentum distribution of the atoms (figure 8a). One finds a momentum distribution with equally spaced peaks, which is expected if the wave function is delocalized over the periodic lattice.

When one increases the amplitude U_0 of the potential, a transition occurs when the tunnel coupling becomes notably smaller than the interaction energy between two atoms occupying the same site. The fluctuations of the number of atoms per site then become costly from an energy point of view, and this favors a ground state of the system with a well-defined number of atoms on each site:

$$|\Psi\rangle \propto \prod_{i=1}^{M} \left(a_i^\dagger\right)^n |0\rangle \,, \tag{8}$$

with n of the order of 1 or 2 in the Munich experiment. The ballistic expansion reveals the transition towards such a state (figure 8b). One gets a momentum distribution with a broad single component. The coherence of the one atom wave function over the whole lattice is lost.

By switching on and off slowly the periodic potential created by the light, the Munich team has shown that one could commute in a reversible way between the insulating and the conducting states. This system, which allows one to accumulate a given number of atoms at a given site of the lattice, opens very promising perspectives for the treatment of quantum information.

4 Quantum degenerate Fermi gases

4.1 Sympathetic cooling of fermions

In contrast to bosons (particles with integer spin) which can undergo the Bose-Einstein condensation, fermions (with half-integer spin) cannot populate the trap quantum states with more than one fermion because of the Pauli exclusion principle. At zero temperature, fermions fill one by one the trap energy states up to an energy called the Fermi energy E_F. Therefore an ultra-cold Fermi gas will exhibit properties which drastically differ from those of a Bose gas. In the quantum regime, Fermi gases can constitute a model system for more complex problems such as electrons in solids and superconductors, atomic nuclei or neutron stars.

In order to cool a Fermi gas confined in a magnetic trap, the now standard evaporative cooling method for bosons must be modified for fermions. For polarized fermions, s-wave collisions are forbidden and, for higher partial waves, the cross-sections are dramatically reduced at low temperature. A new cooling method, sympathetic cooling, is employed. Both the Fermi gas and another species are confined in the same trap and the evaporation is performed selectively over one (or both) of the two gases. Elastic collisions between the two species are then allowed and the mixture is cooled. The second species can be another spin state in the same fermionic atom [82, 83], or a bosonic isotope of the same gas [84, 85], or a bosonic atom of a different species [87, 88].

Today two fermionic species ^{40}K and ^{6}Li have been cooled to quantum degeneracy, with temperatures as low as 0.1 T_F, where T_F is the Fermi temperature. Quantum degeneracy effects become prominent when the gas temperature is significantly below T_F. ^{40}K has been cooled using collisions between two spin states [82], or using collisions with ^{87}Rb [87]. ^{6}Li has been cooled by collisions between two spin states [83, 90], by collisions with ^{7}Li [84, 85], and with ^{23}Na [89].

Consider for instance the ^{6}Li Fermi gas mixed with ^{7}Li bosons (Fig.9). ^{6}Li is confined in state $|F = 3/2, m_F = 3/2\rangle$ and ^{7}Li in state $|F = 2, m_F = 2\rangle$. The system is cooled by evaporating the bosons using a microwave transition which selectively excites ^{7}Li. Starting initially with a number of ^{7}Li atoms much larger (a thousand-fold) than that of ^{6}Li, this method enables to reach simultaneous quantum degeneracy for the two species [84, 85]. The Fermi quantum degeneracy is reached when the gas temperature is below the Fermi temperature T_F. For an harmonic trap, T_F is given by:

$$k_B T_F = \hbar\bar{\omega}(6N)^{1/3} \,, \tag{9}$$

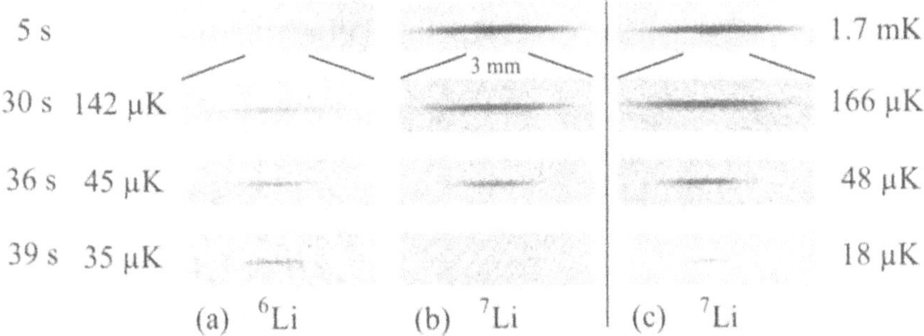

Figure 9: *Sympathetic cooling of fermionic lithium (^{6}Li) by thermal contact with the bosonic isotope (^{7}Li). (a) et (b). Absorption images at various stages of evaporation. The similar sizes of the ^{7}Li and ^{6}Li clouds indicate that thermal contact between the two species is good except at the end of the microwave evaporation ramp, when the number of bosons becomes smaller than the number of fermions (decoupling). For comparison, evaporation images of the single bosonic isotope with the same initial number is shown in (c) [84].*

where $\bar{\omega} = (\omega_1\omega_2\omega_3)^{1/3}$ is the mean trap oscillation frequency and N the number of fermions. In a typical Ioffe-Pritchard trap, $N = 3.2 \times 10^5$, $\bar{\omega} \sim 2\pi \times 2\,\mathrm{kHz}$, and $T_F \sim 11\mu\mathrm{K}$.

4.2   Mixtures of quantum gases

As shown in figure 10, sympathetic cooling enables one to produce mixtures of quantum degenerate gases; ^{7}Li bosons form a Bose-Einstein condensate (narrow peak surrounded by a thermal cloud) while ^{6}Li fermions form a Fermi sea which is much larger because of the Fermi pressure. The lowest measured temperature for this mixture is $T = 0.28\,\mu\mathrm{K} \simeq 0.2(1)\,T_\mathrm{C} = 0.2(1)\,T_\mathrm{F}$, where T_C is the Bose-Einstein condensation temperature and T_F the Fermi temperature. These two gases are in strong interaction in the same magnetic trap [86]. The scattering length $a_\mathrm{BF} = +2\,\mathrm{nm}$ which is proportional to the boson-fermion interaction is positive and much larger than the scattering length $a_\mathrm{BB} = +0.27\,\mathrm{nm}$ describing the boson-boson interaction. After the spectacular results obtained previously with liquid mixtures of helium 4 and helium 3, one can predict that the physics of these cold atomic gas mixtures will be very rich.

4.3   Collapse of a Fermi gas

A spectacular example illustrating the role of the interactions between a Bose condensate and a Fermi gas was recently realized at the University of Firenze [87]. A ^{87}Rb condensate is mixed with a ^{40}K Fermi gas in a magnetic trap at a temperature $T < T_C$ and $T < T_F$. The scattering length a_BB describing the

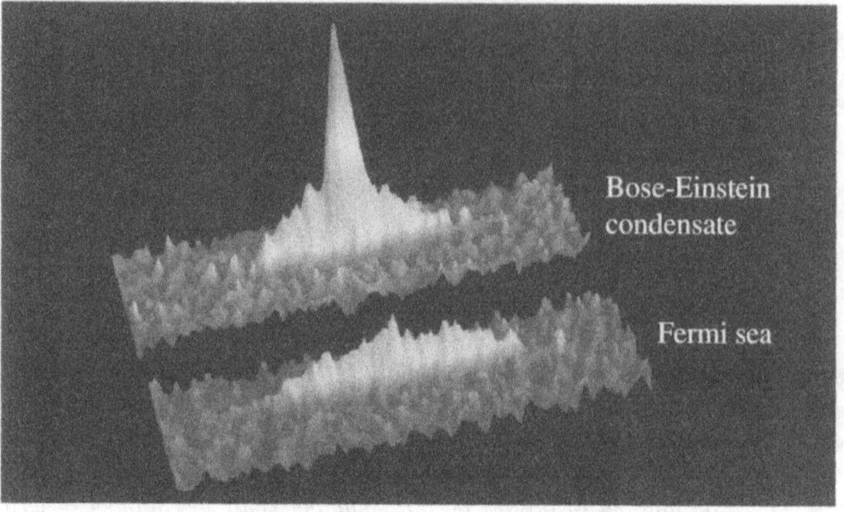

Figure 10: *Mixture of quantum degenerate gases confined in a magnetic trap. A Bose-Einstein condensate of lithium 7 atoms is immersed in a Fermi sea of lithium 6 atoms. The lowest temperature for this mixture is about $0.2\,T_F$ [86].*

interaction between ^{87}Rb bosons is $+5$ nm. This value is much smaller than the scattering length $a_{\mathrm{BF}} = -22\,$nm describing the attractive interaction between bosons and fermions. The authors of ref.[87] observed that, when the number of condensed atoms exceeded 10^5, the number of trapped fermions which initially was on the order of 3×10^4, was abruptly reduced by a factor of ~ 2 (figure 11). The interpretation of this effect is very simple: the attractive mean field between the two species counterbalances the Fermi pressure of the fermionic gas. When it exceeds the Fermi pressure, the system becomes unstable and abruptly collapses in a duration which is less than $50\,$ms. This collapse induces an important increase in fermion losses whereas the Bose gas is almost unaffected.

4.4   Search for superfluidity in an interacting Fermi gas

One of the major goals of this research domain is now to lower the temperature of a two component Fermi gas which exhibits attractive interaction. Well below the Fermi temperature, a phase transition towards a superfluid state is predicted. This transition (called BCS transition for Bardeen, Cooper, Schrieffer) corresponds to the creation of fermion pairs with opposite spins which form a condensate [92]. The transition temperature is given by:

$$T_{\mathrm{BCS}} \simeq 0.3\,T_{\mathrm{F}}\,\exp\left(\frac{-\pi}{2k_{\mathrm{F}}a}\right) \qquad \text{for} \qquad k_{\mathrm{F}}a \ll 1, \qquad \text{i.e.,} \qquad T_{\mathrm{BCS}} \ll T_{\mathrm{F}}\,,$$

$$\tag{10}$$

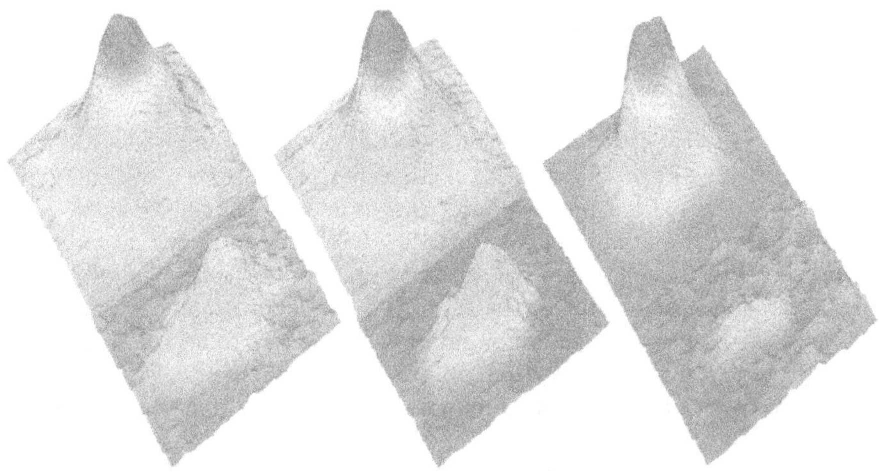

Figure 11: *Density distributions of rubidium atoms (top) and potassium atoms (bottom) after ballistic expansion (19 ms for ^{87}Rb and 4 ms for ^{40}K) at various stages of rubidium evaporation. On the left, the ^{87}Rb condensate begins to form from the thermal cloud and coexists with a relatively large cloud of fermions. At center, when the condensate fraction increases the Fermi gas experiences moderate losses induced by inelastic collisions with the condensate atoms. On the right, when the number of condensed rubidium atoms exceeds 10^5, the Fermi gas abruptly collapses. The vertical scale for fermions is multiplied by 3 [87].*

where k_F is the Fermi momentum and a the scattering length with $a < 0$. A recent theoretical prediction indicates that using a Feshbach resonance might lead to a spectacular increase in the BCS condensation temperature [93]. This transition would occur at a value only 4 times lower than the Fermi temperature. This is accessible to current experiments. In case of success, this research would enable to create a very fruitful bridge with other research domains such as high T_c superconductors, neutron stars and quantum information. In lithium 6, such a Feshbach resonance exists between the two lowest hyperfine energy states and it possesses *a priori* the required properties for the BCS transition. The next question is then: how to detect this superfluid state? Several methods have been proposed as the excitation of specific modes which are sensitive to the superfluid component [94], laser spectroscopy of the BCS gap [95], and study of time of flight images when the trap is made strongly anisotropic [96]. These methods are currently under investigation in a dozen laboratories worldwide.

5   Inertial sensors and atomic clocks

5.1   Why cold atoms?

When atoms are slow, their observation time can be long, giving the opportunity to measure with great precision the transition frequency between two energy levels. Atomic clocks and matter-wave inertial sensors have thus become two of the most important applications of cold atoms [97, 98]. One can show that microwave and optical clocks as well as inertial sensors belong to the same class of atom interferometers [99]. Depending on the experimental geometry, interferometers are sensitive to the frequency of an interrogation field (clocks), to an acceleration field (accelerometers, gravimeters), or to a rotation field (gyrometers). In all of these domains, cold atoms have brought an important gain in sensitivity because of their low velocities.

If T is the duration of coherent excitation between one or several field(s) and N atoms, the resolution of a frequency measurement in a clock is proportional to $N^{-1/2}T^{-1}$. For an interferometric measurement of an acceleration using lasers with wavelength λ, the resolution scales as $N^{-1/2}\lambda T^{-2}$. A rotation measurement for a Sagnac interferometer with length L has a sensitivity scaling as $N^{-1/2}\lambda L^{-1}T^{-1}$.

In a fountain geometry, T can reach one second, bringing a 100 to 1000-fold improvement with respect to experiments with hot atoms. The first atomic fountains have been realized with sodium atoms at Stanford in 1989 [100] and with cesium in 1991 in our laboratory [101]. In a decade, atomic clocks have gained almost two orders of magnitude in accuracy. Today the inaccuracy of the BNM-SYRTE cold atom fountains at Paris Observatory does not exceed 7×10^{-16} corresponding to less than a single second error over 50 million years and prospects for further improvements are still important [102]. Concerning inertial sensors, back in 1991, S. Chu's group in Stanford had demonstrated a sensitivity to changes in the acceleration of gravity of 3×10^{-8} in 40 min averaging time. Today this sensitivity is $\delta g/g \sim 3 \times 10^{-9}$ in 1 min and the accuracy over g is 5×10^{-9} [104]. At Yale University, the rotation sensitivity is currently 6×10^{-10} $\mathrm{rad\,s^{-1}\,Hz^{-1/2}}$ [105], comparable to the best optical gyroscopes after 30 years of development. Finally a very precise measurement of the fine structure constant $\alpha = e^2/4\pi\epsilon_0\hbar c$ can be deduced from a frequency measurement of the recoil energy $E_R = h^2/(2m_{\mathrm{atom}}\lambda^2)$ in a Ramsey-Bordé atom interferometer presented in figure 12 [99]. Indeed in the relation:

$$\alpha^2 = \frac{2R_\infty}{c}\,\frac{m_p}{m_e}\,\frac{m_{\mathrm{atom}}}{m_p}\,\frac{h}{m_{\mathrm{atom}}} \tag{11}$$

all quantities, Rydberg constant R_∞, speed of light c, ratio of the proton mass m_p to electron mass m_e, are known with an uncertainty better than 10^{-9}. Thus a determination at this level of α is possible. The recent Stanford measurements have an uncertainty of 10^{-8}, which are very competitive with respect to other determinations of α with the additional advantage of not being dependent upon quantum electrodynamic calculations [103].

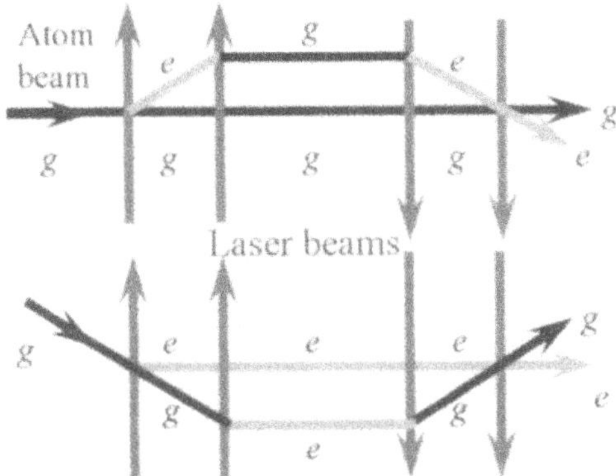

Figure 12: *The two matter-wave interferometers giving access to measurements of the recoil energy in an atom and of the fine structure constant in an optical clock. Atoms are modeled as two-level systems: g (ground state) and e (excited state). Atoms interact with two pairs of counterpropagating laser beams separated in space or in time. In the upper (lower) interferometer atoms are in the ground (excited) state in the central zone between the two beam pairs. For one-photon transitions, the splitting between the two optical resonances is twice the recoil energy $h^2/(m_{\mathrm{atom}}\lambda^2)$. More generally, the 4 interaction zones can utilize multi-photon transitions in order to increase the interferometer sensitivity (image courtesy of Christian Bordé [99]).*

5.2 Example: cold atom clocks

How does an atomic clock operate? Since the 1967 Conférence Générale des Poids et Mesures, "The second is the duration of 9 192 631 770 periods of the radiation corresponding to the transition between the two hyperfine states of the electronic ground state of cesium 133". These two levels correspond to two possible relative orientations (parallel or anti-parallel) of the outer electron magnetic moment and nucleus magnetic moment. A clock simply realizes a resonant spectroscopy experiment. A sample of cold cesium atoms is prepared in their ground state. When sweeping the frequency of a microwave source near $\nu_0 = 9\ 192\ 631\ 770\,\mathrm{Hz}$, the number of atoms transferred by the microwave field into the upper level is measured as a function of detuning from resonance. After locating the peak of the resonance, a servo system maintains the average value of the microwave field frequency locked to the resonance peak. For atoms cooled to a temperature of 1 microKelvin, the thermal velocity is about 1 centimeter par second. On Earth the acceleration of gravity limits the measurement time. Therefore one uses an atomic fountain in which atoms are launched vertically through a microwave cavity where the interrogation field is confined.

Figure 13: *Principle of an atomic fountain. A cloud of cesium atoms is cooled in optical molasses to a temperature of $1\,\mu K$ and launched upwards with a velocity of 4 m/s. Atoms travel upwards and downwards through a cavity which confines a microwave field tuned to a frequency near the cesium hyperfine frequency. Atoms excited by the microwave field are detected by light induced fluorescence. In this Ramsey interrogation method, the duration between the two interactions is about 0.5 s.*

The principle of an atomic fountain is illustrated in figure 13. The atoms are first captured and cooled in a magneto-optical trap or optical molasses. After optical pumping into the ground state g_1 atoms are launched upward through a microwave cavity which put them into a linear superposition of the two clock states g_1 et g_2. This wavefunction then evolves freely during the ballistic flight above the cavity. Gravity turns the atoms back down and they cross the cavity again after a flight time T. Atoms interact a second time with the microwave field and, using light induced fluorescence, atoms which have made the transition to g_2 are counted.

The transition probability $P_{g_1 \rightarrow g_2}(T)$ to be excited from g_1 to g_2 is a function of the resonance detuning and of the flight time T between the two interactions with the microwaves:

$$P_{g_1 \rightarrow g_2}(T) \sim \cos^2\left(\pi(\nu_{12} - \nu)T\right) \tag{12}$$

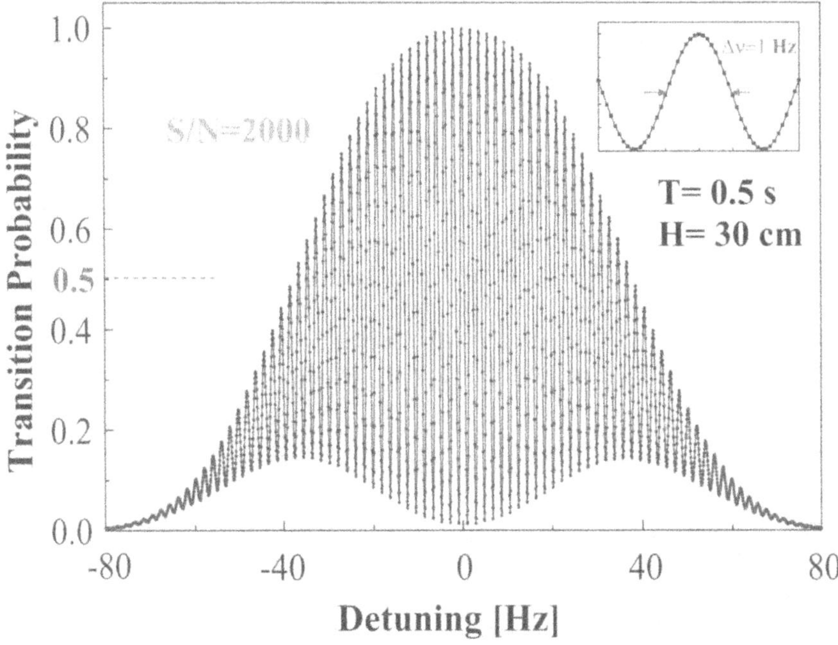

Figure 14: *Ramsey fringes in a cesium atomic fountain: Transition probability versus detuning of the microwave field from the cesium resonance. The fringe width is 1 Hz. Each point corresponds to a single measurement of duration 1 s and the signal to noise ratio per point (2000 at mid-height) is limited by the quantum projection noise [106].*

When the microwave frequency ν is swept, one obtains a modulated signal called *Ramsey fringes* (figure 14). The fringe width at half maximum $\delta\nu$ is $1/2T$. For a fountain height of 30 cm, $T = 0.5$ s and the fringe width is only 1 Hz. This corresponds to a quality factor for this resonance $\delta\nu/\nu = 10^{10}$. In comparison to conventional cesium clocks, this amounts to a 100- to 1000-fold improvement. The signal to noise ratio in figure 14 is not limited by technical noise but by the fundamental noise associated to a quantum measurement. The fluorescence measurement projects the atomic wavefunction which is a linear superposition of g_1 et g_2 in either one of the two states. If the atomic cloud is made of N uncorrelated atoms, then this noise called "quantum projection noise " is simply proportional to $1/\sqrt{N}$. By changing N, it was verified that this noise was the dominant source of noise up to about 10^6 detected atoms [106].

With this method, the accuracy of the cesium fountains developed at the BNM-SYRTE laboratory (Paris Observatory) reaches 7×10^{-16}, currently the best accuracy worldwide. Paradoxically, this accuracy is limited by the strong interactions between cesium atoms at very low temperatures. Of course this was only discovered with the advent of laser cooling and could not have been anticipated in 1967 when cesium was chosen for the definition of the second in the SI unit system! These cold atom interactions are also responsible for the great difficulties encountered to produce a Bose-Einstein condensate of cesium, which were only overcome in 2002 [14]. Other alkalis have a far smaller scattering length and therefore a reduced mean field shift. Thus fountains operating with rubidium atoms exhibit a collisional shift 50 to 100 times smaller than for cesium and are likely to significantly surpass cesium devices [107, 108]. It is remarkable that only five years after the measurement of the collisional shift in rubidium fountains, the master clocks of the US GPS navigation system will be based in 2005 on an ensemble of operational rubidium fountains. The second scientific interest of rubidium fountains is that it enables one to search for a drift of fundamental physical constants.

5.3   Search for a drift of fundamental constants

The search for a drift of fundamental physical constants, such as the fine structure constant α, the electron mass m_e, the proton mass m_p, or the gravitational interaction G, is currently a very active research subject because of the potential impact that the discovery of such a drift would have. It would indeed constitute a violation of Einstein's equivalence principle [109]. Recent models based on string theory do predict a variation of fundamental physical constants at levels which are close to experimentally realizable sensitivities [110]. In the case of atomic clocks, the principle of the test is to perform at various epochs a comparison between the hyperfine energy of two (or more) alkali atoms with different atomic number Z [111]. Because of large relativistic corrections for high Z values, the hyperfine structure of alkali atoms is not simply proportional to α^2 as for hydrogen. It is a more complex function of the product $Z\alpha$. By successive frequency comparisons between a rubidium fountain and several cesium fountains over a duration close to 5 years a new upper limit for the drift of the frequency ratio $\nu(^{87}Rb)/\nu(^{133}Cs)$ was obtained:

$$\frac{d}{dt}\ln\left(\frac{\nu_{\text{Rb}}}{\nu_{\text{Cs}}}\right) = (0.2 \pm 7)\,10^{-16}\,\text{year}^{-1}\;. \tag{13}$$

Within the theoretical description of reference [111], one deduces a new upper limit over the variation of the fine structure constant α at the present epoch:

$$\frac{\dot{\alpha}}{\alpha} = (-0.4 \pm 16)\,10^{-16}\,\text{year}^{-1}\;. \tag{14}$$

With this sensitivity, no drift is detected and this measurement represents a 100-fold improvement over previous laboratory measurements. The method will considerably improve in the future by using a double cesium-rubidium fountain which is under construction at Paris Observatory.

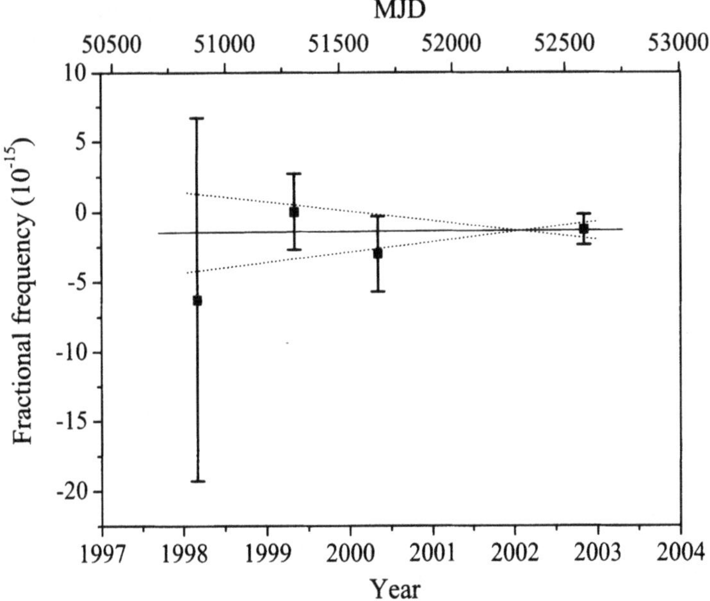

Figure 15: *Hyperfine frequency of ^{87}Rb with respect to cesium over a duration of 57 months. The 1999 value ($\nu_{\mathrm{Rb}}^{1999} = 6\,834\,682\,610.904\,333$ Hz) is conventionally chosen as a reference. A weighted linear fit gives: $\frac{d}{dt}\ln\left(\frac{\nu_{\mathrm{Rb}}}{\nu_{\mathrm{Cs}}}\right) = (0.2 \pm 7.0) \times 10^{-16}\,\mathrm{year}^{-1}$.*

5.4 Space Clocks

In an atomic fountain, the duration between the two interactions with the microwave field in the cavity is limited by the Earth gravity acceleration. In the microgravity environment of a satellite, this duration can be increased by another order of magnitude. This is the goal of the PHARAO project (Projet d'Horloge Atomique par Refroidissement d'Atomes en Orbite). PHARAO is developed by CNES, BNM-SYRTE, and our laboratory. PHARAO was selected by ESA (the European Space Agency) to fly within the ACES European mission on board the International Space Station at the end of 2006 for a duration of 18 months [113]. The various components of the Space hardware are under development in industry. The PHARAO cesium clock will have excellent performance in space, its frequency reference will be accessible to a large number of time and frequency laboratories worldwide, and it will contribute to the realization of the International Atomic Time (TAI). By comparison with other ground clocks, various fundamental physics tests will be performed such as an improved measurement of the Einstein effect (the influence of a gravitational field over the proper time of a clock, or red-shift) and improved tests on drift of fundamental constants. On the application side, satellite navigation systems (GPS, Global Positioning System) or the future European navigation system, GALILEO, rely on very stable atomic clocks on board a constellation of satellites orbiting the Earth at 20 000 km altitude. Improved clocks will improve the precision of the positioning and velocity determination. Thus progress in clocks and time transfer methods will in turn enable new advances in other disciplines, as high rate telecommunication, geodesy and astrophysics.

5.5 Optical Clocks

We currently witness a fast development of cold atom optical clocks and comparison methods between clocks operating in various parts of the electromagnetic spectrum. In the visible domain, the clock rate ν is 10^{15} Hz, instead of 10^{10} Hz in the microwave domain. For a given atomic resonance linewidth $\delta\nu$, the quality factor of the clock $\nu/\delta\nu$ can be considerably increased. Relative frequency stabilities in the range 10^{-17} to 10^{-18} seem accessible to these new optical clocks. Neutral atoms or ions having strongly forbidden lines (therefore very narrow linewidths) in the visible or UV range are used. Ultra-stable lasers with sub-Hertz emission spectrum are required to excite these narrow transitions. Today the highest quality factor has been obtained with laser cooled and trapped Hg^+ ions at NIST (USA) [115] and several other ions (In^+, Yb^+, Ca^+, Sr^+, ...) are under study. In the case of neutral atoms, the transition $1s \rightarrow 2s$ in atomic hydrogen has an ultimate quality factor of 10^{15} and continuous progress towards this goal has been made at the Max Planck Institute in Garching (Germany) [116]. Other atoms (Mg, Ca, Sr, Yb, Ag, Au) have also interesting ultra-narrow lines in the visible domain. Among these, thFermionicic isotope of strontium under study in Japan and at BNM-SYRTE seems a very attractive candidate [117, 118]. In parallel a major breakthrough has been realized in 1999 for comparison methods between frequency standards operating at large frequency differences. The new method uses mode-locked femtosecond lasers [114, 115]. Such lasers produce a frequency comb with very regular spacing between the comb lines. If the spacing between the comb lines is locked to a primary frequency standard, then the absolute frequency of all comb lines is known with respect to the standard. If such a laser is coupled into an optical fiber with strong index gradients, the strong dispersive optical nonlinearity of the fiber increases the bandwidth of the comb beyond one octave in frequency enabling direct microwave to optical frequency comparisons. In addition, this octave in frequency allows by a simple frequency doubling nonlinear crystal ($\nu \rightarrow 2\nu$) to self-reference the frequency comb spacing. Thus stabilizing one of the comb lines on a narrow atomic resonance provides a set of ultra-stable secondary frequency references at each of the comb line. This very simple and performing method has replaced the very cumbersome frequency multiplication chains used formerly. Femtosecond lasers not only link microwave and optical clocks together but also link directly two optical clocks. In summary the era of the cesium atom for the definition of the second may well reach an end soon!

References

[1] S.N. Bose, *Z. Phys.* **26**, 178 (1924).

[2] A. Einstein, *Sitzber. Klg. Preuss. Akad. Wiss.* **261** (1924) ; A. Einstein, *Sitzber. Klg. Preuss. Akad. Wiss.* **3** (1925).

[3] A. Pais, *Subtle is the Lord* (Oxford University Press, Oxford 1982).

[4] P. Kapitza, *Nature* **141**, 74 (1938).

[5] J.F. Allen, A.D. Misener, *Nature* **141**, 75 (1938).

[6] F. London, *Nature* **141**, 643 (1938).

[7] M.H. Anderson, J. Ensher, M. Matthews, C. Wieman, E. Cornell, *Science* **269**, 198 (1995).

[8] D.G. Fried, T. C. Killian, L. Willmann, D. Landhuis, S.C. Moss, D. Kleppner, T.J. Greytak, *Phys. Rev. Lett.* **81**, 3811 (1998).

[9] A. Robert, O. Sirjean, A. Browaeys, J. Poupard, S. Nowak, D. Boiron, C.I. Westbrook, A. Aspect, *Science* **292**, 461 (2001).

[10] F. Pereira Dos Santos, J. Léonard, Junmin Wang, C. J. Barrelet, F. Perales, E. Rasel, C. S. Unnikrishnan, M. Leduc, C. Cohen-Tannoudji, *Phys. Rev. Lett.* **86**, 3459 (2001).

[11] C.C. Bradley, C.A. Sackett, R.G. Hulet, *Phys. Rev. Lett.* **78**, 985 (1997). Voir aussi C.C. Bradley, C.A. Sackett, J.J. Tollett, R. G. Hulet, *Phys. Rev. Lett.* **75**, 1687 (1995).

[12] K.B. Davis, M.O. Mewes, N. Van Druten, D. Durfee, D. Kurn, W. Ketterle, *Phys. Rev. Lett.* **75**, 3969 (1995).

[13] G. Modugno, G. Ferrari, G. Roati, R. J. Brecha, A. Simoni, M. Inguscio, *Science* **294**, 1320–1322 (2001).

[14] T. Weber, J. Herbig, M. Mark, H.-C. Nägerl, and R. Grimm, *Science* **299**, 232 (2003).

[15] Y. Takasu, K. Maki, K. Komori, T. Takano, K. Honda, M. Kumakura, T. Yabuzaki, and Y. Takahashi, *Phys. Rev. Lett.* **91**, 040404 (2003).

[16] J.P. Wolfe, J.L. Lin, D.W. Snoke, 'Bose-Einstein Condensation of a Nearly Ideal Gas: Excitons in Cu_2O'. In *Bose-Einstein Condensation*, ed. by A. Griffin, D.W. Snoke, S. Stringari (Cambridge University Press, Cambridge 1995) pp. 281–329.

[17] A. Mysyrowicz, 'Bose-Einstein Condensation of Excitonic Particles in Semiconductors'. In *Bose-Einstein Condensation*, ed. by A. Griffin, D.W. Snoke, S. Stringari (Cambridge University Press, Cambridge 1995) pp. 330–354.

[18] G. M. Kavoulakis and A. Mysyrowicz, *Phys. Rev. B* **61**, 16619 (2000).

[19] *Coherent Atomic Matter Waves (Les Houches-Ecole D'été de Physique Theorique*, R. Kaiser, C. Westbrook, F. David edts. (Springer-Verlag, 2001).

[20] M. Inguscio, S. Stringari, C.E. Wieman eds, *Proceedings of the International School of Physics Enrico Fermi*, Course CXL, *Bose–Einstein Condensation in Atomic Gases* (IOS Press, Amsterdam 1999).

[21] F. Dalfovo, S. Giorgini, l.P. Pitaevskii, S. Stringari, *Rev. Mod. Phys.* **71**, 463 (1999).

[22] C. J. Pethick and H. Smith, *Bose-Einstein Condensation in Dilute Gases* (Cambridge University Press, 2001).

[23] *Nature insight*, *Nature* **416**, 205–246 (2002).

[24] L. Pitaevskii and S. Stringari, *Bose-Einstein Condensation* (Clarendon Press, Oxford, 2003).

[25] S. Chu, *Rev. Mod. Phys.* **70**, 685 (1998).

[26] C. Cohen-Tannoudji, *Rev. Mod. Phys.* **70**, 707 (1998).

[27] W.D. Phillips, *Rev. Mod. Phys.* **70**, 721 (1998).

[28] W. Ketterle, N.J. Van Druten, *Advances in Atomic, Molecular, and Optical Physics* **37**, 181 (1996).

[29] Y. Castin and R. Dum, *Phys. Rev. Lett.* **77**, 5315 (1996).

[30] Yu. Kagan, E. L. Surkov, and G. V. Shlyapnikov, *Phys. Rev. A* **55**, R18 (1997).

[31] I. Bloch, T.W. Hänsch, T. Esslinger, *Phys. Rev. Lett.* **82**, 3008 (1999).

[32] M.-O. Mewes, M. R. Andrews, D. M. Kurn, D. S. Durfee, C. G. Townsend, and W. Ketterle, *Phys. Rev. Lett.* **78**, 582 (1997).

[33] B. P. Anderson and M. A. Kasevich, *Science* **282**, 1686 (1998).

[34] M. Kozuma, L. Deng, E. W. Hagley, J. Wen, R. Lutwak, K. Helmerson, S. L. Rolston, and W. D. Phillips, *Phys. Rev. Lett.* **82**, 871.

[35] I. Bloch, T. W. Hänsch, T. Esslinger, *Nature* **403**, 166 (2000).

[36] W. Hänsel, P. Hommelhoff, T. W. Hänsch, J. Reichel, *Nature* **413**, 498 (2001).

[37] H. Ott et al., *Phys. Rev. Lett.* **87**, 230401 (2001).

[38] R. Folman *et al.*, *Adv. Atom. Mol. Opt. Phys.* **48**, 263 (2002).

[39] A.E. Leanhardt et al., *Phys. Rev. Lett.* **89**, 040401 (2002).

[40] E.A. Hinds, C.J. Vale and M.G. Boshier, *Phys. Rev. Lett.* **86**, 1426 (2001).

[41] Des expériences similaires sont actuellement en cours au laboratoire Charles Fabry de l'Institut d'Optique.

[42] A. P. Chikkatur, Y. Shin, A. E. Leanhardt, D. Kielpinski, E. Tsikata, T. L. Gustavson, D. E. Pritchard, W. Ketterle, *Science* **296**, 2193 (2002).

[43] P. Cren, C.F. Roos, A. Aclan, J. Dalibard, D. Guéry-Odelin, *Eur. Phys. Jour.* **D20**, 107 (2002).

[44] E. Mandonnet, A. Minguzzi, R. Dum, I. Carusotto, Y. Castin, J. Dalibard, *Eur. Phys. J. D* **10**, 9 (2000).

[45] J. Dalibard: 'Collisional dynamics of ultra-cold atomic gases'. In: *Proceedings of the International School of Physics Enrico Fermi, Course CXL, Bose-Einstein Condensation in Atomic Gases*, ed. by M. Inguscio, S. Stringari, C.E. Wieman (IOS Press, Amsterdam 1999) pp. 321–350.

[46] S. Inouye, M.R. Andrews, J. Stenger, H.-J. Miesner, D.M. Stamper-Kurn, W. Ketterle, *Nature* **392**, 151 (1998).

[47] S. L. Cornish, N. R. Claussen, J. L. Roberts, E. A. Cornell, and C. E. Wieman, *Phys. Rev. Lett.* **85**, 1795 (2000); E. A. Donley, N. R. Claussen, S. L. Cornish, J. L. Roberts, E. A. Cornell, C. E. Wieman, *Nature* **412**, 295 (2001).

[48] L. Khaykovich, F. Schreck, G. Ferrari, T. Bourdel, J. Cubizolles, L. D. Carr, Y. Castin, C. Salomon, *Science* **296**, 1290 (2002).

[49] K. E. Strecker, G. B. Partridge, A. G. Truscott, R. G. Hulet, *Nature* **417**, 150 (2002).

[50] S. Burger et al., *Phys. Rev. Lett.* **83**, 5198 (1999).

[51] J. Denschlag et al., *Science* **287**, 97 (2000).

[52] E.M. Lifshitz, L.P. Pitaevskii, *Statistical Physics, Part 2* (Butterworth-Heinemann, Oxford 1980).

[53] R.J. Donnelly, *Quantized Vortices in Helium II* (Cambridge University Press, Cambridge 1991).

[54] A. Fetter, A Svidzinsky, *J. Phys. Condens. Matter* **13**, 135 (2001).

[55] L. Onsager, *Nuovo Cimento* **6**, suppl. 2, 249 (1949).

[56] R.P. Feynman, 'Application of Quantum Mechanics to Liquid Helium'. In *Progress in Low Temperature Physics*, vol. 1, ed. by C. J. Gorter (North-Holland, Amsterdam, 1955) pp. 17–53.

[57] K.W. Madison, F. Chevy, W. Wohlleben, J. Dalibard, *Phys. Rev. Lett.* **84**, 806 (2000).

[58] J.R. Abo-Shaeer, C. Raman, J.M. Vogels, W. Ketterle, *Science* **292**, 476 (2001).

[59] C. Raman, J.R. Abo-Shaeer, J.M. Vogels, K. Xu, W. Ketterle, *Phys. Rev. Lett.* **87**, 210402 (2001).

[60] P.C. Haljan, I. Coddington, P. Engels, E.A. Cornell, *Phys. Rev. Lett.* **87**, 210403 (2001).

[61] E. Hodby, G. Hechenblaikner, S.A. Hopkins, O.M. Maragò, C.J. Foot, *Phys. Rev. Lett.* **88**, 010405 (2002).

[62] P. Engels, I. Coddington, P. C. Haljan, E. A. Cornell, cond-mat/0204449.

[63] J.J. García-Ripoll, V.M. Pérez-García, *Phys. Rev. A* **63**, 041603 (2001); J.J. García-Ripoll, V.M. Pérez-García, *Phys. Rev. A* **64**, 053611 (2001).

[64] A. Aftalion, T. Riviere, *Phys. Rev. A* **64**, 04s3611 (2001); A. Aftalion, R. L. Jerrard, cond-mat/0204475.

[65] M. Modugno, L. Pricoupenko, and Y. Castin, *Eur. Phys. J. D* **22**, 235 (2003).

[66] P. Rosenbusch, V. Bretin, and J. Dalibard, *Phys. Rev. Lett.* **89**, 200403 (2002).

[67] P. Engels, I. Coddington, P. C. Haljan, and E. A. Cornell, **89**, 100403 (2002).

[68] V. Bretin, P. Rosenbusch, F. Chevy, G.V. Shlyapnikov, and J. Dalibard, cond-mat/0211101.

[69] D. Guéry-Odelin, S. Stringari, *Phys. Rev. Lett.* **83**, 4452 (1999).

[70] O.M. Maragò, S.A. Hopkins, J. Arlt, E. Hodby, G. Hechenblaikner, C. J. Foot, *Phys. Rev. Lett.* **84**, 2056 (2000).

[71] C. Raman, M. Köhl, R. Onofrio, D. S. Durfee, C.E. Kuklewicz, Z. Hadzibabic, W. Ketterle, *Phys. Rev. Lett.* **83**, 2502 (1999).

[72] A.P. Chikkatur, A. Gorlitz, D.M. Stamper-Kurn, S. Inouye, S. Gupta, W. Ketterle, *Phys. Rev. Lett.* **85**, 483 (2000).

[73] R. Onofrio, C. Raman, J.M. Vogels, J. Abo-Shaeer, A.P. Chikkatur, W. Ketterle, *Phys. Rev. Lett.* **85**, 2228 (2000).

[74] M.R. Matthews, B.P. Anderson, P.C. Haljan, D.S. Hall, C.E. Wieman, E.A. Cornell, *Phys. Rev. Lett.* **83**, 2498 (1999).

[75] B.P. Anderson, P.C. Haljan, C.E. Wieman, E.A. Cornell, *Phys. Rev. Lett.* **85**, 2857 (2000).

[76] A. E. Leanhardt, A. Görlitz, A. P. Chikkatur, D. Kielpinski, Y. Shin, D. E. Pritchard, and W. Ketterle, *Phys. Rev. Lett.* **89**, 190403 (2002).

[77] G. Grynberg and C. Robillard, *Phys. Rep.* **355**, 335 (2001).

[78] P. S. Jessen and I. H. Deutsch, *Adnaces in Atomic, Molecular and Optical Physics* **37**, eds. B. Bederson and H. Walther (Academic Press, Cambridge, 1996).

[79] M. P. Fisher, P.B. Weichman, G. Grinstein, D.S. Fisher, *Phys. Rev. B* **40**, 546 (1989).

[80] D. Jaksch, C. Bruder, J. Cirac, C. W. Gardiner, P. Zoller, *Phys. Rev. Lett.* **81**, 3108 (1998).

[81] M. Greiner, O. Mandel, T. Esslinger, T. W. Hänsch, I. Bloch, *Nature* **415**, 39 (2002).

[82] B. DeMarco and D. Jin, *Science* **285**, 1703 (1999).

[83] S. Granade, M. Gehm, K. O'Hara, and J. Thomas, *Phys. Rev. Lett.* **88**, 120405 (2002).

[84] F. Schreck et al., *Phys. Rev. A* **64**, 011402R (2001).

[85] A. G. Truscott *et al.*, *Science* **291**, 2570 (2001).

[86] F. Schreck et al., *Phys. Rev. Lett.* **87**, 080403 (2001).

[87] G. Modugno, G. Roati, F. Riboli, F. Ferlaino, R. Brecha, M. Inguscio, *Science* **297**, 2240 (2002).

[88] Z. Hadzibabic et al., *Phys. Rev. Lett.* **89**, 160401 (2002).

[89] K. Dieckmann et al., *Phys. Rev. Lett.* **89**, 203201 (2002).

[90] K. O'Hara et al., *Science* **298**, 2179 (2002).

[91] H.T.C. Stoof and M. Houbiers, in *Bose-Einstein Condensation in Atomic Gases*, Proc. of the Enrico Fermi school, M. Inguscio, S Stringari and C. Wieman eds., IOS Press (1999).

[92] J. Bardeen, L. Cooper and J. Schrieffer, *Phys. Rev. A* **108**, 1175 (1957).

[93] M. Holland, S. Kokkelmans, M. Chiofalo and R. Walser, *Phys. Rev. Lett.* **87** 120406 (2001).

[94] G. M. Bruun and B. R. Mottelson, *Phys. Rev. Lett.* **87**, 270403 (2001).

[95] G. M. Bruun, P. Törmä, M. Rodriguez, and P. Zoller, *Phys. Rev. A* **64**, 033609 (2001).

[96] C. Menotti, P. Pedri, S. Stringari, *Phys. Rev. Lett.* **89**, 250402 (2002).

[97] Voir par exemple: Proc of the 6th Symposium on Frequency Standards and Metrology, P. Gill ed., World Scientific, (2001).

[98] *Atom Interferometry*, P. Berman ed., Academic Press (1997).

[99] C. Bordé, Metrologia, **39**, 435 (2002).

[100] M. Kasevich, E. Riis, S. Chu and R. de Voe, *Phys. Rev. Lett.* **63**, 612 (1989).

[101] A. Clairon, C. Salomon, S. Guellati and W. D. Phillips, *Europhys. Lett.* **16**, 165 (1991).

[102] P. Lemonde et al., in *Frequency Measurement and control, Topics in Applied Physics* **79**, Springer, 131, (2001).

[103] A. Wicht, J. Hensley, E. Saralic and S. Chu, Proc of the 6th Symposium on Frequency Standards and Metrology, P. Gill ed., World Scientific, 193, (2001).

[104] A. Peters, K. Chung, and S. chu, *Nature*, **400**, 849 (1999).

[105] A. Landragin, T. Gustavson, and M. Kasevich, in *Proc. of the 14th Int. Conf. on Laser Spectroscopy*, R. Blatt, J. Eschner, D. Liebfried, F. Schmidt-Kaler eds., 170, World Scientific, (1999).

[106] G. Santarelli *et al.*, *Phys. Rev. Lett.*, **82**, 4619 (1999).

[107] C. Fertig and K. Gibble, *Phys. Rev. Lett.* **85**, 1622 (2000).

[108] Y. Sortais, S. Bize, C. Nicolas, and A. Clairon, C. Salomon, C. Williams, *Phys. Rev. Lett.* 85 (15), 3117 (2000).

[109] Pour une revue, voir par exemple J. P. Uzan, ArXiv: hep-ph/0205340.

[110] T. Damour, F. Piazza, and G. Veneziano, *Phys. Rev. Lett.* **89**, 081601 (2002).

[111] J. D. Prestage, R. L. Tjoelker and L. Maleki, *Phys. Rev. Lett.* **74**, 3511 (1995).

[112] H. Marion, F. Pereira Dos Santos, M. Abgrall, S. Zhang, Y. Sortais, S. Bize, I. Maksimovic, D. Calonico, J. Grüenert, C. Mandache, P. Lemonde, G. Santarelli, P. Laurent, A. Clairon, and C. Salomon, to appear in Phys.Rev. Lett., Archiv:physics/0212112.

[113] Ch. Salomon, N. Dimarcq, M. Abgrall, A. Clairon, P. Laurent, P. Lemonde, G. Santarelli, P. Uhrich, L.G. Bernier, G. Busca, A. Jornod, P. Thomann, E. Samain, P. Wolf, F. Gonzalez, Ph. Guillemot, S. Léon, F. Nouel, Ch. Sirmain and S. Feltham, C.R. Acad. Sci. Paris, t.2, Série 4, 1313 (2001).

[114] J. Reichert, et al., *Phys. Rev. Lett.* **84**, 3232 (1999).

[115] S. Diddams et al., *Science* **293**, 825 (2001).

[116] M. Niering *et al.*, *Phys. Rev. Lett.* **84**, 5496 (2000).

[117] H. Katori et al., *Phys. Rev. Lett.* **82**, 1116 (1999).

[118] I. Courtillot, A. Quessada, R.P. Kovacich, J-J. Zondy, A. Landragin, A. Clairon, and P. Lemonde, *Opt. Lett.* **28**, p. 468 (2003).

Jean Dalibard and Christophe Salomon
Laboratoire Kastler Brossel
24, rue Lhomond
F-75005 Paris
email: Jean.Dalibard@lkb.ens.fr
email: salomon@lkb.ens.fr

Poincaré Seminar 2003, 85 – 98
© Birkhäuser Verlag, Basel, 2004

Bose-Einstein Condensation and Quantum Coherence: From Superfluidity to Localization, from Fermi Liquids to Superconductors

Philippe Nozières

1 Ideal gas

N bosons of mass m in an unit volume have an equilibrium distribution

$$n_k = \frac{1}{\exp\left[\frac{\varepsilon_k - \mu}{T}\right] - 1} \, ,$$

where μ is the chemical potential. In three dimensions, μ goes to zero for a critical temperature

$$T_c = 3,31 \frac{\hbar^2}{md^2} \, ,$$

where $d = N^{-1/3}$ is the mean distance between particles. Below T_c, a finite number of particles

$$N_o = N \left[1 - \left(\frac{T}{T_c}\right)^{3/2}\right]$$

accumulate in the lowest state: this is Bose-Einstein condensation. But this state is unstable: nothing requires the condensed particles to strictly stay in the same quantum state. This instability becomes clear if we move the condensate at a speed $v = \hbar q/m$, accumulating particles in the state of wave vector q. In the condensate coordinate system, elementary excitations have the kinetic energy $\varepsilon_k = \hbar^2 k^2/2m$. What matters is the energy in the coordinate system of the surfaces which ensure thermalization, $\widetilde{\varepsilon}_k = \varepsilon_k - kv$ (Galilean invariance). To get a stable flow, this energy must be positive ("Landau's criterium"), otherwise a spontaneous excitation creates a dissipation. It is clearly shown on figure 1a that this is not the case for an ideal gas.

2 The essential role of a repulsive interaction

At the lowest order, a repulsion U creates a Hartree energy $E_H = \frac{1}{2}UN^2$, thus a chemical potential at null temperature $\mu = dE_H/dN = UN$, which means, in the end, a finite compressibility $d\mu/dN$. Compressibility evokes the speed of sound c,

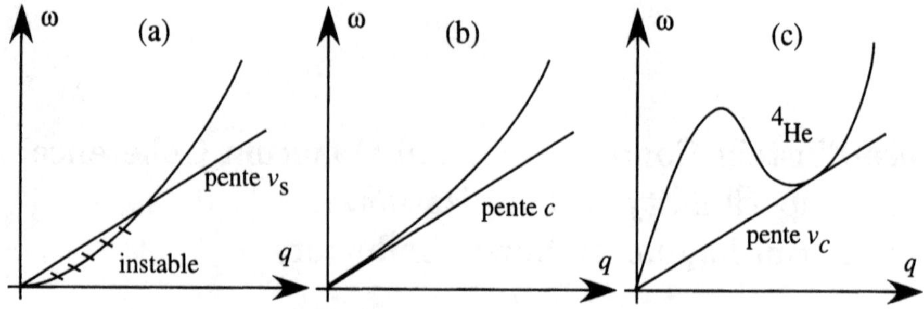

Figure 1: *Excitation spectrum: (a) ideal gas, (b) gas of bosons in weak repulsive interaction, (c) liquid* ^{4}*He.*

thus an excitation spectrum $\varepsilon_k = ck$ linear at high wavelengths. At weak coupling, ε_k interpolates between phonons and free particles, the crossover being at:

$$\frac{\hbar^2 k^2}{2m} \approx NU \approx ck .$$

Landau's criterium then predicts a critical speed for the "superfluid" flow $v_c = c$ (Fig. 1b). The actual reality for ^{4}He is different, because it corresponds to a very strong coupling (Fig. 1c). The main point is the essential role of a repulsive interaction.

We can come to the same conclusion by a different path, interesting because it raises some fundamental questions. Instead of taking a condensate strictly in the lowest state, can we *fragment* it between two states 1 and 2, of arbitrarily close energies in the thermodynamical limit, with populations N_1 and $N_2 = N - N_1$? Only the potential energy can differentiate these two choices! Let's write the interaction with the help of creation and annihilation operators

$$\frac{1}{2} \sum_{qkk'} U_q b_k^* b_{k'}^* b_{k'-q} b_{k+q} .$$

In the non fragmented state, all k's are the same and the energy has a single term $U_o N (N-1)/2$. In the fragmented state, however, the interaction between the components N_1 and N_2 contains two terms, Hartree's energy where operators 1,4 contract on one side, and 2,3 on the other, and *Fock's exchange energy* where operators 1,3 and 2,4 contract. Whereas for fermions the exchange energy is subtracted from Hartree's energy (exchange hole due to the exclusion principle), it is added up for bosons. U_o plays a role in Hartree's term and exactly reproduces the non fragmented condensate's energy. Fragmentation eventually costs Fock's energy $U N_1 N_2$. It's an *extensive* energy, significant in the thermodynamical limit. It's the Coulomb repulsion which prevents fragmentation, thus ensuring the *quantum coherence of the condensate*, which wave function becomes a macroscopic observable quantity.

What happens if the interaction is attractive and not repulsive? This question is only a rethorical one for a uniform system, since an attractive bosons gas spontaneously collapses. Hartree's energy $-\frac{1}{2}|U|N^2$ gives an upper variational bound for the fundamental energy, which thus goes to $-\infty$ for high N. The situation is different for a dilute gas trapped in a harmonic pit, where the energy contains 3 terms, localization, trap measured by the radius of the fundamental harmonic a_o and interaction measured by the diffusion length a:

$$\frac{E}{N} = \frac{\hbar^2}{mR^2} + \frac{m\omega_\mathrm{o}^2 R^2}{2} - \frac{|U|}{R^3}$$

(R is the radius of the "drop" of N bosons). For low enough values of Na/a_o there is a metastability area. What then becomes the coherent condensate? The problem isn't defined for a system at rest, since the fundamental harmonic is not degenerated. However it exists for a system in rotation in an isotropic pit. An approached computation by Gunn *et al.* suggests a spontaneous fragmentation. Although a bit esoteric, this problem is worth a more serious study, because it raises a fundamental question: *is quantum coherence a fatality?*

3 Quantum coherence, phase locking, symmetry breaking

The absolute phase of a wave function is arbitrary, but its variation from one point to another is perfectly defined. As soon as a condensate is in a pure state, this phase gains some stiffness. This happens for example for a system separated into two compartments by a tunnel barrier ("Josephson" junction). This is also the case for an open system traversed by a current: it's the *phase gradient* which is responsible for the current. The energy $\frac{1}{2}\Lambda\left(\nabla S\right)^2$ defines a phase stiffness Λ: the current density is $J = \Lambda\nabla S$. An immediate consequence is the quantification of the flow in a closed circuit: the phase must satisfy the condition $\oint \nabla S \cdot dl = 2\pi p$, where p is an integer. A current thus becomes a topological object, insensitive to the impurities or other defects – thus the superfluidity. At the local level, this quantification acts upon the vorticity: the vortex have discrete energies (which in fact control the critical currents, much lower than the values predicted by Landau's criterium).

It can be easily shown that the wave function's phase is a canonical conjugate quantity of the number of particles. It's a familiar notion in optics: to lock a laser's phase, one must let fluctuate the number of photons. This conjugation implies an uncertainty relation $\Delta N \cdot \Delta S \approx 1$. A pure coherent state locks the phase, and thus must allow high variations ΔN, typically $\approx \sqrt{N}$. Bose-Einstein's traditional condensation is thus a *necessary* condition for superfluidity, but not *sufficient*: the attractive gas example constitutes a proof for that. We note that the phase stiffness corresponds to a breaking of the symmetry, like the existence of a spontaneous magnetization in a ferromagnetic. In the absence of crystalline anisotropy, the magnetization's orientation is arbitrary, but it is the same everywhere. This is the same thing here.

The quantum coherence notion is considerably improved when the bosons are granted a internal degree of freedom α. The condensate may then correspond to an internal pure quantum state, or to a statistical incoherent mixture. The population N_o is replaced by a density matrix

$$\rho_{\alpha\beta} = \langle b_{o\alpha}^* b_{o\beta} \rangle \ .$$

For a pure state, ρ has a single non zero eigenvalue N_o. The consequences of internal coherence are spectacular, since the internal state becomes a macroscopic observable quantity. The standard example is superfluid ^{3}He, where pairs of atoms condensate in the same state $L = 1$, $S = 1$. Because of the coherence, a weak perturbation like the dipolar magnetic coupling between nuclear spins takes place at first order and becomes a determining factor of the magnetic resonance: the resulting behavior is extremely rich. Likewise, the atomic hydrogen possesses a nuclear spin $1/2$: a coherent condensate implies a nuclear ferromagnetism. Another example is the case of the excitons, free states of an electron-hole pair in a semiconductor. If the valence range has symmetry p and the conduction range has symmetry s, the exciton possesses an electric dipolar momentum: the internal coherence aligns these momenta, resulting in superradiance. The internal coherence of a Bose condensate is thus a crucial factor. But it's not clear that this is the rule! In principle, it's the exchange repulsion which goes against fragmentation – but for a scalar interaction, there is no exchange between particles of different spins. Only an interaction using the internal degrees of freedom may play a role. Until now, there is no general argument excluding the possibility of a statistical mix. An example can be found in a forgotten article by Gor'kov and Galitski (1962), who built an isotropic state of a superfluid ^{3}He on these grounds. The experience has invalidated this model – but it's not inconceivable. So this question about the internal coherence remains open: it is an important one at the conceptual level.

4   Dilute gas: correlation length, quantum fluctuations

Interaction in the s channel at low energy can be characterized by a diffusion length a, positive for a repulsion $U \, \delta(r)$, given by:

$$U = \frac{4\pi \hbar^2 a}{m} \ .$$

The strength of the coupling is measured by the dimensionless parameter Na^3. The correlation length ξ is the scale for which the kinetic energy for localization and the interaction energy are comparable:

$$\frac{\hbar^2}{m\xi^2} \approx NU \ .$$

It corresponds to the change from a "phonons" system (small q) to free particles (big q). $\xi \approx \hbar/\sqrt{mNU}$ is thus *big* at low couplings: a mean field approximation is appropriate. We note that a/ξ is $\approx \sqrt{Na^3}$: the true parameter of expansion is the square root of Na^3.

Even at zero temperature, quantum fluctuations are sufficient to potentially excite the particles out of the condensate. At first order, the interaction can take two condensed atoms and create a pair $(q, -q)$: this is called "depletion" of the condensate. Microscopic theory for a dilute system was established by Bogoliubov more than 50 years ago, and can be expressed very easily in a variational scheme. A condensate without depletion is described by the state vector:

$$|\Psi\rangle = \exp\left(\phi b_o^*\right) |\text{vac}\rangle \ ,$$

which incorporates phase locking (via ϕ 's argument) and the related N_o fluctuations. To take into account quantum fluctuations at the lowest order, we can simply write:

$$|\Psi\rangle = \exp\left(\phi b_o^* + \sum_q \lambda_q b_q^* b_{-q}^*\right) |\text{vac}\rangle \ .$$

We can get λ_q by minimizing energy. Elementary Bogoliubov's excitations are then obtained:

$$\varepsilon_q = \hbar q \sqrt{\frac{NU}{m} + \frac{\hbar^2 q^2}{4m^2}} \ ,$$

as is the fundamental energy and the depletion

$$E_o = \frac{2\pi\hbar^2 a N^2}{m} \left[1 + \frac{128}{15\sqrt{\pi}}\sqrt{Na^3}\right] \ ,$$

$$N_o = N\left[1 - \frac{8}{3\sqrt{\pi}}\sqrt{Na^3}\right] \ .$$

Like foreseen, the expansion is in powers of $\sqrt{Na^3}$. For the short story, this result can be compared with a famous calculus by Onsager and Penrose in 1956. Superfluid order is expressed by a non-zero mean $\langle \psi^*(r) \rangle$ of the creation's operator (it's the transition order parameter). N_o is then the limit value of the density matrix $\rho(r, r')$ with big separations $|r - r'|$. It's the first appearance of the central "non diagonal order with high range" concept. This density matrix can be derived from the N bodies wave function $\Psi(r_1, \ldots, r_N)$:

$$\rho(r, r') = \iint dr_2 \ \ldots \ dr_N \ \Psi^*(r, r_2, \ldots, r_N) \ \Psi(r', r_2, \ldots, r_N) \ .$$

If we choose $\Psi = 0$ when two particles overlap and 1 elsewhere, this integral is the same as the partition function for a classic gas, known by the viriel's development. We therefore get

$$\frac{N_o}{N} = 1 - \frac{4\pi}{3}Na^3 + \cdots$$

which disagrees with Bogoliubov's computation: where is the mistake? In the wave function's choice: a discontinuity for $|r - r'| = 2a$ costs an infinite kinetic energy! In fact, the wave function goes to zero at contact, going through a maximum at mid-distance. The depletion is consequently reinforced – thus Bogoliubov's result!

5　Bose-Einstein's condensation at strong coupling

We already observed that the measured excitation's spectrum in ^{4}He (by inelastic diffusion of neutrons), has nothing to do with Bogoliubov's result. The speed of sound is high, ε_q goes through a maximum with a noticeable rotons dip. The correlation's length ξ has an atomic scale, similar to the diffusion's length a: coupling is very strong. The nature of rotons remains mysterious. In Landau's and Feynman's minds, they express an underlying liquid's vorticity. Feynman has the nice sentence "a roton is the ghost of a vanishing vortex ring": attempts to describe the collapse of a small vortex ring whose radius is of magnitude ξ remain unsuccessful for now. I don't even know if the symmetry allows disintegration into one or several rotons. I personally believe that roton is better viewed as a memory of a crystalline order. If this lattice was frozen, we would get a zero energy phonon with Bragg spots. One can imagine that coherent quantum fluctuations restore the translation invariance, blurring the reciprocal lattice. The ε_q energy doesn't go to zero for $q = G$, but it can reveal a pronounced minimum – the roton! Again, this question is a little bit semantic, but it deserves further thought.

At strong coupling, depletion of the condensate N_o/N has to become important. In principle, it can be measured by a Compton diffusion of high energy neutrons (impulse approximation), but interactions in the final state widen the spectra and a precise measure is hard. Roughly, $N_o/N \approx 0,08$ is obtained at vapor's pressure: depletion is huge. Beware that this doesn't invalidate the two fluids model, and the superfluid density at zero temperature remains N! Quantum fluctuations are big, but they are locked to the condensate. If it has a wave vector q, the whole distribution n_k is moved by q. Superfluid current thus remains $J = N\hbar q/m$.

N_o/N goes down with pressure. At thermodynamical equilibrium, solidification's pressure of 25 bars can't be exceeded. But Sébastien Balibar's acoustical techniques allow to reach much higher pressions in transitional systems – thus raising a new question: can depletion reach 100%? Or said in an other way, is there a Bose *normal* liquid, without symmetry breaking? If there is such a state, we don't have any microscopic description for it. But we can understand its significance indirectly, through a very old Matsubara's argument (which goes back to the 40's). This argument concerns a lattice bosons gas, with a hopping amplitude t from a site to its neighbour which provides the kinetic energy. We assume that these bosons have a *hard core*: each site has only two states, $n_i = 0$ or 1. Bose-Einstein statistics imply commutation relations between two sites: this is the exact algebra for a set of spins 1/2. Particles density sets the magnetization M_z. The operators

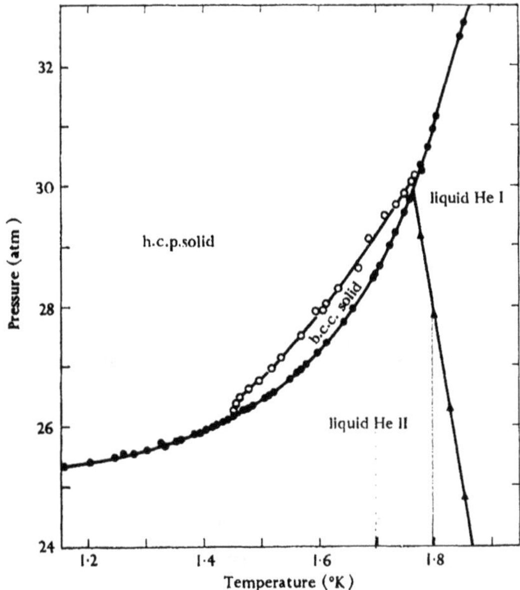

Figure 2: ^4He phase diagram.

b_i and b_i^* are transition operators for the spins S_i^- and S_i^+, kinetic energy $b_i^* b_j$ is a transverse ferromagnetic coupling between two adjacent spins. Let's assume a half-filled lattice ($M_z = 0$). The superfluid state is isomorphic to the transverse ferromagnetism, the order parameter's phase being the magnetization's azimuth. A normal bosons gas should correspond to an isotropic magnetic fundamental, left invariant by a rotation of the spins, a singlet. Such a state can be easily obtained by dimerizing two adjacent spins in a state $S_1 + S_2 = 0$ – but then the translation symmetry is broken, like it would be for a crystalline lattice corresponding to an antiferromagnetic order of the spins ($n_i = 1$ on a sub-lattice, 0 on the other). A state left invariant by translation is what Anderson called the "RVB" state, famous nowadays. Can it be obtained without infinite range correlations? I don't think so! The existence of a normal Bose liquid is thus not an innocent question. Every experimental information about helium under high pressures would be valuable.

6 Bose-Einstein's condensation or localization?

The problem is obvious for ^{4}He: if pressure increases, the superfluid liquid becomes a crystal at 25 bars (Fig. 2). This solid is very plastic, but it has all the attributes of an elastic medium, which resists shear. To set the transition, it is sufficient to notice that superfluidity optimizes kinetic energy at the expense of repulsion energy (non localization implies some proximity), whereas crystallization does the

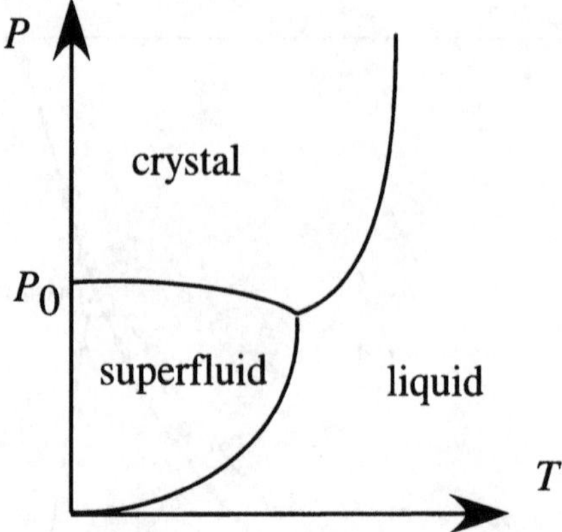

Figure 3: *Phase diagram obtained in Matsubara's model.*

opposite (atoms avoid each other, but we pay for that in terms of localization kinetic energy). The transition must correspond to similar energies:

$$\frac{\hbar^2}{md^2} \approx NU \ ,$$

where d is the average distance between atoms. So the solid must appear if $d \lesssim \xi$, ξ being the correlation's length, thus at strong coupling. But coupling must be very strong since the liquid already corresponds to an important depletion. This competition between superfluidity and localization can also be observed in the hardcore boson model, on a previously considered lattice. The correct interaction energy is then the repulsion between two atoms on adjacent sites: in the isomorphic spins model, it corresponds to an Ising antiferromagnetic coupling:

$$V \left(S_{iz} + \frac{1}{2} \right) \left(S_{jz} + \frac{1}{2} \right) \ .$$

For a half-filled band, V favors a crystalline layout with two sub-lattices, whereas the tunnel hopping t favors phase locking and superfluidity. This is how Matsubara explained, qualitatively, the phase diagram, by a simple mean field approximation. His result is displayed on Fig. 3.

From a theoretical point of view, the transition can be modeled by a Hubbard model for bosons, instead of the usual fermions. N bosons move on a N_L sites lattice, with a hopping amplitude t. They are subjected to a local interaction U. The Hamiltonian is thus:

$$H = -t \sum_{i\delta} b_i^* b_{i+\delta} + \frac{U}{2} \sum_i n_i \left(n_i - 1 \right) \ .$$

If N is an integer, multiple of N_L ($N = pN_L$), the system will clearly be insulating for big U: each site has occupancy p and the "ionization" $(p,p) \rightarrow (p+1, p-1)$

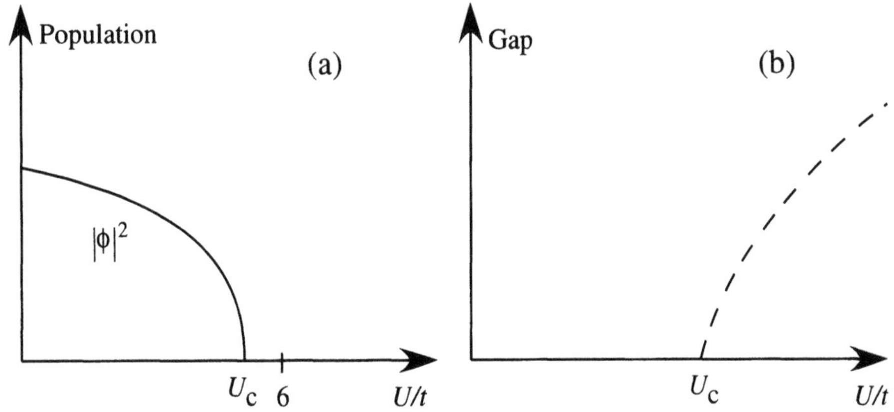

Figure 4: *Hubbard model results for bosons. (a) Condensate's population. (b) Excitations gap.*

costs energy U. If U is small, however, we again obtain superfluidity. It's the boson version of the insulator-metal transition (at partial filling we always have a conductor).

We can attack this problem under two different angles (i) by an adaptation of Gutzwiller's approximation [1] (ii) by an unusual mean field approximation [2]. In the latter case, we write:

$$b_i = \phi + a_i \; ,$$

where $\phi = \overline{b_i}$ is the superfluid order parameter, and we linearize in fluctuation a_i. The Hamiltonian becomes local:

$$H = \frac{U}{2} n_i \left(n_i - 1 \right) - \mu n_i - zt\phi \left(a_i + a_i^* \right) \; ,$$

where μ is the chemical potential and z the coordination. It is enough to minimize with respect to ϕ. We do that numerically by truncating the n_i series to a maximal value $n_{\max}$ that is adjusted to obtain a stable result. Fig. 4a gives the condensate's population $\rho = |\phi|^2$ as a function of U: superfluidity disappears at U_c. Fig. 4b shows the charged particles excitations gap appearing in the insulating stage.

The previous discussion concerns Mott's localization through correlations: an Anderson localization by disorder is also possible, which leads to a glassy state without spatial order, but where atomic motion is frozen (complete filling is not necessary anymore). The theory of this "Bose glass" is much more complex. It has been studied from a renormalization point of view by Fisher *et al.* Competition between Mott and Anderson localizations is discussed by Trivedi *et al.* But it remains very formal, and to this point, nobody is able to understand how superfluidity disappears in a porous medium whose scale is similar to the correlation's length ξ.

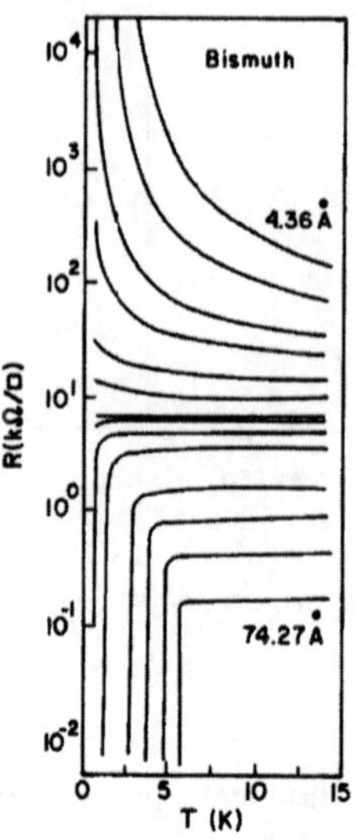

Figure 5: *Resistance of a granular Bismuth film as a function of temperature, for different grain radius R (figure taken from [3]).*

At the experimental level, the most striking example is the direct supra-conductor-insulator transition in granular matter (or in a Josephson junctions network). Below the critical volume's temperature, each grain may be considered as a Bose-Einstein condensate of electrons pairs, containing N pairs. Josephson coupling tends towards locking the order parameter's phase of two adjacent grains. But the electrostatic energy tends to lock the charge fluctuations Q. This capacitive energy is of magnitude $Q^2/2R$ where R is the grain's radius: it gets bigger as the grain gets smaller. If R is big, Josephson wins and we get a supraconductor, whereas if R is small electrostatics win and we get an insulator. This behavior is perfectly illustrated in Fig. 5 which gives the resistance of a granular bismuth film as a function of temperature T for several values of R: the sudden change from a superfluid to an insulator is striking. Bose condensates of trapped atoms have just given birth to an even better example. A trap network is built with a stationary wave of two opposite laser beams. The laser's intensity controls the barriers' height, which means the tunnel amplitude from one pit to another. In the superfluid state, phases are coherent from one pit to another: Bragg spots can be observed when atoms are freed by suppressing the traps. These two experiments give a concrete example of the Hubbard bosonic model.

7 An open problem: condensation kinetics

Let's consider a boson gas at a temperature $T_o > T_c$. We suddenly put the thermostat at a temperature $T < T_c$: how the condensate is going to build up? Here again the answer is quite different for an ideal gas and for a repulsive gas. The thermostat can be modeled by a transition probability $W_{kk'}$ which obeys the detailed balance equation:

$$\frac{W_{kk'}}{W_{k'k}} = \exp\left[\frac{\varepsilon_k - \varepsilon_{k'}}{T}\right] .$$

Distribution n_k is then governed by a master equation of type:

$$\frac{dn_k}{dt} = \sum_{k'} \left[W_{k'k}\, n_{k'} - W_{kk'}\, n_k\right] .$$

The time scale is a collision time $\tau \approx 1/W$.

For an ideal gas, it can be easily shown that a few collisions are enough to prefigure the condensate: N_o particles gather in a tight energy range $\delta(t)$. To obtain a unique state, δ must thus decrease! But then the collisions become totally inefficient because Boltzmann distribution's asymmetry becomes weaker and weaker. It can be shown that $\delta(t)$ decreases as $1/t$: for a macroscopic system, it takes an infinite amount of time to obtain a coherent condensate.

The situation is completely different with an exchange repulsion. As soon as there is a finite N_o population, adding an atom to the previously mentionned condensate changes the exchange energy by a *finite* quantity, destroying the balance between collisions "in" and "out". Relaxation towards equilibrium state will happen in a few collisions. The situation is very similar to the case of a ferromagnetic material, where magnetization's relaxation is the result of collisions reversing the spin, $k_\downarrow \implies k'_\uparrow$. In the unstable paramagnetic state, such a transition is impossible at $T = 0$ (the Fermi levels of the two spins would coincide). However if a small magnetization has been nucleated, the exchange energy separates $\mu_\uparrow$ and $\mu_\downarrow$, a collision window opens and relaxation is then very fast, in a few collisions. In both cases, the only problem is the initial nucleation: we are facing again the classical problem of spin decomposition, common to every phase transition. This nucleation necessarily comes from a localized seed, which implies a surface energy, thus a finite correlation's length. Again interactions play an essential role.

In practice, this condensate's formation seems very fast and the question has hardly any concrete incidence. Nonetheless it would be interesting to understand this problem and to observe these kinetics experimentally.

8 Two binded fermions = one boson?

Existence of supraconductivity and of superfluid ^{3}He definitely shows that the answer is "yes" – to a certain extent! Let's consider two fermions who attract sufficiently each other to form a bound state, for example a spin singlet. Let ψ_k be the internal wave function, of extension a in the real space, $1/a$ in the reciprocal space. The binding energy is $\varepsilon_o \approx \hbar^2/ma^2$. If the center of mass has momentum $q = 0$, this bound pair's creation operator can be written:

$$b_o^* = \sum_k \psi_k \, c_{k\uparrow}^* c_{-k\downarrow}^*$$

(c^* is the creation operator for a fermion). If the distance d between pairs is much larger than a, each pair ignores its neighboring one and can be treated like a specific boson (statistics don't play any role as soon as wave functions don't recover each other). At the lowest order (Hartree and not Bogoliubov!), the fundamental thus has state vector:

$$|\Psi\rangle = \exp\left[\phi b_o^*\right] |\text{vac}\rangle \ .$$

$(k, -k)$ pairs commute, so the exponential can be factorized. Each factor can be expanded in series: thanks to the exclusion principle, only the first two terms remain. After normalization, $|\Psi\rangle$ is of the form:

$$|\Psi\rangle = \prod_k \left[u_k + v_k c_{k\uparrow}^* c_{-k\downarrow}^*\right] |\text{vac}\rangle$$

with $v/u = \phi\psi_k$. We obtain the BCS wave function! The population of the underlying fermion state is v_k^2: the exclusion principle ensures that it is ≤ 1.

In the dilute limit $\phi \approx \sqrt{N}$, $\psi_k \approx a^{3/2}$: v_k^2 is of magnitude $Na^3 \approx a^3/d^3 \ll 1$: thus there is no problem with the exclusion principle. v_k reflects the atomic wave function, the ε_o gap needed to break a pair has nothing to do with Bose-Einstein condensation. If, however, Na^3 goes beyond 1, exchange's repulsion between fermions becomes decisive. The population v_k^2 is distorted as shown in Fig. 6. It can keep a finite gap Δ, without any discontinuity at the Fermi level: it's the BCS supraconducting state, where Δ is a consequence of the superfluid symmetry breaking, as opposed to the dilute case. It can also evolve towards a discontinuous step function: it's the standard Fermi liquid, without any symmetry breaking. For the usual electron gas, supraconductivity exists for an arbitrarily weak attractive interaction (Δ is then exponentially small). But there are more complex situations, for example some excitons where an electron and a hole are paired. In general, the Fermi surfaces of the two bands are anisotropic and differ one from the other: a minimal interaction is thus needed to build a superfluid state (otherwise we get a simple regular plasma of free carriers).

So, the conclusion is that there is a continuous interpolation between Bose-Einstein condensation for preformed pairs (strong attraction) and BCS supraconductivity (low attraction). But the physics is qualitatively different for the two limits. It is especially clear if we look at the critical temperature T_c where the symmetry breaking vanishes. In a BCS state, T_c is controlled by *pair breaking*, of pairs which only exist owing to superfluid order: T_c is thus $\approx \Delta$. In the atomic

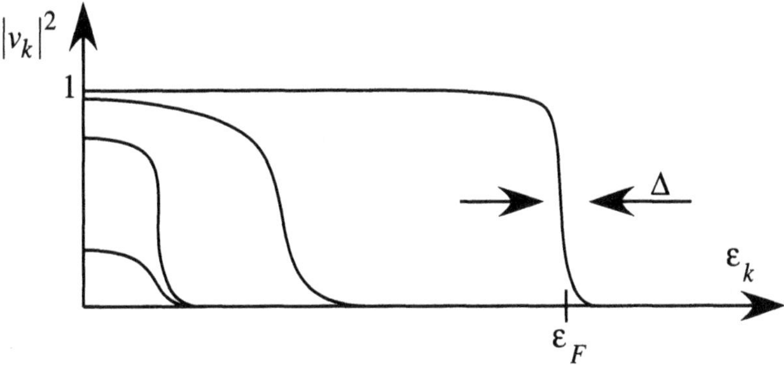

Figure 6: *Population $|v_k|^2$ as a function of energy ϵ_k.*

regime, the binding is strong and pairs don't break: T_c is due to the center of mass motion, which empties the condensate in favor of a normal fluid. If we notice that this normal fluid corresponds to a thermic fluctuation of the order parameter's phase, a new mechanism for the superfluidity destruction appears, entirely free of any mean field description. A precise description of the transition region remains problematic. It can be shown in two dimensions that the two mechanisms give similar T_c's when $\Delta \approx E_F/5$: the notion of free state then no longer makes any sense and current theories are helpless.

A last rhetorical precaution: the previous discussion only concerned exchange repulsion due to the exclusion principle. It ignores another phenomenon equally important, *the screening effect,* which reduces the attraction force. The Debye-Huckel type screening effect is well understood in the dense limit. When we consider the dilute state, it should evolve to a van der Waals attraction. Despite 50 years of efforts, there is no reliable theory for the intermediate region, dealing in a self consistent way with pairing and screening effect. It's one of the great challenges of strong correlations.

9 Conclusion

This short overview shows that Bose-Einstein's condensation is well alive, and has considerably evolved since Bose's initial discovery. We can marvel at Fritz London's exceptional insight, who understood as soon as 1937 the essential quantum nature of the phenomenon, despite some virulent opposition by great scientists such as Landau. This internal quantum coherence makes all the richness of this field. It was barely visible in ^{4}He, but it becomes omnipresent in the ultracold trapped gas. There are many outstanding questions remaining to be clarified: will experiments succeed to answer them?

References

[1] W. Krauth *et al.*, *Phys.Rev. B* **45**, 3137 (1992).

[2] Sheshari *et al.*, *Eur. Phys. Lett.* **22**, 257 (1993).

[3] Haviland *et al.*, *Phys. Rev. Lett.* **62**, 2180 (1989).

Philippe Nozières
CNRS-LEPES
Grenoble
email: nozieres@ill.fr

Part II
Entropy

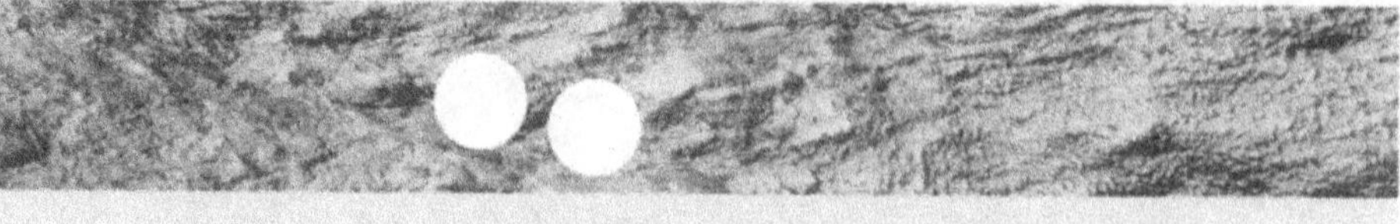

SÉMINAIRE POINCARÉ
Samedi 6 décembre 2003

L'Entropie

Institut Henri Poincaré • Amphi Hermite
11, rue Pierre et Marie Curie • 75005 Paris

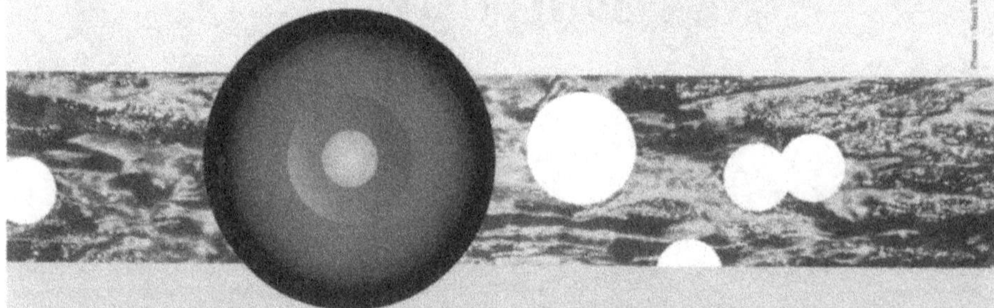

O. Darrigol : La naissance du concept d'entropie • 10h00
R. Balian : L'entropie protéiforme • 11h00
C. Maes : Turning around Entropy • 12h00
F. Ritort : Experiments on Transient Violations
of the Second Law • 15h00
T. Damour : L'entropie des trous noirs • 16h00

www.lpthe.jussieu.fr/poincare • Avec le soutien du CEA (DSM) et du CNRS (SPM)

Poincaré Seminar 2003, 101 – 117
© Birkhäuser Verlag, Basel, 2004

The Origins of the Entropy Concept

Olivier Darrigol

To this day entropy remains a strange, difficult, and multiform concept. Even the great Henri Poincaré renounced precisely defining energy and entropy. In order to justify the success of the two laws of thermodynamics for his students at the Sorbonne, he turned to history:

> Pour expliquer par quelles raisons tous les physiciens ont été amenés à adopter ces deux principes [la conservation de l'énergie et la croissance de l'entropie], je n'ai rien trouvé de mieux que de suivre dans mon exposition la marche historique. Le spectacle des longs tâtonnements par lesquels l'homme arrive à la vérité est d'ailleurs très instructif par lui-même. On remarquera le rôle important joué par diverses idées théoriques ou même métaphysiques, aujourd'hui abandonnées ou regardées comme douteuses. Service singulier que nous a ainsi rendu ce qui est peut-être l'erreur! Les deux principes, appuyés sur de solides expériences, ont survécu à ces fragiles hypothèses, sans lesquelles ils n'auraient peut-être pas encore été découverts. C'est ainsi que l'on débarrasse la voûte de ses cintres quand elle est complètement bâtie.

Following Poincaré's advice, I will show you the scaffolding and seek in the past some clues on the necessity of the entropy concept and on the tensions between its various meanings.[1]

As is well known, there are at least two basic meanings of entropy, one belonging to macroscopic thermodynamics, the other to statistical mechanics. The first derived from studies of the performance of the steam engine, the second from the Maxwell-Boltzmann kinetic theory of gases. I will follow this natural order, beginning with Sadi Carnot's innovative approach to the theory of the steam engine.

1 Classical entropy

Carnot's theorem In his *Traité sur la puissance motrice du feu* of 1824, Sadi Carnot relied on an analogy with hydraulic machines, imitating the forms of reasoning that his father Lazare had introduced in the latter domain. His first fundamental remark was that the production of work by thermal means depended on a temperature

---

[1] Henri Poincaré, Thermodynamique, Sorbonne lectures of 1888–89 ed. by J. Blondin (Paris, 1892), XIII–XIV.

difference that permitted "the fall of caloric." The analogy with hydraulic engines then suggested that in an optimal heat engine there should be no temperature change without a corresponding change of volume of the working substance. Carnot described the simplest cyclic process that meets this criterion, which we now call a Carnot cycle.[2]

Then came Carnot's important theorem: a reversible engine has the highest possible efficiency among all engines that work between two given temperature sources. The proof is based on two axioms: the conservation of the caloric, and the impossibility of a certain form of perpetual motion. Suppose, Carnot reasoned *ab absurdo*, that there exists an engine that is more efficient than the reversible engine. Then the work produced by the hypothetical engine would be superior to the work needed to return the "fallen" caloric to its original level by operating the reversible engine backwards. The simultaneous operation of the two engines would then permit the indefinite production of work without any compensation. In order to avoid this consequence, Carnot's theorem must hold.

As a corollary to this theorem, all reversible bithermal engines must have the same efficiency: the ratio of the produced work to the transferred heat is a function solely of the temperatures of the source and sink. Carnot used this wonderful property to derive what we now call Carnot-Clapeyron relations,[3] that is, relations between the constitutive properties of fluids and the universal efficiency function. Although most of Carnot's results did not survive the later rejection of caloric, his theorem and the style of its derivation and application had a brilliant future. Ideal machines and processes, reversible cycles, *ab absurdo* reasoning, and organizing principles are the gist of modern thermodynamics.

In 1845, William Thomson became aware of Carnot's long-neglected treatise through Emile Clapeyron's derivative memoir of 1834. The young natural philosopher quickly understood the predictive power of Carnot's theory and developed it in a series of publications. At the same time, he admired James Joule's contemporary experiments, which seemed to contradict the conservation of caloric on which Carnot's theorem was based. To face this dilemma, he tentatively assumed that heat could be created but never annihilated by mechanical means. This compromise saved the demonstration of Carnot's theorem, but led to a further paradox:[4]

> When "thermal agency" [a temperature difference]... is spent in conducting heat through a solid, what becomes of the mechanical effect which it might produce? Nothing can be lost in the operations of na-

---

[2] Sadi Carnot, *Réflexions sur la puissance motrice du feu* (Paris, 1824). Cf. Truesdell 1980.

[3] The prototype is the relation $\frac{\partial P}{\partial \theta}|_V = \mu l$ where P is the pressure of the fluid, θ its temperature, V its volume, and l its latent heat of expansion, and $\mu(\theta)d\theta$ is the universal efficiency ratio for a temperature fall $d\theta$.

[4] William Thomson, "An account of Carnot's theory of the motive power of heat, with numerical results deduced from Regnault's experiments on steam," Royal Society of Edinburgh, *Transactions*, 16 (1849), 541–574, also in *Mathematical and physical papers*, 6 vols. (Cambridge, 1882–1911), on 118n–119n.

ture, no energy can be destroyed... It might appear that the difficulty
would be entirely avoided by abandoning Carnot's fundamental axiom
[the conservation of heat during its "fall"]; a view which is strongly
urged by Mr. Joule... If we do so, however, we meet with innumerable
other difficulties, insuperable without farther experimental investigation, and an entire reconstruction of the theory of heat from its foundation. It is in reality to experiment that we must look; either for a verification of Carnot's axiom, and an explanation of the difficulty we have
been considering; or for an entirely new basis of the Theory of Heat.

Clausius's new thermodynamics

With this empiricist attitude Thomson vainly scrutinized contemporary steam
measurements. The true key to the paradox was a modification of Carnot's theoretical reasoning, as Rudolf Clausius explained in his fundamental memoir of 1850
on the motive force of heat. Then a simple *Privatdozent* at the University of Berlin,
Clausius remarked that one could simultaneously assume, in a Carnot engine, the
transfer of heat from the hot source to the cold source and the transformation of
part of this heat into work. Carnot's theorem could then be maintained without
contradicting Joule's statement of the equivalence between heat and work. But the
theorem could no longer be based on the impossibility of perpetual motion. Imitating Carnot's *ab absurdo* reasoning, Clausius supposed the existence of a bithermal
engine with a higher efficiency than a reversible engine. The work produced by
the hypothetical engine could be used to run a reversed Carnot engine between
the same sources. The net result would be a transfer of heat from a cold source
to a hot source without any compensation. The impossibility of such a transfer
implies Carnot's theorem.[5] As this impossibility agreed with "the known behavior
of heat," Clausius made Carnot's theorem his second principle. His first principle
was a statement of Joule's equivalence between heat and work: "In all cases when
work is produced by heat, a quantity of heat is consumed that is proportional to
this work, and reciprocally the same quantity of heat can be produced by con-
suming an equal amount of work."[6] As we may retrospectively judge, Clausius's
memoir provided a complete and secure foundation for thermodynamics. Clausius nonetheless sought a deeper understanding of the principles. In particular, he
tried to compensate for the loss of intuitive understanding implied by the demise
of the caloric. In Carnot's original reasoning the work produced by a steam engine corresponded to the "fall of the caloric," that is, to its transfer from a state
of higher to lower potential energy. What happens to this idea when heat is no
longer conserved?

---

[5] Rudolph Clausius, "Über die Bewegende Kraft der Wärme und die Gesetze, welche sich
daraus für die Wärmelehre selbst ableiten lassen," *Annalen der Physik und der Chemie*, 79
(1850), 368–397, 500–524. Cf. Daub 1971.

[6] Clausius, ref. 5, also in *The mechanical theory of heat, with its applications to the steam-
engine and to the physical properties of bodies* (London, 1867), on 18.

Clausius's transformation value

Clausius answer to this question appeared in a memoir of 1854 entitled "On a modified form of the second principle of the mechanical theory of heat." This is one of the strangest memoirs in the entire history of physics, owing to the odd mixture of the new thermodynamics with older ideas reminiscent of Carnot's theory. In 1850, Clausius had maintained a connection with Carnot's theory by considering that in a cycle of a bithermal machine two simultaneous transformations occurred: a transfer of heat (without loss) from the hot source to the cold source, and a conversion of part of the heat released by the hot source into work. In analogy with Carnot's relation between the fall of heat and the production of work, Clausius declared that for a reversible machine the first transformation was "equivalent" to the second. He further assumed that the equivalence rested on the equality of the "equivalence values" (*Equivalenzwerthe*) of the two transformations, just as in a hydraulic machine the decrease in the potential energy of the water must be equal to the work produced by the machine.[7]

In general for any number of thermal sources and for a given amount of mechanical work produced, Clausius regarded the evolution of the sources as resulting from a combination of two kinds of transformations: heat transfer from one source to another, conversion of heat from one source into work (of course, this decomposition is not unique when there is more than one source). He assumed that the equivalence-value of the product of two transformations was equal to the product of their equivalence-values, and that the equivalence-value of any transformation that could be realized through a reversible cycle of a thermodynamic machine was equal to zero. Noting $Q_{\theta_2}^{\theta_1}$ a transformation of the first kind, and Q_θ a transformation of the second kind, these axioms imply

$$w(Q_\theta) = Q f(\theta), \qquad w(Q_{\theta_2}^{\theta_1}) = Q\left[f(\theta_1) - f(\theta_2)\right] \tag{1}$$

for the equivalence-values w, wherein f is a universal function of the temperature.

Clausius next considered a transformation operated through a bithermal machine taking the heat Q_1 from the source at temperature θ_1 and yielding the heat $-Q_2$ to the source at temperature θ_2. This transformation may be regarded as the combination of the conversion of the heat $Q_1 + Q_2$ from the hot source into work, and the transfer of the heat $-Q_2$ from the hot source to the cold source. If the bithermal machine has performed a reversible cycle, the two axioms lead to

$$(Q_1 + Q_2)f(\theta_1) + (-Q_2)\left[f(\theta_1) - f(\theta_2)\right] = 0 , \tag{2}$$

or

$$Q_1 f(\theta_1) + Q_2 f(\theta_2) = 0 . \tag{3}$$

---

[7]Clausius, "Über eine veränderte Form des zweiten Hauptsatzes der mechanischen Wärmetheorie," *Annalen der Physik und der Chemie*, 93 (1854), 481–506.

As Clausius knew, Thomson had defined the absolute temperature T so that in a Carnot cycle the relation

$$\frac{Q_1}{T_1} + \frac{Q_2}{T_2} = 0 \tag{4}$$

holds. With this definition, $f(\theta)$ must be proportional to $1/T$. Clausius generalized these considerations to an arbitrary number of sources to get the expression

$$w = \oint \frac{\delta Q}{T} \tag{5}$$

of the value of a combination of infinitesimal transformations in which the net heat δQ is released by the source at temperature T. For a transformation that can be realized through a reversible cycle of a thermodynamic machine, the condition

$$\oint \frac{\delta Q}{T} = 0 \tag{6}$$

must hold. Clausius further deduced the inequality

$$\oint \frac{\delta Q}{T} < 0 \tag{7}$$

for a transformation that can be realized through an irreversible cycle of a thermodynamic machine. If the opposite inequality held, the integral could be divided into a vanishing part and a part for which all the δQ's are positive so that work could be produced from heat without compensatory heat fall.

Clausius's entropy
It should be emphasized that Clausius's "transformation value," unlike his later "entropy," concerned the transformations of an environment described as a system of heat sources and a work recipient. Its primary purpose was to analytically express the fact that such a transformation can or cannot be realized through a reversible cycle of a thermodynamic machine. However, toward the end of his memoir Clausius shifted the focus from the environment to the working substance of a machine operating in this environment. For a reversible, quasi-static cycle of the substance, he argued, its successive temperatures are identical with the temperatures of the sources with which it exchanges heat. The transformation value $\oint \delta Q/T$ thus becomes a property of the substance. Since it vanishes for any cycle, $\delta Q/T$ must be an exact differential.

In 1854 Clausius used this remarkable property to ease the derivation of relations of the Carnot-Clapeyron type. Much later, in 1865, he forged the word "entropy" from the Greek $\tau\rho\sigma\pi\acute{\eta}$ (transformation), to denote the integral of this differential from a fixed reference state. He also considered irreversible transitions of the substance from one state of equilibrium to another. After such a transition the substance can be returned to its original state through a reversible transformation. During the resultant cycle, the inequality $\delta Q/T < 0$ must hold. Consequently, the entropy of the global system that includes the substance and all the sources can only increase; while its energy is of course invariable. Clausius concluded with two cosmic laws:[8]

"The energy of the world is constant."

"The entropy of the world tends to a maximum."

Reception

The beautiful symmetry of this statement, and the mathematical advantages of explicitly introducing the integral of the complete differential $\delta Q/T$ did not suffice to impose the entropy concept in early thermodynamics. Clausius himself preferred the now forgotten concept of disgregation,[9] which he intuitively grasped as the dispersion of the molecules of a body. In Britain, the engineer William John Macquorn Rankine had introduced the integral $\int \delta Q/T$ in 1853, and in 1854 Thomson had given $\sum_i Q_i/T_i = 0$ as "the mathematical expression of the second principle" for a reversible cycle of a system exchanging the heats Q_i with a series of sources at the temperatures T_i.[10] Yet the leaders of British thermodynamics judged the entropy concept too abstract and rather reasoned in terms of available and dissipated energy, which directly referred to the human ability to exploit energy sources – and, for Thomson, to God's unwillingness to intervene in the created world. In their treatises on heat James Clerk Maxwell and his friend Peter Guthrie Tait cared so little about Clausius's entropy as to give an erroneous definition of it.[11]

French, German, and American thermodynamicists were more receptive to this notion. In the 1870s, the engineer François Massieu discovered that the entropy function could be used to form thermodynamic potentials from which all thermodynamic properties of a substance resulted by simple derivations; and the American Josiah Willard Gibbs founded the laws of chemical equilibrium on Clausius's entropy law. In the 1880s, Hermann von Helmholtz recovered similar laws thanks to the "free energy" $U - TS$, and his disciple Max Planck reformulated thermodynamics on the basis of the energy and entropy laws. By the end of the century, the entropy concept belonged to the standard equipment of well-educated physicists, although some of them still deplored the abstract character of this concept and rather reasoned in terms of the free energy.[12]

---

[8]Clausius, "Über verschiedene für die Anwendung bequeme Formen der Hauptgleichungen der mechanischen Wärmetheorie," *Annalen der Physik und der Chemie*, 125 (1865), 353–400, also in *Mechanical theory of heat*, ref.6, on 365.

[9]The disgregation Ξ is defined by $-Td\Xi = -d\Omega + \delta W$, where Ω is the (average) potential of the internal forces of the system and δW the work of external forces. If Y denotes the "heat content" $U - \Omega$, the disgregation is related to the entropy by $dS = dY/T + d\Xi$. As remarked in Klein 1969, statistical mechanics leads to a similar decomposition if only the Halmiltonian has two separate, kinetic and potential terms.

[10]Cf. Truesdell 1980.

[11]Cf. Smith and Wise 1989.

[12]Cf. Klein 1972.

2   Statistical entropy

A high level of abstraction was the price to be paid for a powerful thermodynamics based on macroscopic principles only. The founders of thermodynamics, especially Rankine, Clausius, and Maxwell, sought a more intuitive understanding of this science in kinetic-molecular theories that they developed in parallel to the phenomenological approach. In two memoirs of 1860 and 1867 Maxwell developed the statistical description of a gas in terms of the velocity distribution and the relative frequency of various kinds of molecular encounters.[13]

Maxwell's demon
Famously, Maxwell used the kinetic-molecular picture to "pick a hole" in the second law of thermodynamics. In a letter to Tait of December 1867, he argued that a "finite being" who could "see the individual molecules" would be able create a heat flow from a cold to a warm body without expense of work. The being only had to control a diaphragm on the wall between warm and cold gas, and let solely the swiftest molecules of the cold gas pass into the warm gas. In discussions with William Thomson and William Strutt, Maxwell related this exception to the second law of thermodynamics with another obtained by mentally reversing all molecular velocities at a given instant. "The 2nd law of thermodynamics," he wrote in 1870, "has the same degree of truth as the statement that if you throw a tumblerful of water into the sea you cannot get the same tumblerful of water out again." In 1878, he further remarked (probably inspired by Gibbs' paradox) that the dissipation of work (or the mixing entropy in Gibbs' terms) during the interdiffusion of two gases depended on our ability to separate them physically or chemically, and concluded: "The dissipation of energy depends on the extent of our knowledge.... It is only to a being in the intermediate stage, who can lay hold of some forms of energy while others elude his grasp that energy appears to be passing inevitably form the available to the dissipated state."[14]

Boltzmann's first entropy formulas
Maxwell's remarks, insightful as they were, remained purely qualitative and did not directly relate entropy and probability. Roughly speaking, the British disliked entropy too much, and the Germans disliked molecular-probabilistic theories too much to seek such a relation. The Austrian physicist Ludwig Boltzmann had neither of these aversions. In 1866, assuming a periodic motion of the atoms of the system and evoking natural definitions of the heat and work exchanged during

---

[13] James Clerk Maxwell, "Illustrations of the dynamical theory of gases," *Philosophical magazine*, 19 (1860), 19–32; 20 (1860), 21–37; "On the dynamical theory of gases," Royal Society of London, *Philosophical transactions*, 157 (1867), 49–88. Cf. Brush (and Everitt) 1973.

[14] Maxwell to Tait, 11 Dec. 1867, in P. Harman, ed., *The scientific letters and papers of James Clerk Maxwell*, vol. 2 (Cambridge, 1995), 328–334; Maxwell, *Theory of heat* (London, 1871), 328–329; Maxwell to Strutt, 6 Dec. 1870, *Letters and papers*, vol. 2, 582–583; Maxwell, "Concerning demons," undated note to Tait, in C.G. Knott, *Life and scientific work of Peter Guthrie Tait* (Cambridge, 1911) 214–215; Maxwell, "Diffusion," *Encyclopedia Britannica* (1878), also in *Scientific papers*, vol. 2, 625–646, on 646. Cf. Klein 1970b.

slow deformations of the system, he gave a first mechanical interpretation of the entropy concept:

$$S = \sum_i 2 \ln \left(\tau \bar{T}_i \right) , \qquad (8)$$

where τ is the period of the motion and $\bar{T}_i$ the average kinetic energy of the atom i. He thus launched a German trend to produce direct analogies between thermodynamic systems and special (periodic or monocyclic) mechanical systems without probabilistic considerations and ignoring irreversible processes. I skip these developments, because they have left few traces in modern physics save for the theory of adiabatic invariants in mechanics.[15]

In most of his later writings on kinetic-molecular theory, Boltzmann developed the statistical point of view found in Maxwell's kinetic theory of gases of 1867. For the number of collisions between two gas molecules occurring within the time interval δt, with initial velocities $\mathbf{v}_1$ and $\mathbf{v}_2$ (within ranges $d^3 v_1$ and $d^3 v_2$), with a relative impact parameter comprised between b and $b + db$, and with a relative azimuthal angle comprised between ϕ and $\phi + d\phi$, Maxwell gave the natural expression

$$dN = |\mathbf{v}_1 - \mathbf{v}_2| \, \delta t \, b \, db \, d\phi \, f(\mathbf{v}_1) \, d^3 v_1 \, f(\mathbf{v}_2) \, d^3 v_2 , \qquad (9)$$

where $f(\mathbf{v}) d^3 v$ is the number of molecules per unit volume within the element $d^3 v$ around $\mathbf{v}$. He then applied this expression to a derivation of the equilibrium distribution of velocities (Maxwell's law) and to the computation of transport phenomena.[16]

In 1868, at the end of a series of generalizations of Maxwell's law, Boltzmann introduced the distribution $\rho(q_1, q_2, \cdots q_N ; p_1, p_2, \cdots p_N) \, d^N q \, d^N p$ that gives the fraction of time spent by the system in the volume element $d^N q \, d^N p$ of phase-space after a very long time has elapsed. Assuming that the trajectory of the system in phase-space filled the energy shell, he then proved that ρ was uniform over the energy shell. In modern terms, he assumed ergodicity and derived the micro-canonical distribution.[17]

Three years later, Boltzmann showed that any small subsystem of the original system was distributed according to the canonical law

$$\rho = \frac{1}{Z} e^{-\beta H} , \qquad (10)$$

wherein H is the Hamiltonian of the subsystem, β is a constant parameter to be identified to the inverse of temperature, and Z is a normalizing factor. He then

---

[15]Ludwig Boltzmann, "Über die mechanische Bedeutung des zweiten Hauptsatzes der Wärmetheorie," Kaiserliche Academie der Wissenschaften zu Wien, mathematisch-naturwissenschaftliche Klasse, *Sitzungsberichte*, 53 (1866), 195–220. Cf. Klein 1972; Bierhalter 1993.

[16]Maxwell, "Dynamical theory of gases", ref. 13.

[17]Boltzmann, "Studien über das Gleichgewicht der lebendigen Kraft zwischen bewegten materiellen Punkten," *Wien. Ber.*, 58 (1868), 517–560, also in *Abhandlungen*, vol. 1, 49–96. Cf. Klein 1973.

submitted a canonically distributed system to an infinitesimal change of external conditions (corresponding for instance to a macroscopic change of temperature and volume). Identifying the work provided to the system during this change with the canonical average $< dH >$ of the resulting change of the Hamiltonian, and the internal energy with the canonical average $< H >$ of the energy, he obtained the expression

$$\delta Q = d < H > - < dH > \tag{11}$$

of the exchanged heat. The product $\beta \delta Q$ is then easily seen to be the differential of $\beta < H > + \ln Z$. In other words, there exists an entropy function, which Boltzmann soon rewrote as[18]

$$S = - \int \rho \ln \rho \, d^N p \, d^N q \ . \tag{12}$$

This is the first occurrence of a mathematical relation between entropy and probability. Remember that in this case Boltzmann defined the probability ρ as the fraction of time spent by the system around a given point of phase space after a very long time has elapsed. Remember also that his derivation of the expression of this probability depended on the assumption of ergodicity.

The H-theorem

As Boltzmann doubted the truth of this hypothesis, he simultaneously developed another approach to kinetic equilibrium in which he generalized Maxwell's collision formula (9) to molecules of arbitrary complexity, and even to the thermal interaction of two ensembles. Maxwell, however, had only proved that the Maxwell distribution of velocities was invariant under molecular collisions; he had no satisfactory proof for the uniqueness of the equilibrium distribution.[19]

In order to remedy this defect, Boltzmann examined the evolution of an arbitrary velocity distribution under Maxwell's assumption for the collision number, and thus obtained the equation

$$\frac{\partial}{\partial t} f(\mathbf{v}_1, t) = \int \left[f(\mathbf{v'}_1) f(\mathbf{v'}_2) - f(\mathbf{v}_1) f(\mathbf{v}_2) \right] |\mathbf{v}_1 - \mathbf{v}_2| \, b \, db \, d\phi \, d^3 v_2 \ , \tag{13}$$

in which the velocities $\mathbf{v'}_1$ and $\mathbf{v'}_2$ denote the final velocities corresponding to the initial velocities $\mathbf{v}_1$ and $\mathbf{v}_2$ in a collision of the kind (b, ϕ). This equation, now called the Boltzmann equation, was published in 1872 under the banal title: "Further studies on the thermal equilibrium among gas molecules." The Maxwell distribution is clearly invariant through this equation. In order to show that any other distribution evolved toward Maxwell's equilibrium distribution, Boltzmann considered the quantity

$$H = \int f \ln f \, d^3 v \tag{14}$$

---

[18] Boltzmann, "Analytischer Beweis des zweiten Hauptsatzes der mechanischen Wärmetheorie aus den Sätzen über das Gleichgewicht der lebendigne Kraft," *Wien. Ber.*, 63 (1871) 712–732, also in *Abhandlungen*, vol. 1, 288-308.

[19] Boltzmann, "Über das Wärmegleichgewicht zwischen mehratomigen Gasmolekülen," *Wien. Ber.*, 63 (1871), 397–418, also in *Abhanlungen*, vol. 1, 237–258.

(originally noted E), probably by analogy with his earlier entropy formula (12). As a consequence of the Boltzmann equation, H is a strictly decreasing function of time, unless the distribution is Maxwell's. This is the so-called H-theorem. Boltzmann further noted that the value of $-H$ corresponding to Maxwell's distribution was identical to Clausius's entropy. For other distributions, he proposed to regard this quantity as an extension of the entropy concept to states out of equilibrium, since it was an ever increasing function of time.[20]

As can be inferred from earlier publications of his, Boltzmann was aware of exceptions to Maxwell's collision formula and therefore could not possibly believe that the Botzmann equation and the decrease of H applied to every possible microscopic configuration of the molecular system. He nevertheless formulated the H-theorem in absolute terms (H "must necessarily decrease"), presumably because his main purpose was to retrieve macroscopic thermodynamics, not to point to exceptions. In 1876 his Viennese colleague Joseph Loschmidt remarked that the reversibility of the laws of mechanics implied that to every H-decreasing evolution of the gas system corresponded a reverse evolution for which H increased.[21]

To this "extremely pertinent" paradox Boltzmann replied (in the more intuitive case of the spatial distribution of hard spheres): "One cannot prove that for every possible initial positions and velocities of the spheres, their distribution must become more uniform after a very long time; one can only prove that the number of initial states leading to a uniform state is infinitely larger than that of initial states leading to a non-uniform state after a given long time; in the latter case the distribution would again become uniform after an even longer time." Boltzmann's intuition, expressed in the modern terminology of micro- and macro-states, was that the number of microstates compatible with a uniform macrostate was enormously larger than that compatible with a non-uniform macrostate. Consequently, an evolution of the gas leading to increased uniformity was immensely more probable. "Out of the relative number of the various state-distributions," Boltzmann went on, "one could even calculate their probability, which perhaps would lead to an interesting method for the computation of the thermal equilibrium."[22]

The combinatorial entropy

This is precisely what Boltzmann managed to do a few months later. The probability Boltzmann had in mind was proportional to the number of microstates corresponding to a given macrostate. Such a number is ill-defined as long as the configuration of the molecules can vary continuously. Boltzmann, who generally believed in a discrete foundation of analysis, began with a "fiction" wherein the

---

[20]Boltzmann, "Weitere Studien über das Wärmegleichgewicht unter Gasmolekülen," *Wien. Ber.*, 66 (1872), 275–370, also in *Abhandlungen*, vol. 1, 216–402.

[21]Boltzmann, *Abhalungen*, vol. 1, 295 (probability needed), 297 (energy fluctuations), 96 (special initial states), 317 (quote), 344 (quote); Joseph Loschmidt, "Über den Zustand des Wärmegleichgewichtes eines Systemes von Körpern mit Rücksicht auf die Schwerekraft," *Wien. Ber.*, 73 (1876), 128–142.

[22]Boltzmann, "Bemerkungen über einige Probleme der mechanischen Wärmetheorie," *Wien. Ber.*, 75 (1877), 62–100, also in *Abhandlungen*, vol. 2, 112–148, on 117, 120, 121.

energy of a molecule can only be an integral multiple of the finite element ϵ. Then a list of N integers giving the number of elements on each molecule defines the microstate of the gas, or "complexion." The macrostate according to Boltzmann is the discrete version of the energy distribution: it gives, for each possible value $i\epsilon$, of the energy, the number N_i of molecules with this energy. The probability of such a macrostate is proportional to its "permutability" $N!/N_1!N_2!\cdots N_i!\cdots$. For a given value of the total number $\sum_i N_i$ of molecules and of the total energy $\sum_i N_i i\epsilon$ and in the Stirling approximation of factorials, the permutability is a maximum when N_i is proportional to $e^{-\beta i\epsilon}$ (wherein β is the Lagrange multiplier associated to the constraint over the total energy). Boltzmann next replaced the uniform division of the energy axis with a uniform division of the velocity-space, and took the continuous limit of the distribution N_i. This procedure yields Maxwell's velocity distribution. For any distribution N_i, the logarithm of the permutability is $-\sum_i N_i \ln N_i$ in the Stirling approximation (up to a constant), or $-H$ in the continuous version. Hence the entropy $-H$ measures the combinatorial probability of the velocity distribution, as Boltzmann already suspected in his reply to Loschmidt. This is the relation that Max Planck later wrote as

$$S = k \ln W \,, \tag{15}$$

with W for Wahrscheinlichkeit (probability), and k for the so-called Boltzmann constant.[23]

In 1878, Boltzmann used the combinatorial probability to explain the existence of a mixing entropy for two chemically indifferent gases. In 1883, after reading Helmholtz's memoirs on the thermodynamics of chemical processes, he showed how his combinatorics explained the dependence of chemical equilibrium on the entropy of the reaction. In this context, Helmholtz (presumably drawing on Maxwell) distinguished between "ordered motion" that could be completely converted into work, and "disordered motion" that allowed only partial conversion. Accordingly, Boltzmann identified the permutability with a measure of the disorder of a distribution. The mixing entropy thus became the obvious counterpart of increased disorder.[24]

The H-curve

Boltzmann's probabilistic interpretation of the H function failed to silence criticism of the H-theorem. In 1894, British kinetic-theoreticians invited Boltzmann to the annual meeting of the British Association, in part to clarify the meaning

[23]Boltzmann, "Über die Beziehung zwischen dem zweiten Haupsatze der mechanischen Wärmetheorie und der Wahrscheinlichkeitzrechnung respektive den Sätzen über das Wärmegleichgewicht," *Wien. Ber.*, 76 (1877), 373–435, also in *Abhandlungen*, vol. 2, 164–223; M. Planck, "Über das Gesetz der Energieverteilung im Normalspektrum, *Annalen der Physik*, 4 (1901), 553–563. La "probabilité" de Planck n'est en fait qu'un nombre de complexions.

[24]Boltzmann, "Über die Beziehung der Diffusionsphänomene zum zweiten Hauptsatze der mechanischen Wärmetheorie," *Wien. Ber.*, 78 (1878), 733–763, also in *Abhandlungen*, vol. 2, 289–317; "Über das Arbeitsquantum, welches bei chemischen Verbindungen gewonnen werden kann," *Wien. Ber.*, 88 (1883), 861–896.

of this theorem. One of them, Samuel Burbury, offered a terminological innovation: "molecular chaos," defined as the validity condition for Maxwell's collision formula. Intuitively, this assumption corresponds to the exclusion of specially arranged configurations, for instance those in which the velocities of closest neighboring molecules point toward each other. It should not be confused with Helmholtz's molar notion of disorder. As long as the gas remains molecularly disordered, the H function evolves according to the Boltzmann equation. Boltzmann did not entirely exclude ordered microstates. He even indicated that an initially disordered microstate could occasionally pass through ordered microstates leading to entropy-decreasing fluctuations, although he judged such events extremely improbable.[25]

To this view Boltzmann's British interlocutors opposed a refined version of the reversibility paradox. H-decreasing and H-increasing states of an isolated gas should be equally frequent, they reasoned, for they correspond to each other by time-reversal. In order to elucidate this point, Boltzmann discussed the shape of the real H-curve determined by molecular dynamics and its relation with the variations of H given by the Boltzmann equation. The real curve results from the cumulative effect of the rapid succession of collisions in the gas. It therefore has an extremely irregular shape, and does not admit a well-defined derivative in the ordinary sense. The refined paradox of reversibility fails, because it implicitly identifies the decrease of H with the negative sign of its derivative.[26]

Boltzmann then offered the following interpretation of the decrease of H: For an initial macrostate out of equilibrium and for a finite time of evolution, the number of compatible microstates for which H decreases is much higher than the number of compatible microstates for which H increases. This statement is perfectly time-symmetrical. Over a very long time, Boltzmann explained, H is for the most very close to zero, and the frequency of its fluctuations decreases very quickly with their intensity. Hence, any significant value of H is most likely to be very close to a summit of the H-curve. From that point H may increase for some time, but this time is likely to be very short and to be followed by a long-term decrease.[27]

The following year Max Planck's assistant Ernst Zermelo formulated another objection to the H-theorem based on Poincaré's recurrence theorem. According to this theorem, any mechanical system (governed by Hamilton's equations) evolving in a finite space with a finite number of degrees of freedom returns, after a sufficiently long time, as close to its initial configuration as one wishes (except for some singular motions). In Zermelo's and Planck's opinion, the theorem excluded any derivation of the entropy law from a mechanical, molecular model. Boltzmann replied, with obvious weariness, that his description of the H-curve was perfectly

---

[25]See various articles by E.P. Culverwell, G.H. Bryan, and S.H. Burbury in *Nature*, 51 (1895); Boltzmann, "Nochmals das Maxwellsche Verteilungsgesetz der Geschwingigkeiten," *Annalen der Physik*, 55 (1895), 223–224; *Gastheorie*, vol. 1, 20–21. Cf. Brush 1973, 616–626.

[26]Boltzmann, "On certain questions of the theory of gases," *Nature* 51 (1895), 413–415, also in *Abhandlungen*, vol. 3, 535–544. Cf. P. and T. Ehrenfest 1909; Klein 1970a.

[27]Paul Ehrenfest later illustrated this behavior with an urns-and-balls model: cf. Klein 1970a.

compatible with recurrences. There was not any conflict with the second law of thermodynamics, as long as the relevant times were far beyond human accessibility. Through a simple calculation he estimated the recurrence time of a macroscopic gas sample to have some 10^{10} digits when measured in a human scale.[28]

Reception

To summarize, Boltzmann proposed three different relations between entropy and probability:

- the relation $S = -\int \rho \ln \rho \, d^N p \, d^N q$ between the equilibrium entropy of a system and the canonical probability of its phases (understood as a temporal frequency),

- the relation $S = -H = -\int f \ln f \, d^3 v$ for the entropy of a gas out of equilibrium and the velocity distribution f, with generalizations to polyatomic molecules and even to ensembles (a whole system regarded as a giant molecule),

- the relation $S = k \ln W$ between the entropy of a gas out of equilibrium and the combinatorial probability W of its macrostate.

Boltzmann regarded the second relation as most important, because it included non-equilibrium states and because it could be established without the assumption of ergodicity, which he did not trust. In his view the third relation only was a "mathematical illustration" of the second, because the equiprobability of the relevant complexions ultimately depended on assumptions already made in earlier approaches, either ergodicity or generalized *Stosszahlansatz*. Thanks to the Boltzmann equation and the H-theorem, Boltzmann could prove the increase of the entropy defined by the second relation. But he clearly recognized that this derivation only had statistical validity, that improbable entropy-decreasing fluctuations could occur; and he provided insightful answers to the resulting paradoxes of reversibility and recurrence. As an epigraph to his lectures of 1898 on gas theory, he cited Gibbs' pronouncement: "The impossibility of an uncompensated decrease of entropy seems to be reduced to an improbability."[29]

British physicists, including Maxwell, welcomed Boltzmann's theory, which they regarded as a monumental, sometimes impenetrable but always deep extension of Maxwell's kinetic theory of gases. In America, the Yale mathematician Josiah Willard Gibbs developed Boltzmann's and Maxwell's ensemble approach in his supremely elegant, general, and powerful *Statistical mechanics* of 1902. In contrast, the Germans and the French ignored Boltzmann's theory because they

---

[28] Poincaré, "Sur le problème des trois corps et les équations de la dynamique," *Acta mathematica*, 13 (1889), 1–270; E. Zermelo, "Über einen Satz der Dynamik und die mechanische Wärmetheorie," *Annalen der Physik*, 57 (1896), 485–494; Boltzman, "Entgegnung auf die Wärmetheoretischen Betrachtungen des Herrn E. Zermelo, ibid., 57 (1896), 773–784, also in *Abhandlungen*, vol. 3, 567–578; "Über die sogenannte H-Kurve," *Mathematische Annalen*, 50 (1898), 325–332, also in *Abhandlungen*, vol. 3, 629–637. Cf. Brush 1873, vol. 2, 627–639.

[29] Boltzmann, *Vorlesungen über die Gastheorie*, 2 vols. (Leipzig, 1896, 1898), vol. 2, 19.

believed that the ordinary methods of thermodynamics to be self-sufficient. Some of them, including Max Planck and Henri Poincaré even declared the impossibility of a kinetic-molecular understanding of the entropy law. This hostile attitude began to change toward the end of the century owing to three circumstances: the development of ionic or electronic theories of electricity, the rise of a new experimental microphysics of electrons, x rays and radioactivity, and the growing interest in black-body radiation. Boltzmann's and Maxwell's methods soon won important successes in these new fields. J.J. Thomson applied kinetic theory to the recombination of x-ray generated ions. The same Thomson, Paul Drude and Hendrik Lorentz developed the electron-theory of metals. Planck surmounted his original distaste of Boltzmann's theory, and based his famous derivation of the black-body law of December 1900 on the formula $S = k \ln W$, which he was first to write in this form.[30]

This spread of Boltzmann's methods did not necessarily imply a better understanding of their foundations. When a young admirer of these methods, Albert Einstein, began to reflect on them in 1902, he complained that the implied probabilities were ill-defined. The following year, he published his own foundations of statistical thermodynamics, mostly recovering aspects of Boltzmann's theory of which he was not aware, but also innovating in one capital respect, the interpretation of fluctuations. Boltzmann, Gibbs, and Maxwell were aware of the statistical fluctuations of thermodynamic quantities, and even knew out to compute them in the ensemble approach. At the same time, they judged these fluctuations to be so extremely rare to be devoid of physical meaning. Most radically, Planck believed that some unknown feature of the microdynamics prevented fluctuations from occurring at all, and he maintained the absolute validity of the entropy law until very late (about 1914).[31]

On the contrary, Einstein focused on the fluctuations around equilibrium that were negligible for Boltzmann and non–existent for Planck. He interpreted the probability in the Boltzmann relation $S = k \ln W$ as the temporal frequency of the fluctuations of the system around equilibrium, and the constant k as the measure of its thermal stability. In 1905, his analysis of Brownian motion showed how fluctuations could become observable at a mesoscopic scale. Inverting Boltzmann's relation, he derived the density fluctuations of black–body radiation from its empirically known entropy and thus arrived at the light–quantum hypothesis in the same year 1905. Whereas Maxwell and Boltzmann meant to provide a mechanical foundation of thermodynamics, Einstein used statistical mechanics to question this foundation.[32]

---

[30] Cf. Klein 1972a; Darrigol 1992.

[31] Cf. Darrigol 1992.

[32] Cf. Klein 1967; Renn 1997.

Conclusions

In my short history of the entropy concept, I have focused on the early period in which the basic principles of thermodynamics and statistical mechanics were first established. I am aware that this period does not exhaust the variety of meanings of entropy. Yet from Clausius's pioneering considerations to Einstein's re-foundation this concept evolved beyond recognition, from a meaning bound to Carnot's concern with steam-engine efficiency to a meaning designed to explore the microworld. The scene changed from the factory to the world of atoms. The plot turned from absolute laws of the macroworld to statistical laws representing the average behavior of enormous numbers of atoms. The man who most contributed to this stupendous evolution of the entropy concept, Ludwig Boltzmann, found himself under fire. When at the turn of the century a few major physicists began to appreciate his methods, they modified them for their own purposes, thus creating a number of different statistical thermodynamics. The later convergence toward a more uniform statistical mechanics "à la Gibbs" was never complete, as may be appreciated from the variety of expositions found in modern textbooks. Reflection on the meaning of entropy goes on, as the forthcoming conference purposes to illustrate.

To which extent can this reflection draw on nineteenth-century sources? I suppose none of you will be tempted to resurrect the Carnot-Clausius concept of transformation, which dragged along the antiquated concept of heat as a substance. Yet from this origin of entropy it is well worth remembering that the concept has to do with human ability to exploit the energy stored in a given system. This connotation prepares the statistical interpretation, for the inaccessibility of certain forms of energy depends on the impossibility of acting on individual molecules, as emphasized by Maxwell. It is also good to remember that classical thermodynamics provides a consistent framework for defining entropy without any reference to the molecular level or to probabilistic considerations.

Nineteenth-century sources are most inspiring when it comes to the statistical meanings of entropy. Boltzmann was aware of the connections between various meanings of that kind, and offered deep insights into their apparent conflict with the thermodynamic meaning, for instance in his discussion of the H-curve. Unfortunately, his writings are hard to penetrate. Maxwell himself complained: "By the study of Boltzmann I have become unable to understand him. He could not understand me on account of my shortness and his length was and is an equal stumbling block to me."[33] As a result of Boltzmann's style, many of his ideas have been periodically rediscovered or attributed to his followers. At the same time, he could not foresee every modern development of the entropy concept. I am sure he would be glad to attend the forthcoming conference.

---

[33]Maxwell to Tait, circar Aug 1873, in Harman, ref. 14, 915.

References

[1] Günther Bierhalter, "Helmholtz's mechanical foundation of thermodynamics," in David Cahan (ed.), *Hermann von Helmholtz and the foundations of nineteenth-century science* (Berkeley, 1993), 432–458.

[2] Ludwig Boltzmann, *Wissenschaftliche Abhandlungen*, 3 vols. (Leipzig: Barth, 1909).
1896-98 Vorlesungen über die Gastheorie, 2 vols. (Leipzig, 1896).

[3] Stephen Brush, *The kind of motion we call heat: A history of the kinetic theory of gases in the 19th century*, 2 vols. (Amsterdam, 1976).

[4] Donald Cardwell, *From Watt to Clausius* (Ithaca, 1971).

[5] Olivier Darrigol, *From c-numbers to q-numbers: The classical analogy in the history of quantum theory* (Berkeley, 1992).
"Thermodynamics," to be published in the *Enciclopedia Italiana*.

[6] Olivier Darrigol and Jürgen Renn, "The emergence of statistical mechanics," preprint 139 of the Max-Planck-Institut für Wissenschaftsgeschichte (2000), to be published in the *Enciclopedia Italiana*.

[7] Edward Daub, "Rudolf Clausius," in *Dictionary of scientific biography* **vol. 3**, 303–311 (1971).

[8] Paul and Tatiana Ehrenfest, "Begriffliche Grundlagen der statistischen Auffassung in der Mechanik," *Encyklopädie der mathematischen Wissenschaften* **IV/2.II**, 1–90 (1909).

[9] C.W.F. Everitt, "Maxwell, James Clerk," in *Dictionary of scientific biography*, 16 vols. (New York, 1970-80), **vol. 9**, 198–230 (1975).

[10] Robert Fox, *The caloric theory of gases from Lavoisier to Regnault*, (Oxford, 1971).

[11] Josiah Willard Gibbs, *Elementary principles in statistical mechanics developed with especial reference to the rational foundation of thermodynamics*, (New York: Scribner's sons, 1902).

[12] Peter Harman, *Energy, force, and matter: The conceptual development of nineteenth century physics*, (Cambridge, 1982).

[13] Martin Klein, "Thermodynamics in Einstein's thought," *Science* **157**, 509–516 (1967).
"Gibbs on Clausius," *Historical studies in physical science* **1**, 127–149 (1969).
Paul Ehrenfest (Amsterdam: North-Holland, 1970).

"Maxwell, his demon, and the second law of thermodynamics," *American scientist* **58**, 84–97 (1970).

"Gibbs, Josiah Willard," *Dictionary of scientific biography* **vol. 5**, 386–393 (1972).

"Mechanical explanation at the end of the nineteenth century," *Centaurus* **17**, 58–82 (1972).

"The development of Boltzmann's statistical ideas," in E.G.D. Cohen and W. Thirring, eds., "The Boltzmann equation: Theory and applications" (Vienna: Springer, 1973), 53–106.

[14] James Clerk Maxwell, *The Scientific papers of James Clerk Maxwell*, 2 vols. (Cambridge: Cambridge University Press, 1890).
Maxwell on molecules and gases, ed. E. Garber, S.G. Brush, and C.W.F. Everitt (Cambridge: MIT Press, 1986).

[15] J. von Plato, "Boltzmann's ergodic hypothesis," *Archive for the history of exact sciences* **42**, 71–89 (1991).
Creating modern probability. Its mathematics, physics and philosophy in historical perspective, (Cambridge: Cambridge University Press, 1994).

[16] Jürgen Renn, "Einstein's controversy with Drude and the origin of statistical mechanics: A new glimpse from the 'love letters'," *Archive for the history of exact sciences* **51**, 315–354 (1997).

[17] L. Sklar, *Physics and chance: Philosophical issues in the foundations of statistical mechanics*, (Cambridge: Cambridge University Press, 1993).

[18] Crosbie Smith and Norton Wise, *Energy and empire: A biographical study of Lord Kelvin*, (Cambridge, 1989).

[19] Clifford Truesdell, *The tragicomic history of thermodynamics*, 1822–1854 (New York, 1980).

Olivier Darrigol
CNRS: Rehseis
83, rue Broca
F-75013 Paris
email: darrigol@paris7.jussieu.fr

Poincaré Seminar 2003, 119 – 144
© Birkhäuser Verlag, Basel, 2004

Entropy, a Protean Concept

Roger Balian

Abstract. We review at a tutorial level the many aspects of the concept of entropy and their interrelations, in thermodynamics, information theory, probability theory and statistical physics. The consideration of relevant entropies and the identification of entropy with missing information enlighten the paradoxes of irreversibility and of Maxwell's demon.

The concept of entropy, invented one and a half century ago, has given rise to an immense literature. Under various guises it appears in many branches of physics, of mathematics, and even of most other sciences. Like the Greek divinity Proteus, it uses to change its shape so as to escape anyone who tries to grasp it, and it also presides prevision and deceit. To catch its meaning, we need to recognize it through its metamorphoses. We shall therefore review with an introductory scope some of its aspects, focusing on those which are relevant to statistical physics.

1 Macrophysics

1.1 Entropy in thermostatics

Entropy has first been introduced as a mathematical tool in the framework of thermodynamics. This science, born in the first half of the XIXth century, deals with the general laws that govern the transformations of systems. Actually, what is usually called "the laws of thermo*dynamics*" are physical constraints about the transformations allowed by physics that lead from one equilibrium state to another. These laws do not pertain to the dynamics of the processes, that is, their time-dependence, but only refer to their *initial and final state*. We therefore prefer here to speak of "thermostatics".

In the modern formulation of thermostatics [1], a physical or mechanical or chemical isolated system is analyzed into homogeneous subsystems referred to by the index a. The equilibrium state of each one is characterized by a set of *extensive variables* A_k such as volume, energy, number of constitutive particles. These quantities can be transformed or transferred but are *conserved* (the First Law and its extensions). For the composite system, the state variables are denoted as A_i, where $i = k, a$ is a double index indicating the nature k of the variable and the subsystem a. In the initial state the exchanges between subsystems are blocked and the system lies in a constrained equilibrium state with fixed A_i's. If we release some of these blockings by letting the subsystems interact, only some constraints on the set $\{A_i\}$ remain, in particular those imposed by the conservation laws. After some time, we reach a global equilibrium state. The extensive variables of each subsystem take new values, and the *Second Law*, in the following formulation, determines them. There exists for each subsystem a at equilibrium a function $S_{\mathrm{th}a}(\{A_{ka}\})$ of the extensive variables that characterize its state, its *thermodynamic entropy*. It is concave, extensive and additive: the entropy of the whole system is the sum of those of its subsystems. The overall equilibrium state is then the one that *maximizes the total entropy* $S_{\mathrm{th}}(\{A_i\})$, $i = k, a$, subject to the constraints imposed on the variables A_i by the initial state, the conservation laws and the allowed exchanges.

By stating that the entropy of an isolated system cannot decrease when it goes from a constrained equilibrium state to a less constrained one, the Second Law expresses the *irreversibility* of macroscopic processes. It also provides an upper bound for the efficiency of thermal machines, as was first shown by Carnot (1824). However its field of application covers not only thermal processes (as suggested by the word "*thermo*dynamics") but also other types of transfers, for instance, mixing of substances or chemical reactions: any such spontaneous process raises the entropy. The connection between entropy and irreversibility is directly exhibited by an alternative approach [2] to the Second Law, initiated by Carathéodory. It relies on a comparison hypothesis between any pair of states X, Y of a system: either X is adiabatically accessible from Y, or Y is adiabatically accessible from X. This property is sufficient to ensure the construction of the entropy as function of the macroscopic state variables.

As a consequence of the Second Law, the partial derivative γ_i of the entropy $S_{\mathrm{th}}(\{A_i\})$ with respect to one of its variables A_i $(i = k, a)$ takes the same value for two systems a and b that are in relative equilibrium as regards the considered quantity k: if the two *intensive variables* $\gamma_{ka} = \gamma_{kb}$ are equalized, no transfer of this quantity occurs although exchanges between a and b are allowed. The differential

$$dS_{\mathrm{th}} = \sum_i \gamma_i \, dA_i \tag{1}$$

exhibits γ_i and A_i as conjugate variables with respect to the entropy. If A_i is an energy, the corresponding γ_i is the inverse temperature $\beta \equiv T^{-1}$ of the subsystem; if A_i is a particle number, γ_i is $-\mu/T$ where μ is the chemical potential. In particular, since S_{th} is maximum at equilibrium, $\beta = \partial S_{\mathrm{th}}/\partial E$ takes the same value for two systems which have been brought to relative equilibrium by thermal contact. The *Zeroth Law* follows: relative thermal equilibrium between pairs of systems is a relation of equivalence, implemented by the existence of relative temperatures, which are any functions of β. In particular $\beta^{-1} = T$ is the absolute temperature.

1.2 Entropy in thermodynamics

Thermodynamics proper describes the time-dependence of physical, mechanical or chemical processes, in conditions of local quasi-equilibrium [3], that is, when the macroscopic state at each time is fully characterized by the same conservative variables as in thermostatics, for instance the local energy, momentum or particle density. As above the system is analyzed into a set of subsystems a, each of which is thus at any time nearly at equilibrium. The *conservation laws* are expressed, for each physical quantity k and each subsystem a, by an equation

$$\frac{dA_{ka}}{dt} + \sum_b \Phi_k(b \to a) = 0 \;, \tag{2}$$

which involves fluxes of k from the neighbouring subsystems b.

For continuous media, the subsystems a may be infinitesimal; more precisely each one should be a volume element large on the molecular scale but sufficiently small so that the densities of conserved extensive quantities (energy, particle, or momentum densities) are nearly constant within it. The instantaneous state is thus characterized, for continuous media, by the densities $\rho_k(\mathbf{r}, t)$ of the conserved quantities A_k. These variables are labelled by a continuous index $\mathbf{r}$ and a discrete index k. The conservation laws (2) involve fluxes Φ of local current densities $\mathbf{J}_k(\mathbf{r}, t)$ for each A_k, and take for each volume element the form

$$\frac{\partial \rho_k}{\partial t} + \mathrm{div}\mathbf{J}_k = 0 \;. \tag{3}$$

The index k denotes again the conserved quantities, energy, particle number and, for a fluid, momentum $\mathbf{P}$.

Entropy is defined at each time as in thermostatics. For each subsystem, its time-dependence is obtained from (1) and (2) as

$$\frac{dS_{\mathrm{th}a}}{dt} + \sum_{k,b} \gamma_{ka}\Phi_k(a \to b) = 0 \;,$$

or equivalently, separating symmetric from antisymmetric contributions,

$$\frac{dS_{\mathrm{th}a}}{dt} + \sum_{k,b} \frac{\gamma_{ka} + \gamma_{kb}}{2}\Phi_k(a \to b) = \frac{1}{2}\sum_{kb}(\gamma_{kb} - \gamma_{ka})\Phi_k(a \to b) \;, \tag{4}$$

which exhibits the *fluxes* Φ_k and the *affinities* $\gamma_{kb} - \gamma_{ka}$ as conjugate variables with respect to the *dissipation* dS_{th}/dt. The continuous version of (4) is

$$\frac{\partial \rho_S}{\partial t} + \mathrm{div}\mathbf{J}_S = \sum_k \nabla\gamma_k \cdot (\mathbf{J}_k - \mathbf{J}_k^{(0)}) \;, \tag{5}$$

where the *current density of entropy* is defined as

$$\mathbf{J}_S - \sum_k \gamma_k\mathbf{J}_k + \frac{\mathcal{P}}{T}\mathbf{u} \tag{6}$$

in terms of the hydrodynamic velocity $\mathbf{u} = -T\partial S_{\text{th}}/\partial \mathbf{P}$ and of the pressure $\mathcal{P} = T\rho_S - \sum_k \gamma_k \rho_k$. In the right-hand side of (5), $\mathbf{J}_k^{(0)}$ denotes the current densities that exist in equilibrium due to the motion.

One of the laws of the thermodynamics of non-equilibrium processes is the *Clausius–Duhem inequality*, which expresses the fact that entropy cannot be destroyed in any circumstance, or equivalently that the dissipation, given by the right-hand side of (4) or (5), is never negative. This inequality sets constraints on the response equations that relate the fluxes Φ_k or $\mathbf{J}_k$ to the affinities $\gamma_{kb} - \gamma_{ka}$ or $\nabla \gamma_k$, and hence through (2) or (3) on the dynamical equations for the variables A_{ka} or the densities ρ_k.

The Clausius–Duhem inequality should not be confused with the Second Law although both express an entropy increase. This inequality holds *locally* and at *each time*, but it applies only to systems evolving sufficiently slowly, in a local equilibrium regime. The Second Law compares merely the *global* entropies of the *initial* and the *final* state, both being in equilibrium (but with more constraints in the initial state); during intermediate times, the system may well be far from equilibrium, for instance, if an explosive chemical reaction takes place, and its entropy need not be defined.

2   Information and probability

2.1   Entropy in communication theory

The theory of communication, founded in 1948 by Shannon and Weaver, aims at improving the transmission of signals. At first sight, its only analogy with thermodynamics seems to be the search for an optimum efficiency for the considered process, but we shall see that the entropies introduced in the two fields of science are actually related to each other. The basic idea, here, is that the amount of information transmitted by a message can be measured. One can then compare the performance of different transmission devices or of different coding systems, so as to minimize the duration of a transmission or the size of a memory.

Information is a concept related to *probability*. Indeed the information brought along by some message to a receiver is meaningful only if this message is extracted from a set of messages m that might a priori have been emitted. Before transmission the receiver ascribes to each message m a probability p_m. His *surprisal*, that is, the amount I_m of information that he gains by getting knowledge of some message m among the possible set should be a decreasing function $I_m = I(p_m)$ of the probability p_m: we gain very little information when being told about something we were practically certain of, whereas nearly unexpected messages are very informative. Moreover, a compound message mn consisting in two uncorrelated submessages m and n with respective probabilities p_m and q_n should carry an amount of information equal to the sum $I_m + I_n$. Since probabilities are then multiplicative while information is additive, the continuity and the decrease of the function $I(p_m)$ imply

$$I_m = -\log p_m \ , \tag{7}$$

within a positive factor which defines the unit of information. This unit is the bit if the logarithm in (7) has base 2.

While I_m measures the amount of information *gained* by reception of the message m, Shannon's entropy measures the amount of information which is *missing before reception*, and which on average will be gained through reception. Since each message m has the probability p_m to reach the receiver, Shannon's entropy is defined by weighting I_m by the probability p_m:

$$S_{\mathrm{Sh}}(\{p_m\}) = \sum_m p_m I_m = -\sum_m p_m \log p_m \ . \tag{8}$$

For a number W of equally probable messages, for which $p_m = 1/W$ ($m = 1, 2, \cdots W$), Shannon's entropy reduces to the celebrated expression of Boltzmann within a multiplicative factor:

$$S_{\mathrm{Sh}} = \log W \ . \tag{9}$$

Shannon's entropy thus characterizes the perplexity of the receiver before transmission of some message among the set $\{m\}$, or the average missing information.

A direct proof of (8) was given by Shannon who postulated a strong additivity property of $S_{\mathrm{Sh}}(\{p_m\})$: its expression should be additive, not only for compound events mn such that m and n are uncorrelated, but also if information is gained by steps. Suppose the events m are grouped in bunches, and suppose we are first informed about the occurrence of one among these bunches. The entropy should contain a corresponding first contribution, which is a function of the probabilities of the bunches. In addition, it should contain, weighted by the probability of each bunch, contributions associated with the conditional probability of each event among the considered bunch. The identity thus satisfied by $S_{\mathrm{Sh}}(\{p_m\})$ for different numbers of events, which results from the simplest identity

$$S_{\mathrm{Sh}}(\{p_1, p_2, p_3, \cdots p_m\}) = S_{\mathrm{Sh}}(\{p_1 + p_2, p_3, \cdots p_m\})$$
$$+ (p_1 + p_2) S_{\mathrm{Sh}}(\{\frac{p_1}{p_1 + p_2}, \frac{p_2}{p_1 + p_2}\}) \ ,$$

together with continuity and symmetry requirements, implies the form (8).

Shannon's entropy is a powerful tool for optimizing the amount of information involved in the transmission of messages or the storing of data. In both cases, the messages should be coded. Shannon and Weaver proved the existence of an *optimum coding*, which depends on the probabilities p_m and on a possible noise that may destroy part of the messages.

2.2 Entropy of a probability distribution

Shannon's expression (8) associates with any discrete probability set $\{p_m\}$ a number S, whether the index m labels messages or any other set of events. In the latter case S characterizes the *uncertainty* associated with this probability law, or its *spreading*. If the weights p_m are concentrated on some events, the uncertainty is lower than when they are spread. If all events are equally probable the uncertainty increases with their number. A quantitative evaluation of such an uncertainty is provided by (8) and (9).

Many properties of the function (8) enforce this interpretation of entropy as a measure of uncertainty. For a given number W of events, it is *minimum* and equal to 0 when one event is certain while all other ones have probability zero. It reaches its *maximum* (9) in the most random situation where all probabilities p_m are equal (to $1/W$). For compound events mn with joint probabilities P_{mn}, the separate probabilities of the m's and the n's are $p_m = \sum_n P_{mn}$ and $q_n = \sum_m P_{mn}$, respectively. The entropy is then *additive* for independent events, *subadditive* for correlated ones:

$$S(\{p_m\}) + S(\{q_n\}) = S(\{p_m q_n\}) \geq S(\{P_{mn}\}) . \tag{10}$$

(The equality holds only when $P_{mn} = p_m q_n$.) This inequality expresses that correlations carry some information. As another property of the entropy, consider a single set of events m, to which two different sets of probabilities p_m and q_m can be ascribed for two different statistical ensembles. If these two ensembles are mixed with non-vanishing weights λ and $(1 - \lambda)$ into a single one, the new ensemble is characterized by probabilities $P_m = \lambda p_m + (1 - \lambda) q_m$. The *concavity* property of entropy,

$$S(\{P_m\}) > \lambda S(\{p_m\}) + (1 - \lambda) S(\{q_m\}) , \tag{11}$$

ensures that the uncertainty, as measured by S, is raised by mixing of populations.

2.3 Continuous probabilities

A difficulty arises for continuous distributions of probability. Consider a random real variable x, governed by a continuous probability density $p(x)$. In order to extend to this situation the definition (8), we split the x-axis into intervals x_m, $x_m + \Delta_m \equiv x_{m+1}$ and define p_m as the probability for x to lie between x_m and x_{m+1}. It would be natural to define $S(\{p(x)\})$ as the limit of $S(\{p_m\})$ when all Δ_m's tend to zero, but this quantity diverges. However, if all Δ_m's are equal, adding the constant $\log \Delta$ to $S(\{p_m\})$ provides a finite limit

$$S(\{p(x)\}) = - \int dx \, p(x) \log \, p(x) , \tag{12}$$

which defines the entropy of the probability distribution $p(x)$.

This quantity is not invariant under a change of the variable x. Whereas a linear transformation simply adds a constant to it, a non-linear transformation or equivalently a choice of unequal Δ_m's may modify (12) arbitrarily. Additional hypotheses are therefore needed to define unambiguously the entropy associated with a continuous distribution. For instance, translational invariance over x of the phenomenon characterized by a probability law $p(x)$ justifies the choice (12), which arises from a uniform partition $\Delta_m = \Delta$.

More generally, for continuous variables x lying on some manifold, the existence of an *invariance group* or a metric is necessary to define unambiguously $S(\{p(x)\})$. Actually such a group already existed implicitly for the discrete probabilities p_m involved in $S(\{p_m\})$, since all the events m were treated on the same footing: the very construction of Shannon's entropy implies that it is invariant under permutation of theses events.

3 Statistical physics

3.1 Von Neumann's entropy

Twenty years before Shannon, von Neumann introduced a similar expression, in the quite different context of quantum theory. There, probabilities are unavoidable as exemplified by the Heisenberg uncertainty relations. This irreducible intrusion of probabilities arises from the non-commutative nature of the observables, the algebraic objects that represent the physical quantities. To this replacement of usual random variables $A(m)$ by *non-commuting observables* $\hat{A}$ corresponds the replacement of probability distributions p by *density operators* $\hat{D}$, which are represented by matrices in the Hilbert space associated with the considered system. The expectation value of $\hat{A}$ is given by

$$\hat{A} \;\mapsto\; <\hat{A}> \;=\; \mathrm{Tr}\ \hat{A}\hat{D}\ , \tag{13}$$

where the trace, taken on the Hilbert space, replaces the summation over the elementary events m.

Since any quantity at a given time can be represented in quantum mechanics under the form (13), our whole information at the considered time is represented in a probabilistic way by the density matrix $\hat{D}$. Anticipating the idea of Shannon, who associated the missing information (8) with the set of probabilities p_m, von Neumann associated the entropy

$$S_{\mathrm{vN}}(\hat{D}) = -\mathrm{Tr}\ \hat{D}\ln\hat{D} \tag{14}$$

with the density operator $\hat{D}$. We thus interpret (14) as a measure of the uncertainty associated with the description by $\hat{D}$ of the state of the system. When written in terms of the eigenvalues p_m of $\hat{D}$, the expression (14) is identical with (8). It can be constructed directly, starting from some natural axioms [4]. The invariance of S_{Sh} under the group of permutations of the events m is replaced here by the unitary invariance in the Hilbert space: all representations by matrices of quantum mechanics, deduced from one another by unitary transformations, are equivalent.

Through its diagonalization, $\hat{D}$ behaves, for a finite quantum system, more like a discrete than like a continuous probability distribution, in spite of the continuity of the underlying unitary group. The definition (14) of its entropy thus does not involve the difficulties of (12).

The properties of von Neumann's entropy are similar to those of Shannon's entropy: additivity, subadditivity, concavity. They enforce the interpretation of (14) as a measure of the uncertainty associated with the density operator $\hat{D}$. The minimum, $S_{vN} = 0$ of (14) is reached when $\hat{D}$ is a pure state, that is, a projection onto a single wavefunction. Contrary to what happens for discrete probabilities, such pure states still involve uncertainties although $S_{vN} = 0$, but are the best defined states allowed by quantum mechanics.

States $\hat{D}$ with non zero (positive) entropy occur in *quantum statistical physics* because the wave functions of systems made of a large number of particles cannot be fully determined. The entropy $S_{vN}(\hat{D})$ then measures the uncertainty about such a state. The von Neumann entropy is also of interest in the framework of *quantum measurement theory*. Even for a quantum system having very few degrees of freedom, the measurement of some of its observables implies interaction with a macroscopic apparatus, which can be described only by means of quantum statistical mechanics. The equivalence between negentropy and information that we shall discuss in § 5.2 explains why von Neumann's entropy allows us both to measure the dispersion of a state $\hat{D}$, which will be identified with the thermodynamic entropy, and to evaluate the amount of information gained through a quantum measurement.

3.2 Entropy in classical statistical mechanics

In classical statistical physics, a state is described by the *density in phase* $D(\mathbf{r}_1, \mathbf{p}_1, \cdots \mathbf{r}_N, \mathbf{p}_N, t)$, which is a density of probability in the phase space of the N considered particles. If we were to define for the entropy of such a state an expression similar to (12) with integration over the $3N$-dimensional phase space, this entropy would depend on the measure of integration and thus would not be defined unambiguously. However, by regarding a density in phase as a limit of a state $\hat{D}$ of quantum statistical physics, one can show that the trace in (13) and (14) tends to an integral over phase space (as would be directly introduced in classical statistical mechanics), but with the well-defined measure

$$\frac{1}{N!} \prod_{n=1}^{N} \frac{d^3\mathbf{r}_n \, d^3\mathbf{p}_n}{h^3} , \tag{15}$$

where h is Planck's constant, and where the factor $1/N!$ arises from Pauli's principle about indistinguishability $\hat{D}$ of particles.

As we shall see in § 4.2 the entropy S_{th} of thermostatics can be identified with von Neumann's entropy for equilibrium states. Two problems arising in classical statistical mechanics are then solved. On the one hand the *Gibbs paradox*, according to which the entropy of classical statistical mechanics does not seem extensive for a set of identical particles, is elucidated owing to the factor $1/N!$ in (15). On the other hand the limiting process which starts from quantum statistical mechanics generates the *absolute entropy*, without any additive constant, which satisfies the *Third Law*, or Nernst Law: the limit towards the zero absolute temperature of the absolute entropy vanishes. The occurrence of Planck's constant in (15) is essential in this respect.

3.3   The entropy as measure of disorder

Both Shannon's entropy (8) for a probability set and von Neumann's entropy (14) have been introduced as a *measure of missing information*. They do not appear as properties of the object under study in itself, but rather characterize the knowledge about it of its observers, who describe it by means of probabilities. These entropies thus have a partly subjective character, since they numerically characterize the *uncertainty of the observers*. Such a concept fits with the subjective interpretation of probabilities [5]. They should be regarded as mathematical tools for making consistent predictions, starting from the available information. The entropy (8) measures the quality of such predictions. In fact, since all observers placed in the same conditions should attribute the same probabilities to a set of possible events, probabilities are intersubjective rather than subjective.

Likewise, in statistical mechanics, a state represented by a density operator collects our information on some system. It does not describe this system in itself, but as a sample chosen among an *ensemble* of systems all prepared by the same procedure. This ensemble may be real, or may just be a set of thought similar copies, not completely identical but all having the same known features.

We may alternatively interpret the von Neumann entropy as a *measure of disorder* of the state described by the (probabilistic) density operator $\hat{D}$. Actually the concept of disorder should be identified with that of uncertainty: when we say that a fully mixed pack of cards is disordered, it only means that we know nothing about their ordering; for a conjurer who is aware of this ordering, there is no disorder in the pack. Maxwell already wrote: "Confusion, like the correlative term order, is not a property of material things in themselves, but only in relation to the mind who perceives them." Entropy allows us to make this idea quantitative.

4   Maximum statistical entropy and applications

4.1   The maximum entropy criterion

According to this interpretation, the assignment of probabilities to a set of events should depend on the data available to the observer, but should be made in a consistent way so that any other observer makes the same inferences starting from the same data. In statistical physics on which we focus, the problem is the same: which density operator $\hat{D}$ should we assign to describe the state of a system belonging to some ensemble characterized by a set of macroscopic data?

When nothing is known but the list of the W possible events m, it is natural to resort to Laplace's *principle of indifference* or of insufficient reason, and to assign the same probability $p_m = 1/W$ to all these events, as is done in the theory of games. Likewise, if no information is available about the spin $\frac{1}{2}$ of a particle, the obviously unbiased choice for its density operator is $\hat{D} = \frac{1}{2}\hat{I}$. This state describes an unpolarized spin, the expectation value of which vanishes in any direction. This principle relies on the idea that any other probability distribution would introduce bias, by favourizing without any reason the prediction of some events to which larger probabilities would have been assigned.

The maximum entropy principle [6] extends this idea to situations where some probabilistic information is given, in the form of expectation values. Suppose that, for instance in statistical mechanics the expectation values $A_i \equiv <\hat{A}_i>$ of some observables $\hat{A}_i$, that we shall refer to as the the *relevant observables*, are known. We shall call these data A_i the *relevant variables*. From this information we wish to infer the expectation values of other quantities. To this aim we need to assign a density operator $\hat{D}$ to the system. It should of course satisfy the constraints

$$\mathrm{Tr}\ \hat{A}_i\hat{D} = A_i \tag{16}$$

about the known quantities, but they do not suffice to determine $\hat{D}$. Whithin this allowed set, consider two density operators $\hat{D}_1$ and $\hat{D}_2$ such that $S_{\mathrm{vN}}(\hat{D}_1) < S_{\mathrm{vN}}(\hat{D}_2)$. This inequality means that $\hat{D}_1$ contains more information that $\hat{D}_2$, by an amount $S_{\mathrm{vN}}(\hat{D}_2) - S_{\mathrm{vN}}(\hat{D}_1)$. However the density operator that should describe the state of the considered system (or rather of the considered ensemble of systems) should not carry more information than what is contained in the relevant variables $\{A_i\}$. Thus $\hat{D}_1$, which contains more information than $\hat{D}_2$, is certainly biased. Hence we are led to select for $\hat{D}$ the density operator that maximizes $S_{\mathrm{vN}}(\hat{D})$, subject to the constraints (16) for the set $\{\hat{A}_i\}$. This maximum entropy criterion means that we describe the situation by means of the *least biased* density operator, or the *most uncertain*, or the *most disordered*, among the set compatible with the available data.

The result of this procedure is found by introducing *Lagrange multipliers* γ_i for each constraint on A_i and ζ for the normalization. We thus have to express that

$$\delta\left[S_{\mathrm{vN}}(\hat{D}) - \sum_i \gamma_i \mathrm{Tr}\ \hat{A}_i\hat{D} - \zeta\,\mathrm{Tr}\ \hat{D}\right]$$

vanishes for any Hermitian variation $\delta\hat{D}$ of $\hat{D}$. Letting $\zeta = \Psi - 1$ and regarding Ψ as a function of the variables γ_i, we thus obtain for $\hat{D}$ the generalized Gibbsian distribution

$$\hat{D}_{\mathrm{R}} = \exp\left(-\Psi - \sum_i \gamma_i\hat{A}_i\right), \qquad \Psi(\{\gamma_i\}) \equiv \ln \mathrm{Tr} \exp\left(-\sum_i \gamma_i\hat{A}_i\right), \tag{17}$$

that we shall call the *reduced density operator* associated with the relevant variables A_i. The multipliers $\{\gamma_i\}$ are related to the data $\{A_i\}$ through

$$A_i = -\frac{\partial \Psi}{\partial \gamma_i} , \tag{18}$$

a consequence of (16) and (17).

The value $S_{\text{vN}}(\hat{D}_{\text{R}})$ of the von Neumann entropy (13) of the state (17) is larger than the entropy associated with any other $\hat{D}$ compatible with the data $\{A_i\}$, that is, satisfying (16). It is given by

$$S_{\text{R}}(\{A_i\}) = \max_{\hat{D}} S_{\text{vN}}(\hat{D}) = \Psi + \sum_i \gamma_i A_i . \tag{19}$$

We refer to it as the *relevant entropy* associated with knowledge of the variables $\{A_i\}$. Altogether, if the von Neumann or the Shannon entropy, which measures uncertainty or disorder, is also regarded as a measure of bias, the *least biased* inferences based on the knowledge of the set $\{A_i\}$ should rely on the probability distribution (17), (18). The unicity of this distribution is ensured by the concavity of entropy, the maximum of which is (19).

The validity of the maximum entropy criterion has been questioned. Is it legitimate to identify the missing information as a measure of bias? For Shannon's entropy, its use can be justified by direct approaches where requirements on the consistency of the inference procedure [7] lead to the same result as the maximization of S_{Sh}, provided we deal with discrete events m. In quantum statistical physics, an alternative justification for (17) was given, based on the idea that the expectation value A_i of $\hat{A}_i$, expressed by (16) in terms of the density operator $\hat{D}$ of the considered system, may be identified with the mean value $\sum_\alpha A_i^\alpha / \mathcal{N}$ over an ensemble of $\mathcal{N}$ analogous systems $\alpha = 1, 2, \ldots \mathcal{N}$ described by $\hat{D}$, in the limit of large $\mathcal{N}$ [4, 8]. This identification is consistent with the equivalence of the two interpretations of probabilities, either a tool for predictions or a set of frequencies [5].

The maximum entropy criterion is currently used in various contexts. We resume below some of its outcomes in statistical physics.

4.2   Equilibrium statistical physics

Equilibrium statistical physics underlies thermostatics at the microscopic level. It allows in particular to derive the Laws of thermostatics as statistical consequences of microphysics. It is based on the choice, as relevant observables $\{\hat{A}_i\}$, of the same *conserved* quantities as in thermodynamics. The macroscopic variables $\{A_i\}$ are identified, at the microscopic, statistical level, with the expectation values (16). The relative fluctuations are small as $1/\sqrt{N}$ where N is the number of elementary constituents, so that the statistical nature of the microscopic description does not appear in thermostatics. The equilibrium state of thermostatics is described by means of the density operator $\hat{D}_{\mathrm{R}}$ constructed as in (17), (18) from these macroscopic equilibrium data $\{A_i\}$, which remain fixed in time. We do not regard equilibrium as the final stage of a dynamical process, but rather as a state statistically characterized by the relevant variables A_i.

Consider first a homogeneous piece of material contained in a volume V. Its macroscopic equilibrium states are characterized by its number N of particles (we assume for simplicity that they are all of the same kind) and by their energy. At the microscopic level these two quantities are the only data (16); they are identified with the expectation values of the particle number operator $\hat{N}$, and of the Hamiltonian $\hat{H}$, respectively. We therefore describe this piece of material by the grand canonical density operator (17), where the observables $\hat{A}_i$ are here $\hat{N}$ and $\hat{H}$.

Consider now more generally a compound system as for the Second Law of thermostatics, but described from the viewpoint of statistical physics. In the *initial* state the data $\{A_i\}$ (where $i = k, a$ now denotes both the nature k of the variable and the subsystem a) are the extensive conserved variables of each subsystem; they may take arbitrary values since the exchanges are blocked. The overall equilibrium density operator, which maximizes the uncertainty for the given set $\{A_i\}$, has the form (17). It factorizes as a product of contributions associated with each subsystem, e.g., grand canonical distributions, with multipliers γ_i $(i = k, a)$ taking independent values for each subsystem a. Hence its relevant entropy is the sum of the entropies (19) of all parts. The determination of the *final* equilibrium density operator may then be performed by maximizing S_{vN} in *two steps*. The first step, that we just described, leads to the function $S_{\mathrm{R}}(\{A_i\})$ of the extensive conservative variables of all the subsystems, which we have identified with those entering the entropy S_{th} of thermostatics. Then the second step is exactly the same as in the Second Law of thermostatics; to wit, this function $S_{\mathrm{R}}(\{A_i\})$ is maximized as function of the variables A_i, subject to the remaining constraints. Since the entropy of thermostatics S_{th} is defined in a unique fashion within a multiplicative constant, we can altogether identify it with the sum of the relevant entropies of the subsystems, each one being evaluated at equilibrium.

The Second Law thus appears simply as this second step of the maximum entropy criterion. The *entropy of thermostatics* (and of thermodynamics) is therefore the macroscopic manifestation of the *relevant entropy* associated with the conserved variables $A_i = A_{ka}$:

$$S_{\mathrm{th}}(\{A_{ka}\}) = kS_{\mathrm{R}}(\{A_{ka}\}) \ . \tag{20}$$

The multiplicative constant k depends on the choice of units for S_{th}. The natural choice $k = 1$, where $S_{\mathrm{th}}(\{A_i\}) = S_{\mathrm{R}}(\{A_i\}) = S_{\mathrm{vN}}(\{A_i\})$ is dimensionless, leads to temperatures measured in energy units and taking very small values, whereas entropies are very large since the uncertainty S_{vN} increases as the number N of elementary constituents of the material. On the other hand, measuring temperatures in kelvin, a unit defined from the triple point of water, provides for $k \simeq 1.38 \times 10^{-23} \mathrm{JK}^{-1}$ the Boltzmann's constant.

The Legendre transformation (18), (19) implies that

$$\frac{\partial\, S_{\mathrm{R}}(\{A_i\})}{\partial\, A_i} = \gamma_i \; . \tag{21}$$

The identification (20) of the macroscopic and statistical entropies and the comparison of (1) with (21) justify (for $k = 1$) the identification of the *Lagrange multipliers* γ_j in (17) with the *intensive variables* defined by (1). In particular the multiplier associated with $\hat{H}$, noted β, is interpreted as the inverse of the absolute temperature.

Actually, the Second Law alone defines the entropy S_{th} not only within a multiplicative constant, but also within an additive constant. This arbitrariness is lifted by the Third Law. The above definition $S_{\mathrm{R}}(\{A_i\})$ of entropy issued from statistical physics involves no additive constant, and it *implies the Third Law* since (19) tends to zero (for usual materials) in the zero-temperature limit $\beta \to \infty$.

In the above reduction of thermostatics to equilibrium statistical mechanics, we had first to single out, among the whole set of microscopic variables, the conserved variables $\{A_i\}$ that characterize an equilibrium state at the macroscopic scale. It should be noted that their choice cannot always be made a priori, since the microscopic equations of motion do not necessarily exhibit all the conserved or nearly conserved variables. Consider, for instance, a ferromagnetic piece of material. If it is not submitted to an external magnetic field, taking the energy and the density as variables $\{A_i\}$ is sufficient at high temperature. However, below the Curie point, we cannot understand its properties (even purely thermal) without introducing among the set $\{A_i\}$ the spontaneous magnetization. At those temperatures, the rotational invariance is spontaneously broken, and a new conserved variable, the magnetization, which is the order parameter, emerges. Likewise, in Bose condensation or in superconductivity, an order parameter of quantum nature has to be introduced among the conserved variables $\{A_i\}$ below the critical temperature in the macroscopic description of the material at equilibrium.

The idea of reducing thermostatics to equilibrium statistical mechanics, and of interpreting the entropy as a measure of the disorder or the uncertainty at the microscopic scale after introduction of probabilities, can be traced back to Boltzmann who dealt with the kinetic theory of classical gases. As we already noted, the advent of quantum mechanics has paradoxically brought up simplifications: Owing to the introduction of discreteness, S_{vN} is defined unambiguously. Accordingly, even for a classical system like a monatomic gas, Planck's constant enters through (15) the expression of the thermostatic entropy.

4.3 Reduced states and relevant entropies

A typical problem of non-equilibrium statistical physics is the prediction of the expectation values of specified quantities at time t from a set of initial data. The density operator $\hat{D}(t)$, which characterizes the values (13) of all physical quantities at all times, evolves according to the Liouville-von Neumann equation

$$i\hbar\frac{d\hat{D}}{dt} = [\hat{H}, \hat{D}] \; , \tag{22}$$

and its knowledge at the time t provides the required results. In principle, it can be found by solving (22) with the initial condition $\hat{D}(t_0)$, which is assigned from the initial data as in (17). The observables $\{\hat{A}_i\}$ controlled at the initial time t_0 are, however, no longer constants of the motion as in thermostatics, so that $\hat{D}(t_0)$ does not commute with the Hamiltonian $\hat{H}$ and the evolution (22) leads to a state $\hat{D}(t)$, the construction of which solves the problem.

However, the detailed description by means of $\hat{D}(t)$ involves a huge amount of variables, both without interest and unpracticable. We therefore select some set of *relevant observables* $\{\hat{A}_i\}$ which are expected to govern the evolution. Their choice is guided by phenomenological macroscopic approaches. We should include, apart from the variables of thermodynamics, other macroscopic variables the dynamics of which is of interest. For instance, in nuclear magnetic resonance, we consider the three components of the total magnetic moment, which evolve under the influence of an external magnetic field and of interactions with the other degrees of freedom of the material; the dynamics of this magnetic moment will involve Larmor precession and relaxation, an irreversible process. In nuclear physics, we include in the relevant set collective variables such as the deformation of the nucleus. We give two other examples below. Most often they are the slowest variables, whereas the remaining irrelevant variables evolve on a much shorter time scale. The observables which are initially controlled and the observables about which we want to make predictions belong to the relevant set $\{\hat{A}_i\}$. It is therefore sufficient to follow the evolution of the relevant variables $\{A_i(t)\}$, by eliminating the irrelevant observables which are coupled by (22) to the dynamics of the relevant set. This elimination may be performed, at least formally, by the projection method of Nakajima and Zwanzig [9] that we sketch below.

The relevant variables $\{A_i(t)\}$ are deduced at each time from $\hat{D}(t)$ through (16). Since the observables initially given belong to the relevant set $\{\hat{A}_i\}$, the initial density operator $\hat{D}(t_0)$ has the form (17); however, if the set $\{\hat{A}_i\}$ contains observables whose initial value $A_i(t_0)$ is not specified, the corresponding multipliers vanish. At later times, the solution $\hat{D}(t)$ of eq. (22) has no reason to retain the same exponential form, involving only the relevant observables $\{\hat{A}_i\}$. Considering then the set $A_i(t) = \mathrm{Tr}\,\hat{A}_i\hat{D}(t)$ of relevant variables at the time t, we can associate with them by means of the maximum entropy criterion a *reduced density operator* $\hat{D}_{\mathrm{R}}(t)$ of the form (17), where the multipliers $\gamma_i(t)$ depend on time. Their value is determined by the conditions $A_i(t) = \mathrm{Tr}\,\hat{A}_i\hat{D}_{\mathrm{R}}(t)$ where the $\{A_i(t)\}$ in the left-hand side are deduced from (22) and $A_i(t) = \mathrm{Tr}\hat{A}_i\hat{D}(t)$, and where the right-hand side satisfies eq. (18). Thus $\hat{D}(t)$ and $\hat{D}_{\mathrm{R}}(t)$ are equivalent as regards the relevant variables, but $\hat{D}_{\mathrm{R}}(t)$ has the *maximum entropy*, given by (19) in terms of the set $\{A_i(t)\}$. In other words both states $\hat{D}(t)$ and $\hat{D}_{\mathrm{R}}(t)$ account for the information $\{A_i(t)\}$, but $\hat{D}_{\mathrm{R}}(t)$ involves no further information. The difference $S_{\mathrm{vN}}[\hat{D}_0(t)] -$

$S_{vN}[\hat{D}(t)]$ measures an extra amount of information about the irrelevant variables, which is included in $\hat{D}(t)$ that keeps full track of our knowledge of the initial data $\{A_i(t_0)\}$.

The quantity $S_{vN}[\hat{D}_R(t)]$ defines the *relevant entropy* $S_R(\{A_i\})$ associated with the quantities $\{A_i(t)\}$. It depends on the chosen set $\{\hat{A}_i\}$ of observables, and characterizes the uncertainty at the time t when one follows only the evolution of their expectation values. By construction we have $S_R(\{A_i\}) \geq S_{vN}[\hat{D}(t)]$. On the other hand, we find from (14) and (22) that

$$i\hbar \frac{dS_{vN}(\hat{D})}{dt} = -\mathrm{Tr}\,[\hat{H}, \hat{D}] \ln \hat{D} = 0 \;, \tag{23}$$

and hence that $S_{vN}[\hat{D}(t)]$ is constant. Thus the relevant entropy satisfies $S_R(\{A_i(t)\}) \geq S_R(\{A_i(t_0)\})$, which means that some information about the relevant variables is *lost during the evolution*. On the other hand the constancy of $S_{vN}[\hat{D}(t)]$ implied by (23) means that our information about *all* possible observables, which is described by $\hat{D}(t)$, is conserved by the Hamiltonian evolution (22).

4.4 Projection method and dissipation

From a geometric viewpoint, the *reduction of the description*, which associates with $\hat{D}(t)$ the less detailed distribution $\hat{D}_R(t)$, is in the space of states $\hat{D}$ a *projection* onto the *manifold of reduced states* (17), in the direction of constant relevant variables $\{A_i\}$ (fig. 1). Algebraically, in order to implement this idea, the space of states $\hat{D}$ is regarded as a vector space. The whole set of observables $\{\hat{A}\}$ then appears as its dual vector space, with a scalar product defined by

$$(\hat{A}; \hat{D}) \equiv \mathrm{Tr}\,\hat{A}\hat{D} = <\hat{A}>_{\hat{D}} \;. \tag{24}$$

The relevant observables $\{\hat{A}_i\}$ define a subspace of $\{\hat{A}\}$. Linear transformations in either vector spaces $\{\hat{D}\}$ or $\{\hat{A}\}$ are then represented by "superoperators"; for instance, regarding $\hat{D}$ as a vector leads to rewrite (22) in the form

$$\frac{d\hat{D}}{dt} = \mathcal{L}\hat{D} \tag{25}$$

in terms of the Liouvillian superoperator which transforms $\hat{D}$ into $\mathcal{L}\hat{D} \equiv [\hat{H}, \hat{D}]/i\hbar$. (In the Heisenberg picture, the observables would evolve likewise as $d\hat{A}/dt = \hat{A}\mathcal{L}$.)

The operators of the form (17) define in the vector space $\{\hat{D}\}$ the manifold of relevant states $\{\hat{D}_R\}$, and a superoperator $\mathcal{P}$ describing projection on the relevant states or observables is associated with each $\{\hat{D}_R\}$. It has the form

$$\mathcal{P} = \hat{D}_R \otimes \hat{I} + \sum_i \frac{\partial \hat{D}_R}{\partial A_i} \otimes (\hat{A}_i - A_i\hat{I}) \;, \tag{26}$$

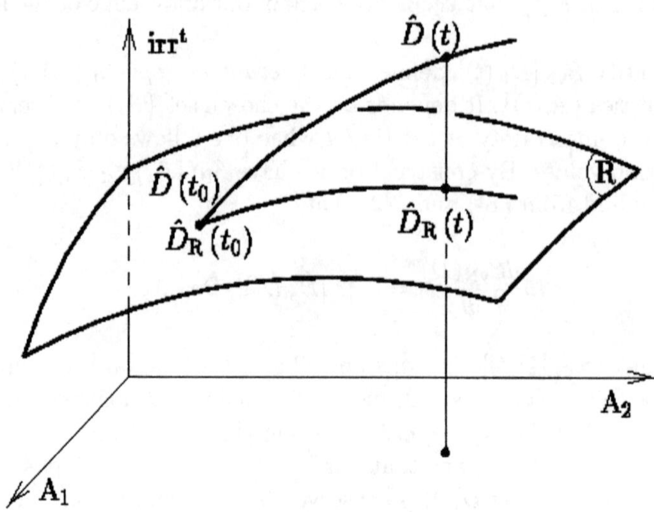

Figure 1: The reduction of the description. A state $\hat{D}$ of quantum statistical physics is represented by a point in a vector space. Here the first two axes schematize the relevant variables $A_i = \mathrm{Tr}\,\hat{A}_i\hat{D}$, regarded as coordinates of $\hat{D}$, while the third axis stands for its many irrelevant coordinates. The trajectory of $\hat{D}(t)$ is governed by the Liouville-von Neumann equation (22), (25). The reduced state $\hat{D}_{\mathrm{R}}(t)$, equivalent to $\hat{D}(t)$ as regards the relevant variables $\{A_i\}$ but maximizing the von Neumann entropy, is the intersection of the plane $A_i(t) = \mathrm{Tr}\,\hat{A}_i\hat{D}$ (represented here as a line) and the surface R, parametrized by the set $\{\gamma_i\}$ according to (17). The trajectory of $\hat{D}_{\mathrm{R}}(t)$ is obtained by projecting that of $\hat{D}(t)$ on R. It starts with the initial point $\hat{D}_{\mathrm{R}}(t_0) = \hat{D}(t_0)$ and is governed by eq.(28) or its approximation (31). The von Neumann entropy is constant along the line $\hat{D}(t)$, whereas for $\hat{D}_{\mathrm{R}}(t)$ information is lost towards the irrelevant degrees of freedom.

where $\hat{I}$ is the unit operator in the space $\{\hat{A}\}$, where $A_i = (\hat{A}_i\,;\,\hat{D}_i)$, and where $\hat{D}_{\mathrm{R}}$ expressed by (17) is regarded as a function of the variables A_i through the variables γ_i and the equations (18) or (21). The complementary superoperator is $\mathcal{Q} = \mathcal{I} - \mathcal{P}$ where $\mathcal{I}$ is the unit superoperator. The reduced state $\hat{D}_{\mathrm{R}}(t)$ associated with $\hat{D}(t)$ then appears as the projection $\hat{D}_{\mathrm{R}} = \mathcal{P}\hat{D}$ of $\hat{D}$.

Actually, the space of states $\hat{D}$ has not only a structure of vector space, but can also be regarded as a Riemannian space, where a natural *metric generated by entropy* [9] according to

$$ds^2 = -d^2 S_{\mathrm{vN}} \tag{27}$$

allows to define distances between states and angles. The projection $\mathcal{P}$ from $\hat{D}(t)$ to $\hat{D}_{\mathrm{R}}(t)$ then appears as an *orthogonal projection*.

Our dynamical problem of non-equilibrium statistical mechanics, deducing the set $\{A_i(t)\}$ from $\{A_i(t_0)\}$, then amounts to the search for the projection $\hat{D}_{\mathrm{R}}(t)$ of the detailed trajectory of $\hat{D}(t)$, which is generated by (22) and by the initial condition $\hat{D}(t_0) = \hat{D}_{\mathrm{R}}(t_0)$. Indeed, at each time, it is equivalent to know the set $\{A_i(t)\}$, the corresponding set $\{\gamma_i(t)\}$ given by (18) or (21), or the reduced state $\hat{D}_{\mathrm{R}}(t)$ given by (17). The *projection method* provides for $\hat{D}_{\mathrm{R}}(t)$ an integro-differential equation that relates $d\hat{D}_{\mathrm{R}}/dt$ to $\hat{D}_{\mathrm{R}}$ at the same time and also at earlier times. This equation, of the form

$$\frac{d\hat{D}_{\mathrm{R}}}{dt} = \mathcal{P}\,\mathcal{L}\,\hat{D}_{\mathrm{R}} + \int_{t_0}^{t} dt'\,\mathcal{M}(t,t')\hat{D}_{\mathrm{R}}(t') \;, \tag{28}$$

is obtained from (25) and the initial condition $\hat{D}_{\mathrm{R}}(t_0)$ by starting from the time-dependent equation $\hat{D}_{\mathrm{R}} = \mathcal{P}\hat{D}$ and taking advantage of the properties $\mathcal{P}\partial\hat{D}_{\mathrm{R}}/\partial A_i = \partial\hat{D}_{\mathrm{R}}/\partial A_i$, $\hat{I}\mathcal{P} = \hat{I}$, $\hat{A}_i\mathcal{P} = \hat{A}_i$ and $(d\mathcal{P}/dt)\hat{D} = 0$ of the projection. It involves a *memory kernel* $\mathcal{M}(t,t')$ defined by

$$\mathcal{M}(t,t') = \mathcal{P}(t)\mathcal{L}\mathcal{W}(t,t')\mathcal{L}\mathcal{P}(t') \tag{29}$$

in terms of the evolution superoperator $\mathcal{W}(t,t')$ in the irrelevant space, which is characterized by the equation

$$\left[\frac{d}{dt} - \mathcal{Q}(t)\mathcal{L}\right]\mathcal{W}(t,t') = \mathcal{Q}(t)\delta(t-t') \;. \tag{30}$$

The equation (28) is equivalent to an *equation of motion for the relevant variables* $\{A_i(t)\}$. Whereas its instantaneous term describes a direct coupling among the relevant variables, the last, *retarded term* describes a *memory effect* resulting from the *elimination of the irrelevant variables* and expressing $\hat{D}_{\mathrm{R}}(t)$ in terms of its past history.

If the set $\{\hat{A}_i\}$ has been suitably chosen, the memory time is short compared to the time-scale for $\{\hat{A}_i(t)\}$, and $\hat{D}_{\mathrm{R}}(t')$ can then be replaced by $\hat{D}_{\mathrm{R}}(t)$ in (28). Then $\hat{D}_{\mathrm{R}}(t)$ is governed within a good approximation by the mere differential equation

$$\frac{d\hat{D}_{\mathrm{R}}}{dt} = \mathcal{P}\,\mathcal{L}\,\hat{D}_{\mathrm{R}} + \mathcal{K}\,\hat{D}_{\mathrm{R}}(t) \;, \tag{31}$$

where the superoperator $\mathcal{K}$ is defined by

$$\mathcal{K}(t) = \int_{-\infty}^{t} dt'\,\mathcal{W}(t,t') \;. \tag{32}$$

The relevant variables (18) or $A_i = (\hat{A}_i\,;\hat{D}_{\mathrm{R}})$ thus evolve autonomously according to a set of differential equations of the same type as usually encountered in phenomenological approaches. The memory term is replaced by a *dissipative term*, which prevents $d\hat{D}_{\mathrm{R}}/dt$ from being generated by an effective Hamiltonian, and which thus allows for a time-dependence of the relevant entropy $S_{\mathrm{R}}(\{A_i\})$. Indeed we find from (14), (17) and (28) that the *dissipation* is given in terms of the memory kernel by

$$\frac{dS_{\mathrm{R}}}{dt} = \sum_i \gamma_i(t)\, (\hat{A}_i\,;\, \int_{t_0}^{t} dt' \mathcal{M}(t,t')\hat{D}_{\mathrm{R}}(t')) \,. \tag{33}$$

Only the coupling of the set $\{A_i\}$ with the irrelevant variables contributes to (33). Dissipation is a retardation effect, and becomes instantaneous in the short-memory approximation.

In physical situations where the selection of relevant variables thus leads to a differential equation for $\hat{D}_{\mathrm{R}}(t)$, $\hat{D}_{\mathrm{R}}(t+\Delta t)$ depends only on $\hat{D}_{\mathrm{R}}(t)$, for Δt small but large compared to the memory time associated with the evolution of the irrelevant variables. We would therefore have found the same $\hat{D}_{\mathrm{R}}(t+\Delta t)$ by starting the evolution from $\hat{D}(t) = \hat{D}_{\mathrm{R}}(t)$, letting $\hat{D}$ evolve according to (25), then projecting $\hat{D}(t+\Delta t)$ on the set of relevant states. Hence we have $S(\hat{D}_{\mathrm{R}}(t+\Delta t)) \geq S(\hat{D}(t+\Delta t)) = S(\hat{D}(t)) = S(\hat{D}_{\mathrm{R}}(t))$. Altogether, for a choice of the set $\{\hat{A}_i\}$ such that the memory time is short, we find $S_{\mathrm{R}}(\{A_i(t+\Delta t)\}) \geq S_{\mathrm{R}}(\{A_i(t)\})$; this inequality holds along the motion of $\hat{D}_{\mathrm{R}}(t)$: the *relevant entropy cannot decrease*, whereas the von Neumann entropy remains constant. This means that *information about the relevant observables is continuously lost* towards the irrelevant ones during the evolution, and this loss is *irretrievable*. In other words the *dissipation* $dS_{\mathrm{R}}(\{A_i\})/dt \geq 0$ measures the rate at which a *loss of order* takes place in the relevant variables $\{A_i(t)\}$. The *entropy production* characterizes an *irreversible flow of uncertainty* from the irrelevant to the relevant variables.

We illustrate below this general approach by two examples.

4.5 The thermodynamic entropy as a relevant entropy

For the processes described by non-equilibrium thermodynamics, the variables A_i governed by the macroscopic equations are the conservative variables of each subsystem or of each volume element. For instance, for a fluid, they are the densities $\rho_k(\mathbf{r})$ of particles, of energy and of momentum at each point. They are identified at the microscopic level with the relevant variables A_i $(i = k, \mathbf{r})$ of the projection method. One can then derive from the microscopic conservation laws the macroscopic ones (3), after microscopic identification of the current densities $\mathbf{J}_k$. It is the conservative nature of the variables $\{A_i\}$ which ensures that they evolve over much longer time scales than the other variables which are discarded. Indeed, without couplings between subsystems or volume elements, the thermodynamic variables $\{A_i\}$ or $\rho_k(\mathbf{r})$ would remain constant. For sufficiently weak couplings and for small affinities $\nabla\gamma_k$, the currents $\mathbf{J}_k$ are weak and from (3) it appears that the motion of the variables $\{A_i\}$ is slow.

The parameters $\{\gamma_i\}$ of the reduced density operator $\hat{D}_{\mathrm{R}}(t)$ given by (17) are then identified at each time with the local intensive variables $\gamma_j(\mathbf{r})$ $(i = j, \mathbf{r})$

associated with the set $\rho_j(\mathbf{r})$, which are directly related to the local temperature, chemical potential and hydrodynamic velocity.

In the short-memory, weak-coupling limit, the equation of motion (31), which governs the dynamics of the the macroscopic variables $\rho_k(\mathbf{r})$ or $\gamma_k(\mathbf{r})$, yields the equations of thermodynamics. Moreover it provides microscopic expressions for the transport coefficients. For instance, for a fluid, we find thus the Navier–Stokes and Fourier equations, together with microscopic expressions for the viscosities and the heat conductivity.

Since $\hat{D}_{\mathrm{R}}$ is factorized into a product of contributions from each subsystem, the relevant entropy associated with it is the sum over all subsystems of the entropy of equilibrium statistical mechanics, that we already identified with the entropy of thermostatics. The relevant entropy relative to the present choice of variables A_i is therefore the same as the entropy of macroscopic non-equilibrium thermodynamics. Its increase gives a microscopic justification for the Clausius–Duhem inequality (5) if the time-scale of the thermodynamic variables is longer than that of the microscopic variables that have been projected out.

4.6 Boltzmann's entropy

Boltzmann's equation for *gases* describes their dynamics in terms of the density of particles $f(\mathbf{r}, \mathbf{p}, t)$ in the *single-particle* phase space: $f(\mathbf{r}, \mathbf{p}, t) \, d^3\mathbf{r} \, d^3\mathbf{p}$ is the expectation value of the number of particles lying in the volume element $d^3\mathbf{r} \, d^3\mathbf{p}$ of this phase space at time t. This quantity evolves according to Boltzmann's equation

$$\frac{\partial f(\mathbf{r}, \mathbf{p}, t)}{\partial t} + \frac{\mathbf{p}}{m} \cdot \nabla_{\mathbf{r}} f(\mathbf{r}, \mathbf{p}, t) = \mathcal{I}\{f\} \, , \tag{34}$$

where the left-hand side accounts for the drift of f due to the free motion of the particles (with mass m). The right-hand side is the collision integral; it is quadratic in f and describes the change in f due to interparticle collisions. Since the collisions are nearly local in space and in time, $\mathcal{I}$ involves $f(\mathbf{r}, \mathbf{p}', t)$ at the same point and time as in the left-hand side, but involves integrations over the two momenta entering the two f's. The establishment of (34) requires that $f(\mathbf{r}, \mathbf{p}, t)$ as function of $\mathbf{r}$ and t varies slowly compared to the sizes of the particles and the duration of the collisions.

Boltzmann proved that his equation obeys the *H-theorem*: the quantity

$$H(t) \equiv \int d^3\mathbf{r} \, d^3\mathbf{p} \, f(\mathbf{r}, \mathbf{p}, t) \ln f(\mathbf{r}, \mathbf{p}, t) \tag{35}$$

is a decreasing function of time.

In order to recover Boltzmann's equation and H-theorem from microphysics, we should start from the most detailed description, in terms of the *density in phase*, that is the probability density $D(\mathbf{r}_1, \mathbf{p}_1, \cdots \mathbf{r}_N, \mathbf{p}_N, t)$ in the N-particle phase space. Its time-dependence is generated by the Liouville equation, the classical limit of (25), where collisions are governed by the interparticle potential.

We take as the set of relevant variables the values of the single-particle density $f(\mathbf{r}, \mathbf{p}, t)$, obtained by integrating $D(\mathbf{r}_1, \mathbf{p}_1, \cdots \mathbf{r}_N, \mathbf{p}_N, t)$ over the phase space of $N - 1$ particles. The index i stands here for $\mathbf{r}, \mathbf{p}$. In the reduced density $D_{\mathrm{R}}(\mathbf{r}_1, \mathbf{p}_1, \cdots \mathbf{r}_N, \mathbf{p}_N, t) \propto f(\mathbf{r}_1, \mathbf{p}_1, t) \times \cdots \times f(\mathbf{r}_N, \mathbf{p}_N, t)$, all correlations that possibly exist in D are eliminated. These correlations are generated at each collision. In Boltzmann's theory they are discarded, but are present in the exact equations of motion (25). The projection method then produces an equation of motion (28) for D_{R} or equivalently for $f(\mathbf{r}, \mathbf{p}, t)$, in which the memory-time is the duration of a collision. The neat separation of time scales allows us to neglect this duration, and f is thus governed by a differential equation (31) that is identified to Boltzmann's equation with its instantaneous collision term.

The relevant entropy S_1 associated with the relevant variables $f(\mathbf{r}, \mathbf{p}, t)$ is identified with Boltzmann's entropy, which is easily shown to equal $-H(t)$ within a multiplicative and an additive constant. The H-theorem is recovered as a special case of the increase of relevant entropies in regimes where the memory is short. Here the increase of the Boltzmann entropy S_1 is interpreted as a loss of information which results from the fact that, although correlations are created by each collision, these correlations have no effect on the subsequent evolution because two particles which underwent a collision have little chance to meet again. The evolution of $f(\mathbf{r}, \mathbf{p}, t)$ is the same as if all these *correlations are forgotten*.

Boltzmann's entropy S_1 related to (35) should not be confused with the thermodynamic entropy of subsection 4.5: Even for a gas of non-interacting particles, S_1 coincides with S_{th} only when at each point $f(\mathbf{r}, \mathbf{p}, t)$ behaves as a Gaussian in $\mathbf{p}$; otherwise S_{th} is larger than S_1 for the same values of the thermodynamic variables $\rho_k(\mathbf{r}, t)$. Nor should the H-theorem be confused with the Second Law or with the Clausius–Duhem inequality. On the one hand, Boltzmann's equation deals with gases only. On the other hand, it holds beyond local equilibrium, in ballistic regimes that cannot be described in terms of the thermodynamic variables. Indeed, in a thermodynamic or hydrodynamic regime, local equilibrium implies that $f(\mathbf{r}, \mathbf{p}, t)$ has at each point $\mathbf{r}$ a Maxwellian form, that is, behaves as a Gaussian in $\mathbf{p}$. The reduced description provided by $f(\mathbf{r}, \mathbf{p}, t)$, a function of 6 variables, is more detailed than the reduced hydrodynamic description in terms of 5 functions of 3 variables, the densities of energy, particles and momentum in ordinary space (or equivalently the local temperature, chemical potential and hydrodynamic velocity).

Boltzmann's entropy accounts for the uncertainty associated with the sole knowledge at each time of the single-particle density $f(\mathbf{r}, \mathbf{p}, t)$ of a gas enclosed in a vessel. For sufficiently large times, a local equilibrium, then the global equilibrium corresponding to the initial values of the total particle number and energy are attained. At these stages the Boltzmann entropy grows so as to reach the entropy of non-equilibrium thermodynamics, then of thermostatics. Before the thermodynamic regime is settled, starting from the full density in phase $D(\mathbf{r}_1, \mathbf{p}_1, \cdots, \mathbf{r}_N, t)$ we can derive at the microscopic scale two-particle, three-particle, ... correlation functions. More and more detailed reduced descriptions are thus obtained by dropping all the correlations of more than n particles [10]. Boltzmann's description corresponds to $n = 1$, with Boltzmann's entropy S_1, and we find a *hierarchy of relevant entropies* $S_2, \cdots S_n \cdots$ such that

$$S_{\text{th}} \geq S_1 \geq S_2 \geq \cdots \geq S_n \geq \cdots \geq S_{\text{vN}}(\hat{D}) \, . \tag{36}$$

Starting all from $S_{\text{vN}}(\hat{D}(t_0)) = S_{\text{vN}}(\hat{D}(t))$ at the initial time, the entropies S_N all increase as a function of time, but later and later; all of them finally reach the thermostatic entropy, which is the upper bound of the whole set (fig. 2).

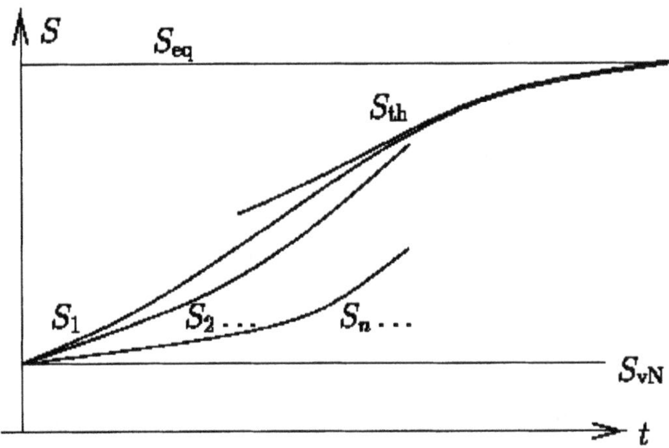

Figure 2: Hierarchy of entropies in a gas. Larger and larger entropies are associated with less and less detailed descriptions. Boltzmann's entropy S_1 increases (H-theorem) from the von Neumann entropy S_{vN} to the equilibrium thermostatic entropy S_{eq}, evaluated for the initial values of the energy, particle number and density of the gas. It is smaller than the thermodynamic entropy S_{th}, a function of the local densities of energy, particles and momentum, but both remain nearly equal after the distribution $f(\mathbf{r}, \mathbf{p}, t)$ has become gaussian in $\mathbf{p}$ at any point $\mathbf{r}$ (local equilibrium regime). The entropies $S_2, S_3 \cdots S_n$ accounting for 2-, 3-, $\cdots n$-particle correlations are at each time smaller and smaller and increase later and later.

5 Two paradoxes

5.1 The paradox of irreversibility

Soon after the birth of the kinetic theory of gases, Loschmidt, Zermelo and Poincaré pointed out the paradox of irreversibility: although the microscopic evolution (22) is invariant under time-reversal, the macroscopic evolutions are not. As regards thermodynamic variables, viscosity and thermal conduction are irreversible phenomena. In the dynamics of gases, the H-theorem also exhibits and *"arrow of time"*. We gave a general explanation to such irreversible behaviours: whenever the memory-time is short, the dissipative term in the reduced description generates an increase of the relevant entropy, the signature of irreversibility. This increase is a *statistical* property; it measures a loss of information towards the irrelevant variables. The explanation of irreversibility also involves the choice of $D_{\mathrm{R}}(t_0)$ as the initial state.

This argument relies on the irretrievable nature of this loss. In principle, the equation of motion (22) does not prevent, for a finite system, some order hidden within the irrelevant degrees of freedom to surge back into the relevant ones, and to show off as a decrease of the relevant entropy. If we forget about the observer and consider an initial microstate *completely defined* at the initial time, it remains completely defined at all times so that nothing seems to prevent this complete order from showing off. For instance, if the N particles of a gas are all grouped at the initial time in the left half of a container, they will fill the full vessel after some delay. However the equations of motion are reversible and allow the converse evolution, where the particles occupying the full container spontaneously come together in its left half. Why do we never observe such a behaviour in practice? The reason why it does not occur is that we deal with systems composed with an extremely large number N of particles. Poincaré's *recurrence time*, after which a system governed by an equation such as (22) returns in the vicinity of its initial states, is then immensely large, even compared to the age of the Universe.

Boltzmann's explanation of the paradox of irreversibility is again probabilistic although we consider here the single trajectory of a well-defined configuration. It relies on an analysis of the initial state. In the above example of the expansion of a gas, let us discretize the positions and momenta of the particles so as to count the configurations (in agreement with quantum mechanics). Denote as W the number of compressed states, such that the N particles lie in the left half of the vessel. The total number of states in the full vessel is $2^N W$, and in nearly all of them the particles are spread all over. Among the latter states, those for which, after some time τ, the particles are gathered in the left half are in one-to-one correspondence with the W compressed states. Hence only a very tiny proportion 2^{-N} of the spread states gives rise to the anomalous grouping process. Since N is of the order of the Avogadro number 6×10^{23}, we have no chance whatsoever to observe such a process.

More generally, consider an irreversible thermodynamic process for which the entropy increases by ΔS_{th}. In the above example $\Delta S_{\mathrm{th}} = kN \ln 2$. The corresponding increase of the relevant von Neumann entropy is $\Delta S_{\mathrm{th}}/k$ where k is Boltzmann's constant $1.38 \times 10^{23} \mathrm{JK}^{-1}$ and where ΔS_{th} has the order of JK^{-1}. Noting from (9) that the ratio between the numbers W of initial and final microstates is of the order of $e^{-\Delta S_{\mathrm{th}}/k}$, we see that anomalous processes with reversed time are not forbidden in principle but that the *initial configurations* from which they might arise *are completely improbable*, as an exponential of -10^{23}, among the whole set of configurations involved in the most disordered thermodynamic state.

Nevertheless, experiments exist in which many microscopic variables that appear as irrelevant at first sight can actually be controlled. The most celebrated ones are *spin echo* experiments, in which only the total magnetic moment of a material containing N spins is observed, and only an applied magnetic field can be controlled. Under normal circumstances, a relaxation of the total magnetic moment occurs; the magnitude of this moment decreases, so that the relevant disorder increases as expected. However, some time after the total moment has vanished, it is possible through suitable pulses of the applied field to manipulate the individual spins in such a way that the order hidden in their correlations manifests itself by an increase of the magnetic moment. The conceptual interest of such experiments (which also have practical applications in NMR) is to demonstrate that some initial information which is apparently lost within microscopic degrees of freedom may in some exceptional cases be retrieved, and that the choice of relevant variables should depend on the circumstances.

5.2   Equivalence of information and negentropy

Another paradox has fed for one century many discussions about the meaning of entropy, the thought experiment of *Maxwell's demon* (1867). Two vessels A and B with the same volume are filled with a gas, initially at the same density and temperature. They communicate through a hole that the demon may open or close at will with negligible work. Whenever a particle arrives from A towards B, the demon lets it pass, but he stops the particles arriving from B towards A. The density thus increases on the side A, that eventually all the N particles reach. The Second Law seems violated, since the entropy S_{th} has decreased by $kN \ln 2$.

However, in order to operate, the demon must *know* on which side each particle lies. He must therefore have gained an amount of information equal according to (7) or (9) to $S_{Sh} = N \log 2$. If entropy and information are measured in the same unit, it can be shown [11] that information may be transformed into negentropy, with possible losses. One may let the entropy of a system decrease by some amount, provided one uses to this aim at least the same amount of information.

Conversely, how is this information acquired? It can be shown that the measurements required in this purpose involve macroscopic physical devices which undergo observable and irreversible transformations, and that in these transformations the entropy must increase by an amount at least equal to the information gained. We indicated that the changes of the entropy (14) during measurement processes were among the incentives of von Neumann when he introduced this expression. The equivalence between negentropy and information which is exhibited by transformations in either direction enforces the interpretation (20) of the thermodynamic entropy in terms of S_{vN} issued from the maximum entropy criterion.

Thus, altogether, if the demon is automatized and if we do not consider the gain of information in the intermediate steps of the process, the whole system including the vessels, the mechanism of gate opening or closing, and the measuring device that governs this operation, obeys the Second Law. The decrease of entropy of the gas is compensated for by at least the same increase of entropy in the measuring device. Proteus has exorcized Maxwell's demon, owing to his two shapes, entropy and information.

6   Other entropies

We list below various other entropies [4] which, in one way or another, measure disorder or missing information.

6.1   Relative entropy

The relative entropy of Kullback and Leibler,

$$S(p|q) = \sum_m p_m \log\left(\frac{p_m}{q_m}\right) , \qquad (37)$$

characterizes the gain of information when prior probabilities q_m are replaced by an actual probability set $\{p_m\}$. Its introduction is natural in the continuum limit when no invariance exists to support (12); in this case a prior distribution $q(x)$ replaces the translationally invariant measure of integration.

Its quantum equivalent

$$S(\hat{D}_2|\hat{D}_1) = \mathrm{Tr}\ \hat{D}_2(\ln \hat{D}_2 - \ln \hat{D}_1) \qquad (38)$$

is currently used to build mean-field approximations in which the exact $\hat{D}_1$ is replaced by a simpler density operator $\hat{D}_2$. Minimization of (38), which is positive for $\hat{D}_1 \neq \hat{D}_2$, provides the best $\hat{D}_2$, closest to $\hat{D}_1$.

6.2   Rényi's entropy

By releasing some among the conditions that characterize Shannon's entropy, other entropies can be defined. Rényi's α-entropies

$$S_\alpha(\{p_m\}) = \frac{1}{1-\alpha} \log \sum_m p_m^\alpha \qquad (39)$$

are additive for a pair of uncorrelated events, but not subadditive (eq.(10) is violated) except in the limit as $\alpha \to 1$ where Shannon's entropy (8) is recovered from (39). They are useful in the context of fractality. The many attempts to build non extensive thermodynamics from the Tsallis entropy

$$S_q(\{\hat{D}\}) = \frac{1}{1-q}[\mathrm{Tr}\ \hat{D}^q - 1] , \qquad (40)$$

directly related to (39), run counter to the Zeroth and Second Laws, because maximization of (39) or (40) introduces correlations between non-interacting subsystems [12].

6.3 Quantum information

Many recent works have been devoted to quantum information theory, in which a bit of information taking values 0 or 1 is replaced by a q-bit, that is, by the quantum state of a two-level system. In this context, it is useful to evaluate the *entanglement* of two systems, that is the purely quantum contribution to their order of correlations, for which several expressions are being proposed. The von Neumann entropy cannot be used since it takes into account both ordinary and quantums correlations. Moreover, when the two entangled systems are an emitter and a receiver, they may be manipulated separately but not together, and the unitary invariance that underlies von Neumann's definition is no longer a requirement here.

6.4 Kolmogorov-Sinai entropy

In the theory of *deterministic chaos*, the evolution of a system is characterized by a set of non-linear differential equations which generates a flow in the space of dynamical variables. The entropy of Kolmogorov and Sinai is a measure of the more or less disordered character of this *flow*, while all the entropies considered above referred to the probabilistic description of a state at a given time. It is defined by partitioning the space of dynamical variables into cells, by discretizing the time, and by analyzing how the cells map on one another after many time steps.

Chaotic dynamics have been proposed to explain, even for systems with few degrees of freedom, the irreversibility paradox. Indeed, prediction becomes hazardous if the motion is chaotic. However, the loss of information that the equations of motion (17) entail in statistical physics is due to the complexity associated with the large number of variables, not to non-linearity.

6.5 Algorithmic complexity

All the types of entropy reviewed so far characterize statistical ensembles of systems, or for Kolmogorov's entropy the full family of trajectories. The concept of algorithmic complexity has been proposed to measure the disorder existing in an *individual* message or in the configuration of a *single system*. The idea is the following. The considered configuration is first represented numerically by coding all its specific features. One then imagines how the resulting number can be constructed by means of algebraic operations in a universal Turing machine, that is, in an ideal computer. The algebraic complexity is the logarithm of the number of steps of the shortest program needed to realize this task. For a family of messages or of systems, its average can be identified with Shannon's missing information. This is still another face of the concept of entropy.

I wish to thank B. Duplantier for his careful reading.

The literature about entropy is immense, and specific searches should be made for each of its very many aspects. We quote below only a few books or articles, either for their tutorial nature or because they contain extensive bibliographies.

References

[1] H.B. Callen, Thermodynamics (Wiley, 1975).

[2] E.H. Lieb, J. Yngvason, *Phys. Rep.* **310**, 1 (1999).

[3] S.R. de Groot and P. Mazur, Thermodynamics of irreversible processes (North Holland, 1962). For an elementary introduction, see http://e2phy.in2p3.fr/2003.

[4] W. Thirring, Quantum mechanics of large systems (Springer, 1983).

[5] R.T. Cox, *Amer. J. Phys.* **14**, 1 (1946); B. de Finetti, Theory of probability (Wiley, 1974).

[6] E.T. Jaynes, *Phys. Rev.* **106**, 620 (1957); **108**, 171 (1957). An introduction aimed at a general audience is included in R. Balian, Université de Tous les Savoirs, vol. 4 (O. Jacob, 2001) p. 947.

[7] J. Uffink, *Studies in History and Phil. of Modern Phys.* **26B**, 223 (1995).

[8] R. Balian, N.L. Balazs, *Ann. Phys. (NY)* **179**, 97 (1987).

[9] A tutoral introduction to this subject is given in R. Balian, *Amer. J. Phys.* **67**, 1078 (1999). Many references can be found in R. Balian, Y. Alhassid and H. Reinhardt, *Phys. Reports* **131**, 1 (1986), where the entropy-based metric is introduced, and in J. Rau and B. Müller, *Phys. Reports* **272**, 1 (1996).

[10] J.E. Mayer and M.G. Mayer, Statistical mechanics, 2nd edition (Wiley, 1977) pp. 145–154.

[11] L. Brillouin, Science and information theory, (Academic Press, 1956).

[12] M. Nauenberg, *Phys. Rev. E* **67**, 036114 (2003) and references therein.

Roger Balian
Service de Physique Théorique
CEA – Saclay
F-91191 Gif-sur-Yvette Cedex
email: balian@spht.saclay.cea.fr

Poincaré Seminar 2003, 145 – 191
© Birkhäuser Verlag, Basel, 2004

On the Origin and the Use of Fluctuation Relations for the Entropy

Christian Maes*

Abstract. Since about a decade various fluctuation relations for the entropy production have been derived and analyzed. These relations deal with symmetries of the entropy production under time-reversal and have been proposed as a non-perturbative generalization of fluctuation–dissipation relations. I describe a unifying framework for understanding these relations and I present an algorithm to derive them. The fluctuation relations all follow from the main observation that in great generality the path-dependent entropy production is the source-term of time-reversal breaking in the Lagrangian over space-time histories. That is illustrated via a number of examples as well as via a general theoretical argument. I move these relations away from the strict dynamical background in which they originated and take them back to the context of statistical mechanics where entropy is understood in the sense of Boltzmann, as measuring the typicality of a manifest condition. I discuss how a relation between work and free energy is naturally put in that framework and how the transient and steady state fluctuation theorems are simple consequences. The fact that fluctuation symmetries for the entropy production are in general only valid asymptotically for large times, makes them mostly inaccessible for experimental verification, in contrast with a recent claim that they would usefully quantify second law violations. Part of the interest in the resulting fluctuation symmetries is that they are so universally valid, a rare occasion in nonequilibrium statistical mechanics. However they do not provide a systematic perturbation expansion for response functions. For that one needs to go back to the full Lagrangian and also consider the nonequilibrium modifications to its time-symmetric part.

1   Scales

Our first look at the world does not invite to think of nature as one. Depending on scales of length, speed, energy and time, new worlds appear that to a large extent can be explored independently. As a matter of fact, at least when sufficiently excited, man feels quite decoupled from underlying microscopic laws and realities. It is also largely due to that decoupling between and the apparent independence of various scales of description that progress in different scientific domains, in particular in physics, is at all possible. Ignoring microscopic details offers often fast and reliable access to questions belonging to the mesoscopic or macroscopic domain. Nevertheless, part of scientific progress is also in the unification of phenomena and in the convergence of explanations into a limited set of more elementary mechanisms. The extent to which a microscopic theory corresponds to reality depends on

---

*http://itf.fys.kuleuven.ac.be/~christ/

the success of that application of reductionism to a variety of natural phenomena of which we all share the same experiences. From these successes we have learnt to speak not of independent scales but rather of a hierarchy of scales with interconnected theories through which we can move and derive one description from another finer level.

Statistical mechanics plays the role of a transfer mechanics between different levels of description. The microscopic laws get connected with the macroscopic behavior. Obviously, the microscopic dynamics is important; perhaps not all details but many features of it, like symmetries and conservation laws, remain visible even on much larger scales. Additional considerations come from counting; these are the statistical aspects. It is there that entropy appears. Entropy throws a bridge between the microscopic motion and the thermal phenomena by introducing the notion of typicality: what can one reasonably expect from a system containing lots of particles and how big are the fluctuations. Not only energy considerations but also entropy considerations make the world work. In the words of Boltzmann: *The general struggle for existence of living creatures is therefore not one for the basic elements – the elements of all organisms are present in abundance in air, water, and the soil – or for energy, which is unfortunately contained in abundance in any body in the form of unconvertible heat, but it is a struggle for the entropy that is available through the passage of energy from the hot Sun to the cold Earth.*

In equilibrium statistical mechanics, one usually forgets about the microscopic dynamics and the central object is the microscopic Hamiltonian. The Gibbs ensembles are constructed and one shows how thermodynamics is obtained from the resulting formalism. Entropy gets translated in various other thermodynamic potentials or free energies. The Gibbs formalism also allows studies of fluctuation and response relations and numerous techniques, computational, perturbative or variational allow detailed studies of the equilibrium system. The equilibrium phases with the resulting phase diagram are constructed from weighing the various possible macroscopic regimes on the energy-entropy balance.

The power of the Gibbs formalism has so far no real counterpart in nonequilibrium physics which to all appearances, shows an even greater variety of phenomena. Most people would agree that there is yet no such thing as nonequilibrium statistical mechanics. In recent years however we have witnessed a renewed interest both in foundational and computational issues from which attempts have been formulated and results have been obtained which are not restricted to the close to equilibrium regime. One guiding idea of the present contribution is to exploit the Gibbsian aspects of the space-time distribution for nonequilibrium systems. The main object of study is now the Lagrangian defined on time-dependent reduced variables and the entropy production will be seen to coincide with the source term for time-reversal breaking. From there it presides over the system's fluctuations and response behavior.

In the next section I specify the key-message of the paper. Sections 3 to 6 contain my personal look at textbook material regarding entropies and the second law. Sections 4.4 and 6 are more original but except perhaps for some notation, much of Sections 3–6 can be skipped when one enjoys running. The rest of the paper substantiates and illustrates the main observation, which is next.

2   Main observations

Here I summarize what is going to come. To get a quick start, I avoid here the many definitions and notations and I write about quantities yet to be defined more precisely later.

There are two main observations.
The first one is that in very great generality the physical entropy production can be identified with the source term of time-reversal breaking in the action governing the space-time distribution. By that I mean the following. Given some type of dynamics on phase space, we can be interested in the probability $P_{\rho_0}(\omega)$ of a trajectory ω of reduced variables. Think about a time-sequence of values that some macroscopic observable can take, or more generally, about some type of contracted description of the system's evolution. The subscript ρ_0 gives the initial distribution on phase space. There is a natural notion of time-reversal which transforms ω into a new trajectory $\Theta\omega$. At some later time τ, the state of the system is described by the distribution ρ_τ, its time-reversal is denoted by $\rho_\tau\pi$, and we are interested in the logarithmic ratio

$$R(\omega) = k_B \, \log \frac{P_{\rho_0}(\omega)}{P_{\rho_\tau\pi}(\Theta\omega)} \tag{2.1}$$

The constant k_B is Boltzmann's constant which I will forget when convenient. I claim that, whenever a physical interpretation is possible, $R(\omega)$ is the physical entropy production over the time-interval $[0, \tau]$, up to a total time-difference. In other words, writing formally

$$P_{\rho_0}(\omega) = \rho_0(\omega_0)\, e^{-\mathcal{L}(\omega)} \tag{2.2}$$

under time-reversal, the antisymmetric term in the Lagrangian $\mathcal{L}$ gives

$$S(\omega) = \mathcal{L}(\Theta\omega) - \mathcal{L}(\omega) = R(\omega) + \log \rho_t(\omega_t) - \log \rho_0(\omega_0) \tag{2.3}$$

and is the variable entropy production modulo a temporal boundary term. That is true for time-dependent or homogeneous dynamics, stochastic or deterministic, in the transient or in the steady state regime (for which $\rho_0 = \rho_\tau$) but of course we need a physical context to verify it.
The statement is non-trivial and should be argued. I do that first by giving four examples in the next subsection. From it, the reader will also better understand what is meant by the statement. A general argument takes more space and some of it will be provided in Section 7 and further.

The second main observation is that the first observation is useful, basically because of two reasons. First, from that general relation (2.1) between time-reversal and entropy production, a unification follows for a variety of recently obtained results concerning nonequilibrium fluctuations. I have in mind the transient and stationary fluctuation theorems that have appeared first in the study of dynamical systems, but also the so called Jarzynski relation relating equilibrium free energies with irreversible work done. I will discuss these in Sections 8–9. An algorithm appears for deriving fluctuation symmetries and they reformulate (2.1) in terms of probabilities and expectations.

A second reason why (2.1) can be useful is that it gives a general way of departure for studying response relations and for obtaining extensions of fluctuation-dissipation relations. Section 10 gives a glimpse at it. We will see there how the observation (2.1) and hence the fluctuation symmetries, are insufficient to go beyond linear order in perturbation theory. One needs also information about how the time-symmetric term in the Lagrangian gets modified under nonequilibrium conditions. The broader perspective is thus that of a Lagrangian statistical mechanics, see below, which would also include insights about the thermodynamic nature of the time-symmetric part of the Lagrangian.

2.1 Examples

2.1.1 Heat conduction

Consider a finite graph $G = (V, \sim)$ with vertex set V. Every site $i \in V$ carries a momentum and position coordinate $(p_i, q_i) \in \mathbf{R}^2$. These oscillators are coupled for a Hamiltonian

$$H(p, q) = \sum_{i \in V} \frac{p_i^2}{2} + U(q) \qquad (2.4)$$

for some symmetric nearest neighbor potential

$$U(q) = \sum_i U_i(q_i) + \sum_{i \sim j} \lambda_{ij} \Phi(q_i - q_j) \qquad (2.5)$$

where $\lambda_{ij} = \lambda_{ji} \neq 0$ whenever $i \sim j$ and Φ is even. I refer to [62] for a recent review.

Select a non-empty subset $\partial V \subset V$ of boundary sites, that are imagined connected to thermal baths at possibly different temperatures. The dynamics is Hamiltonian except at the boundary ∂V where the interaction with the reservoirs has taken the form of Langevin forces as expressed by the Itô stochastic differential equations

$$\begin{aligned}
dq_i &= p_i dt, \quad i \in V \\[1mm]
dp_i &= -\frac{\partial U}{\partial q_i}(q) dt, \quad i \in V \setminus \partial V \\[1mm]
dp_i &= -\frac{\partial U}{\partial q_i}(q) dt - \gamma p_i + \sqrt{\frac{2\gamma}{\beta_i}} dW_i(t), \quad i \in \partial V
\end{aligned} \qquad (2.6)$$

The β_i are the inverse temperatures of the heat baths coupled to the boundary sites $i \in \partial V$; $dW_i(t)$ are mutually independent, one-dimensional white noises. Consider a time-interval $[0, \tau]$; we start the dynamics from the distribution ρ_0 on the (p_i, q_i) obtaining a distribution ρ_τ at time τ. The kinematical time-reversal is $\pi(p, q) = (-p, q)$. A trajectory $\omega = (p_i(t), q_i(t))$ has time-reversal $\Theta\omega = (-p_i(\tau - t), q_i(\tau - t))$. The density (2.1) was computed in [47] and was found to be (with $k_B = 1$)

$$R(\omega) = \sum_{i \in \partial V} \beta_i J_i(\omega) + \log \rho_0(p(0), q(0)) - \log \rho_\tau(p(\tau), q(\tau)) \qquad (2.7)$$

where the J_i are the time-integrated heat currents at the boundary. The definition of these heat currents follows from applying a global energy balance to the equations (2.6). The change of entropy in the reservoirs corresponds to that energy dissipated in the environment divided by the temperature of the reservoirs: $S(\omega) = \sum_i \beta_i J_i(\omega)$.

When we take the average of R in the steady state $\rho_0 = \rho_\tau$ we immediately have from (2.1) that

$$\langle e^{-R} \rangle = 1 \text{ and hence } \langle R \rangle \geq 0$$

or

$$\sum_i \beta_i \langle J_i \rangle \geq 0$$

If the volume V is a linear chain $[0, N]$ with $\partial V = \{0, N\}$ the left and right endpoint of the chain, then by the steady state condition $\langle J_N \rangle = -\langle J_0 \rangle$. Therefore

$$(\beta_0 - \beta_N) \langle J_0 \rangle \geq 0$$

which shows that the heat current into the coldest reservoir is positive. By applying the inequalities at the end of Section 9.2.4 strict inequalities can be obtained. I know of no other method which shows so easily and in such great generality that thermodynamically elementary result.

The computation that leads to (2.7) is the generalization of a Langevin-type calculation starting from:

$$dx(t) = -F(x(t)) \, dt + dW(t)$$

for which the path-space measure (2.2) is formally given by

$$P(\omega) = \exp -\frac{1}{2} \int_0^\tau dt \left[\left(\dot{x}(t) + F(x(t)) \right)^2 - \nabla F(x(t)) \right]$$

The antisymmetric part under time-reversal in the action $\log P$ indeed gives the dissipated power in terms of a stochastic Stratonovich integral of the "force" $F(x(t))$ times the "velocity" $\dot{x}(t)$.

2.1.2   Asymmetric exclusion process

I now look at a bulk driven diffusive lattice gas where charged particles, subject to an on-site exclusion, hop on a ring in the presence of an electric field and in contact with a heat bath at inverse temperature β. Write $\xi(i) = 0$ or 1 depending on whether the site $i \in \mathcal{T}$ is empty or occupied; $\mathcal{T} = \{1, \ldots, \ell\}$ with periodic boundary conditions. The electric field does work on the system and the 'probability per unit time' to interchange the occupations of i and $i+1$ is given by the exchange rate

$$c(i, i+1, \xi) = e^{\beta E/2}\xi(i)(1 - \xi(i+1)) + e^{-\beta E/2}\xi(i+1)(1 - \xi(i)) \qquad (2.8)$$

Consider a path ω of the process in which at a certain time, when the configuration is ξ, a particle hops from site i to $i+1$, obtaining the new configuration $\xi^{i,i+1}$. Then, the time-reversed trajectory shows a particle jumping from $i+1$ to i. Therefore, our (2.1) is given by summing over all jump times, contributions of the form

$$\log \frac{c(i, i+1, \xi)}{c(i, i+1, \xi^{i,i+1})} = \beta E \left[\xi(i)(1 - \xi(i+1)) - \xi(i+1)(1 - \xi(i)) \right] \qquad (2.9)$$

The right-hand side reconstructs the particle current because we get $+1$ or -1 depending on whether the particle has jumped at time t in the direction of E or opposite to it:

$$S(\omega) = \beta E \sum_{i,t} J_{i,t}(\omega)$$

with $J_{i,t}(\omega) = \pm 1$ depending on whether a particle has passed $i \to i+1$ or $i+1 \to i$. Multiplying the current with the force E (with charge and distance taken to be 1) we obtain the variable Joule heating.

The above is not surprising as it is already implicit in the very definition of the model. The same structure will always be recovered for Markov chains. Here is how it goes:

Suppose that K is a finite set and let $(x(t), t \in [0, \tau])$ be a K–valued stationary Markov process. The transition rate to go from a to b is denoted by $k(a, b), a, b \in K$ and we assume that $k(b, a) = 0$ whenever $k(a, b) = 0$ (dynamic reversibility). Suppose now that

$$\frac{k(a, b)}{k(b, a)} = \frac{k_o(a, b)}{k_o(b, a)} e^{E\, J(a,b)} \qquad (2.10)$$

with the detailed balance condition

$$\frac{k_o(a, b)}{k_o(b, a)} = e^{U(b) - U(a)}$$

for the reference (equilibrium) process with (driving) $E = 0$. The time-reversal of a trajectory $\omega = (x(t))$ is $\Theta\omega = (x(\tau - t))$. We can compute (2.1) or (2.3) directly via a so called Girsanov formula:

$$S(\omega) = U(x(\tau)) - U(x(0)) + E \sum_t J(x(t), x(t^+))$$

where we sum over all the jump times t at which $x(t) \to x(t^+)$. If the model has a physical interpretation with (2.10) expressing local detailed balance for a driving force E and a current J, then $S(\omega)$ indeed reproduces the entropy production. One can extend this to spatially extended systems, so called interacting particle systems, see [49, 50]. One shows there can be no current without heat, meaning that detailed balance is equivalent with zero mean entropy production, see also [60, 61, 68].

2.1.3 Strongly chaotic dynamical systems

Here there is *a priori* no notion of physical entropy production but sometimes, making analogies with systems described via effective thermostated dynamics, one thinks of the phase space contraction as entropy production, see also [18, 1]. The trajectory ω is given by an orbit $(x, \varphi(x), \ldots, \varphi_\tau(x))$ in phase space. For so called Anosov diffeomorphisms φ, there is a natural stationary distribution $P(x)$ which turns out to be a Gibbs measure for the potential

$$U(x) = -\log \|D\varphi|_{E^u(x)}\|$$

where $E^u(x)$ is the unstable subspace of the tangent space at the point x ($U(x)$ is the sum of the positive local Lyapunov exponents). Over a trajectory, the Lagrangian equals

$$\mathcal{L}(\omega) = \sum_{t=0}^{\tau} U(\varphi_t x)$$

There is a time-reversal π for which φ is dynamically reversible: $\pi \circ \varphi = \varphi^{-1} \circ \pi$ and we write $\Theta = \pi \circ \varphi, \Theta^2 = \mathrm{id}$.
For (2.3) we need

$$U(\Theta x) = \log \|D\varphi|_{E^s(x)}\|$$

where the stable subspace $E^s(x)$ of the tangent space at the point x appears. The expansions and contractions in the unstable and stable directions give together rise to a net contraction rate (positive or negative) $U(\Theta x) - U(x)$. Therefore,

$$S(\omega) \simeq \sum_0^{\tau} J(\varphi_t x)$$

with $J(x) = -\log \|D\varphi(x)\|$, the phase space contraction rate. That is again a version of (2.1) but now obtaining the phase space contraction rate for S. I refer to [18, 66, 52] for more details and for an extension, see also further under Section 7.3.

2.1.4   Diffusion in local thermodynamic equilibrium

I consider diffusion processes whose space-time fluctuations in the history ω of the particle density $n_t(r)$ at space-time (r, t) in the volume V are governed by a Lagrangian

$$\mathcal{L}(\omega) = \frac{1}{2} \int_0^\tau dt \int_V dr \left(\vec{w}, \frac{1}{\chi(n_s(r))} \vec{w}\right) \tag{2.11}$$

where $(\cdot, \cdot)$ denotes the scalar product in $\boldsymbol{R}^3$ and

$$\vec{w} = \nabla^{-1}\left(\frac{\partial n_t(r)}{\partial t} - \frac{1}{2}\nabla \cdot (D(n_t(r))\nabla n_t(r))\right)$$

is the vector whose divergence equals the difference between left-hand side and right-hand side in the hydrodynamic equation

$$\frac{\partial n_t(r)}{\partial t} = \nabla \cdot J_r(n_t), \quad J_r(n_t) = \frac{1}{2}\chi(n_t(r))\nabla(-s'(n_t(r)))$$

$$= \frac{1}{2}D(n_t(r))\nabla n_t(r)$$

$D(n_t(r))$ is the diffusion matrix, connected with the mobility matrix χ via

$$\chi(n_t(r))^{-1}D(n_t(r)) = -s''(n_t(r)) \text{ Id}$$

for the identity matrix Id and the local thermodynamic entropy s.
Such a quadratic form for $\mathcal{L}$ can be derived for (2.2) for a number of stochastic dynamics but it is believed to be very general for diffusion processes, see, e.g., [3]. Clearly, (2.3) equals

$$S(\omega) = \int_0^\tau dt \int_V dr \left(\nabla^{-1}\left(\frac{\partial n_t(r)}{\partial t}\right), \chi(n_t(r))^{-1}D(n_t(r))\nabla n_t(r)\right)$$

or

$$S(\omega) = -\int_0^\tau dt \int_V dr \left(\nabla^{-1}\left(\frac{\partial n_t(r)}{\partial t}\right), \nabla s'(n_t(r))\right)$$

That is of the form, current $\nabla^{-1}(\partial_t n_t(r))$ times a gradient in the local chemical potential $-s'(n_t(r))$ as we are used to find in expressions of entropy production for local thermodynamic equilibrium. If there is no particle current in or out of the system, we can integrate by parts to find

$$S(\omega) = \int_V dr \left[s(n_\tau(r)) - s(n_0(r))\right] = S(\tau) - S(0)$$

the total change of entropy.
Apart from the nonlinearities in (2.11) that is fully in line with the Onsager-Machlup set-up of 50 years ago, [58].

2.2 Lagrangian statistical mechanics

In the above examples, the Lagrangian is constructed from a model dynamics. These dynamics need not be fully microscopic; they are valid, effective and useful in some regimes, under some approximations and for some purposes. The thought arises therefore whether one could not as well start the business from specifying the Lagrangian, leaving aside its detailed dynamical origin. After all, also in equilibrium statistical mechanics one constructs models and theories based on approximate Hamiltonians, not worrying all the time about the deeply hidden origins. In that Lagrangian statistical mechanics, far or close to equilibrium, the Gibbs formalism again applies but now for distributions of space-time histories. One advantage is that the step to quantum mechanical processes is much easier to make. Another advantage is that the stationary distribution, the projection to the temporal hyperplane, becomes available for perturbation theory as it is the exponential of Hamilton's principle function, see [3, 36].

At this moment, one requirement for the Lagrangian stands out: that the antisymmetric part under time-reversal be given by the entropy production. The Lagrangian is however more than just its time-antisymmetric part. For perturbation theory beyond linear order, also the symmetric part matters, see Section 10.2. As the fluctuation symmetries of Section 9 will just be reformulations of the main observation (2.1), they cannot be the starting point of a systematic perturbation expansion. For that we need the full Lagrangian, see, e.g., [53].

3 The second law in thermodynamics

In the beginning of the 19th century Sadi Carnot followed his father Lazare in meditating over perfect machines and efficiency, see, e.g., [23, 12, 34]. That abstraction led Sadi Carnot to think of a generalized heat engine which draws heat from a source and delivers useful work. To operate continuously, the engine requires also a cold reservoir to which heat can be disposed. The new idea of Carnot was to think of a reversible machine, one that can be run in the opposite direction thereby restoring its original thermodynamic state. He realized that no heat engine can be more efficient than a reversible one operating between the same temperatures. These pure thoughts (*Réflexions sur la puissance motrices du feu*, 1824) imply practically useful insights, such as that one need not worry about working substances other than water in the design of steam engines.

Still room was left for further theoretical reflections. For example, since the reversible efficiency is a universal function of the reservoir temperatures, by inverting the reasoning, we can define temperature on a universal scale independent of the particular substance. To that, Kelvin and Joule added their insights that constitute the first law of thermodynamics, the conservation of energy in thermal processes as elevated from the principle of conservation of mechanical energy. Heat and work, naturally appearing in expressions of efficiency, can now be written in the same units and Carnot's principle takes the form

$$\frac{J_1}{T_1} + \frac{J_2}{T_2} \geq 0$$

where $J_1(J_2)$ is the heat given by the engine to the first (second) reservoir at absolute temperature $T_1(T_2)$. Its extension to more reservoirs was given by Kelvin in 1854 but it was Clausius who came with integrals:

$$\oint \frac{dJ}{T} \geq 0 \tag{3.1}$$

where one now imagines a cyclic process through which the engine makes contact with reservoirs at temperature T. If the process is reversible, the equality applies in (3.1) and T is also the (instantaneous) temperature of the system. But since then

$$\oint \frac{dJ}{T} = 0$$

for every cycle, we get the path-independence of the line integral

$$\int_{\alpha}^{\alpha'} \frac{dJ}{T} = S(\alpha') - S(\alpha) \tag{3.2}$$

where we integrate over a reversible path (equilibrium states). Thus Clausius discovered in 1865 a new function of the thermodynamic state. He called S the entropy (from the Greek $\tau\rho o\pi\eta$, turn, a turning point). As a direct consequence of (3.1)–(3.2), by closing the cycle,

$$\int_{\alpha'}^{\alpha} \frac{dJ}{T} \geq S(\alpha') - S(\alpha) \tag{3.3}$$

for all paths between macrostate α' and α of the system and processes where the system is in contact with a reservoir at temperature T and gives it heat dJ. The left-hand side is the total heat dissipated into the reservoirs and hence, is the change of entropy in the rest of the universe. Putting everything at the same side, we thus see the familiar

$$S(\text{final}) - S(\text{initial}) \geq 0 \tag{3.4}$$

for the total entropy of the universe, or, as interpreted by Clausius, *Die Entropie der Welt strebt einem Maximum zu.*

The equality (3.2) is the operational definition of entropy. Entropy is only defined here for certain thermodynamic states, what are called equilibrium states under specified restraints on composition, energy, etc. They are interconnected via reversible processes.

One does not need to remain with energy exchanges only. If we have an equilibrium system with energy E and particle numbers N_i in a volume V, the entropy $S(E, N, V)$ is defined as in (3.2) from the exact differential

$$dS = \frac{1}{T}(dE + pdV - \sum \mu_i dN_i)$$

where T is the absolute temperature, p is the pressure and μ_i are the chemical potentials.

That definition of thermodynamic entropy can be extended to systems in local thermodynamic equilibrium. Thermodynamics teaches that entropy is additive and when surface effects can be neglected, we also get extensivity: $S(E, N, V) = Vs(e, n)$. We then introduce local energy densities $e(r)$, particle densities $n(r)$ (for one component) and a velocity profile $u(r)$ for which $\int_V n(r)u(r)\, dr = 0$ to write

$$S(e, n, u) = \int_V s\big(e(r) - \frac{1}{2}mn(r)v^2(r)\, , \, n(r)\big)\, dr \tag{3.5}$$

One imagines here the system as composed of microscopically very large domains that are still much smaller than the thermodynamic size of the system.

The inequality (3.4) describes the net result, nothing intermediate, of a process that starts and ends in equilibrium. One can also apply it to (3.5) with the understanding that during the process the system remains in local thermodynamic equilibrium. One must then add the macroscopic evolution equations for the quantities $e_t(r), n_t(r)$ and $u_t(r)$ and introduce intensive quantities like local temperature to obtain the entropy balance equation. For example, if the system is isolated with local temperature $T(r)$ and constant $n(r)$ and $u(r) = 0$, the changes in energy satisfy the conservation law $de_t(r)/dt + \mathrm{div}\, J(r) = 0$ with J the heat current and we have by partial integration

$$\frac{d}{dt}\int_V s(e(r))\, dr = \int_V J \cdot \nabla\frac{1}{T}\, dr$$

The linear transport law (here, Fourier's law) defines the coefficient λ for which $J = -\lambda/T^2\, \nabla T$ and the entropy production rate equals

$$\frac{d}{dt}\int_V s(e(r))\, dr = \int_V \lambda\, (\nabla\frac{1}{T})^2\, dr \geq 0$$

requiring $\lambda \geq 0$. These manipulations serve many different scenario's, for open or for closed systems, in a stationary or in a transient regime, and are typical for, or, what is worse, are restricted to irreversible thermodynamics close to equilibrium.

The above summary contains the standard propositions of macroscopic phenomenology and can be found in many textbooks, e.g., [2]. Yet, it is not particularly illuminating when asked what entropy really is and how to extend it to truly nonequilibrium situations. The microscopic understanding of entropy and the second law was first present in the works of Maxwell, Boltzmann and Gibbs. It is in fact the start of statistical mechanics.

4 Second law from microscopic considerations

The relation (3.2) can be studied in quite some detail for an ideal gas. A simple calculation shows that there the (equilibrium) entropy $S(E)$ can be related to the total phase volume W. More specifically, up to an additive constant,

$$S = k_B \log W \tag{4.1}$$

where, for an ideal gas of N atoms in volume V, for which the energy lies in $(E, E + dE)$,

$$W \equiv \int_{\sum p_i^2/2m \simeq E, q_i \in V} dq_1 \ldots dq_N \, dp_1 \ldots dp_N$$

The formula (4.1) appears on the tombstone of Boltzmann in Vienna but it was first written down in that form (and criticized) by Planck who introduced the W as thermodynamic *Wahrscheinlichkeit*. It guided the young Einstein to propose experimental verifications of the corpuscular nature of fluids and of radiation. Already by its simplicity we cannot but feel that the first mysteries regarding entropy should now evaporate. Boltzmann had truly realized that counting is essential for predicting macroscopic behavior and that entropy was just that.

4.1 Counting

The formula (4.1) can be read and generalized as follows. An immense number of microstates belong to one and the same macrostate (manifest condition). The entropy counts them.

Counting seems to refer to something discrete and introducing discreteness in classical phase space has something arbitrary. One can of course refer to quantum mechanics but that introduces again other problems (more about that later). I give more precise definitions later but in fact, not much is necessary. Most important however is to treat all microstates, that is every assignment of values of positions and momenta of the particles, as equivalent. That is the microcanonical distribution and its relevance is mostly derived from it being left invariant by the Hamiltonian equations of motion (Liouville's theorem).

Denote by Ω the phase space of a perfectly closed and isolated mechanical system. The elements $x \in \Omega$ represent the microstates, i.e., $x = (q_1, \ldots, q_N, p_1, \ldots, p_N)$ gives the canonical variables for a classical system of N particles. The equations of Hamilton produce the flow $x \mapsto \varphi_t(x)$ on Ω and preserve the phase space volume: $|d\varphi_t(x)/dx| = 1$ for each t; the Liouville measure dx is time-invariant. We think of a large volume in which the many particles are conserved and the dynamics also conserves the total energy. We introduce therefore the state space $\Omega_E \equiv \Gamma$, the energy shell, corresponding to energies within $(E, E + dE)$ and denote by $|B|$ the phase space volume of a region $B \subset \Gamma$ given by the projection of the Liouville measure into Γ.

Now we change scales. For comparison with experimental situations, we look at variables that summarize the manifest image of the system. We have in mind a collection of macroscopic variables $A_r(x)$, mostly approximately additive and locally conserved functions on phase space. We obtain a subdivision of Γ by cutting it up in phase cells defined by $\alpha_r < A_r(x) < \alpha_r + \Delta\alpha_r$ (with some tolerance $\Delta\alpha_r$). As a result, to every $x \in \Gamma$ is associated a region $M(x)$ in phase space consisting of all microstates that share with x the same manifest image. Boltzmann defines then the entropy as

$$S(x) = k_B \log |M(x)| \tag{4.2}$$

where we leave out some irrelevant constants that appear when the number of particles can change. That is the meaning of (4.1). Most important to remember in scrutinizing definition (4.2) is the great disparity in sizes of $W = |M(x)|$ and the smallness of Boltzmann's constant $k_B = 1.381 \times 10^{-23}$ J/K. An entropy difference $S' - S$ of about 0.1 millicalorie at room temperature corresponds to a phase volume ratio of

$$\frac{W}{W'} = \exp -(S' - S)/k_B = e^{-10^{20}}$$

Since entropy is extensive in the number of particles, any visible change in the entropy per particle (as measured in units of k_B) corresponds to a ratio of phase volumes that is exponential in the number of particles.

4.2   The hand-waiving part

Equilibrium is the state of maximal entropy; it has the overwhelmingly largest volume in phase space which makes it the typical endpoint for about every evolution. Equilibrium can thus be described via a variational principle, maximizing the entropy while keeping some constraints fixed. If a constraint is lifted, even in a time-dependent way, new macrostates can be explored and the entropy will increase until a new equilibrium is installed. In that sense, the very definition of equilibrium as solution of a variational principle is already incorporating the second law of thermodynamics.

The microscopic definition (4.2) of entropy allows to move beyond equilibrium thermodynamics. We can now follow a microstate and its entropy as time runs. Accepting the counting procedure above and realizing the great difference in scales, that entropy will increase needs no further explanation. Aside from great conspiracies, it is expected that the dynamics takes the system into macrostates with a larger number of compatible microstates, [24, 4, 41]. If in a large collection of black and white balls the number of white balls is about a billion times larger than the number of black balls, then, when picking a ball by whatever which procedure that treats the balls equivalently, we would be surprised not to have picked a white one.

Violations of that microscopic version of the second law are now absolutely possible and in fact ought to be expected when the number of particles and/or the lapse of time over which the system is monitored gets small. It is only because of the huge amount of particles in one mole of a gas that we never witness Poincaré recurrencies. That is the answer to the Zermelo paradox.

The second law appears very rigid when applied in the laboratory situation or in the understanding of macroscopic behavior. After all that is why it is called a law. Nevertheless, its microscopic roots are statistical. In the end it is a probability statement, a *moral certainty* as Maxwell put it. Or, with Gibbs: *the impossibility of an uncompensated decrease of entropy seems to be reduced to an improbability.* Even if we have a large number of particles, one can easily imagine initial conditions for which the evolution breaks the second law. The second law only applies to typical initial data, microstates randomly selected from within the phase volume that corresponds to the initial macrostate.

4.3   Time-(a)symmetry

It is in the last sentence of the previous section that appears the truly hard part about the second law. To state that more precisely let me first recomfort the reader that so far no arrow of time has appeared. Think indeed of a dynamically reversible time-evolution. By reversing all molecular velocities, the *same* equations of motion carry the system back along their very *same* trajectories to their initial positions where they end up with their velocities reversed. That is the reversibility that is present in the microscopic dynamics of a mechanical system. Everything we have said before to make the second law understandable applies to both directions of time. No *a priori* sign of the microscopic time needs to be selected and the change of entropy is then expected to be equal and positive in both time-directions. Furthermore, for every initial microstate from which the entropy increases, there is also a microstate from which the Boltzmann entropy decreases. Nevertheless, that Loschmidt construction does not give rise to a paradox because the second law applies to a typical initial condition and the microstates that are obtained after evolution (with reversed molecular velocities) make only a very small fraction of the microstates that are actually corresponding to the final macrostate.

The deeper question is however why we can have nonequilibrium states in the first place. The second law can only come in action if the system is initially prepared in a macrostate with a much smaller phase volume than is reachable by the evolution. That has obviously been done by putting constraints of some sort but all the same it is again the endpoint of a previous evolution for a bigger system, perhaps including now ourselves. Iterating this argument an arbitrary number of times, we soon reach questions of physical cosmology. A healthy attitude, here and elsewhere, seems to be that instead of trying to (dis)prove the second law, one should use it. In fact, the second law now guides us to conclude within a particular theory as to the nature of the initial condition of the universe; it must allow for the huge entropy increases that have occurred ever since. Obviously, the theory of general relativity and hence gravity play there an all-important role. Within the standard theory of physical cosmology, that leads to precise mathematical estimates on various aspects of the initial singularity. We refer to Chapter 7 in [59] and to [35] for a first look at such derivations.

4.4 Shadowing the macrostate evolution

I am adding here some more mathematical relations to the discussion above. The advantage is not so much precision but generality and recognizing what is less and more essential.

Let Γ be the phase space of a dynamical system. We ask that Γ is equipped as a probability space with some probability measure ρ. Let φ_t be a map on Γ that leaves ρ invariant.
Since we plan to do statistical mechanics and have in mind systems composed of a huge number of particles, we specify a number of macroscopic variables

$$A_r(x) = \frac{1}{N} \sum_{k=1}^{N} a_r^k(x) \tag{4.3}$$

with a_r^k a function that only depends say on the state of the $k-$th particle. For example, we could divide the six-dimensional one-particle phase space in cells C_r and let a_r^k be one or zero depending on whether the state of the $k-$th particle $x_k \in C_r$ or not. Or, the index r can have a purely spatial interpretation in which case $A_r(x)$ gives the profile of some one-particle observable. We can of course go beyond one-particle variables and r can then indicate various other macrovariables but that is not essential here. In short, a macrostate $\alpha = (\alpha_r)$ is realized by the microstate x when $A_r(x) = \alpha_r$ plus/minus some tolerance. I write $\alpha(x)$ for it.

The macrostate α can be represented in different mostly equivalent ways. The most fundamental way appears to be the microcanonical formulation. We then think of Γ as the constant energy surface and ρ is the projection of the Liouville measure. We associate with α the phase volume M_α of all microstates x for which $\alpha(x) = \alpha$ and the entropy of such an x is then, following (4.2), $S(x) = \log|M_\alpha|$.
It is however often more convenient to imagine that the macrostate α is not given via a phase volume but via a distribution function ρ_α. I thus prefer to go to other Gibbs ensembles and we associate to α a probability measure ρ_α on Γ having a density with respect to ρ. The probability ρ is for example the Gibbs measure at some inverse temperature β with respect to some microscopic interaction and ρ_α has a density

$$\frac{d\rho_\alpha}{d\rho}(x) = \exp[-N \sum_{r=1}^{n} \lambda_r A_r(x) - \ln \frac{Z_\lambda}{Z}] \tag{4.4}$$

with respect to ρ. The Lagrange multipliers λ_r, conjugate to the A_r, are determined from requiring the ρ_α-expectations $\langle A_r \rangle_u = \alpha_r$ and ρ_α is a constrained or, depending on the interpretation of the index r, a local equilibrium measure in the canonical set-up; Z_λ and Z are the partition functions corresponding to ρ_α and ρ respectively.

It is important to select the pure phases for which $A_r(x) = \alpha_r$ with overwhelming ρ_α-probability. These satisfy a law of large numbers. In all cases, the negative logarithm of the density

$$S(x) = -\log \frac{d\rho_{\alpha(x)}}{d\rho}(x) \tag{4.5}$$

corresponds to the variable entropy as in (4.2). Indeed, the negative logarithm of (4.4) corresponds exactly to the entropy governing the (equilibrium) fluctuations of ρ. Write $W_N(\alpha)$ as the ρ-probability to see $A_r(x) = \alpha_r$. Then, approximately for large N,

$$W_N(\alpha) = \exp[N \sum_{r=1}^{n} \lambda_r \alpha_r + \log \frac{Z_\lambda}{Z}], \quad W_N(\alpha(x)) = e^{S(x)} \tag{4.6}$$

The precise formulation (and proof) of (4.6) belongs to the theory of large deviations, see, e.g., [71, 39, 22], but that is just an elaboration of the older Boltzmann-Planck-Einstein relation (4.1): the entropy governing the macroscopic fluctuations.

We have thus physical interest in considering (4.5) and we would hope that $S(\varphi_t(x)) \geq S(x)$ at least for most of the microstates x that are drawn out of a particular macrostate, represented by ρ_α. Expectations with respect to ρ_α are written as $\langle \cdot \rangle_\alpha$ and without subscript, $\langle \cdot \rangle$ is an expectation in ρ. I now claim that

$$\langle e^{-S(\varphi_t x) + S(x)} \rangle_\alpha = 1 \tag{4.7}$$

The reason is simple. For almost all x, under ρ_α, $\alpha(x) = \alpha$ and hence

$$\langle e^{-S(\varphi_t x) + S(x)} \rangle_\alpha = \langle e^{-S(\varphi_t x)} \rangle = \langle e^{-S(x)} \rangle = 1$$

by definition and by the invariance of ρ under φ_t. The equality (4.7) implies that

$$\langle S(\varphi_t x) - S(x) \rangle_\alpha \geq 0 \tag{4.8}$$

with strict inequality in all non-trivial cases (if $S(\varphi_t x) - S(x)$ is really variable, not constant equal to zero).

Observe that for microstates x,

$$\langle S \rangle_{\alpha(x)} = S(x)$$

On the other hand, if we define $F(\alpha) = \alpha'$ from $\alpha'_r \equiv \langle A_r \circ \varphi_t \rangle_\alpha$, i.e., the macrostate after evolution with φ_t, then

$$\langle S \circ \varphi_t \rangle_{\alpha(x)} = S(x')$$

for every x' with $\alpha(x') = F(\alpha(x))$. As a conclusion, the inequality (4.8) is for the change in entropies

$$S(x') - S(x) \geq 0$$

for all x' that realize the macrostate that comes out from evolving the macrostate to which x belongs. The word macrostate is here understood in the Gibbsian sense, i.e., in terms of the probability distributions ρ_α introduced before. That implies that we have not succeeded yet in deriving the microscopic picture and that the entropies in (4.8) remain equilibrium Gibbs entropies.

5 H-theorem

At the end of the previous paragraph, we got somehow an evolution from one macrostate to another one, both equilibrium, almost by definition and not because the dynamics is like that. Not surprisingly, we did not end up with the stronger Boltzmann formulation of the second law. Yet, we can obtain more than the second law by requiring that the evolution of macroscopic variables is autonomous.

We think again about the microcanonical ensemble. In that case, ρ is the Liouville measure on a constant energy surface Γ. We make a partition of Γ according to the possible values of a collection of macroscopic variables $A_r(x)$ as in (4.3). The $M(x)$ of (4.2) is the phase volume containing all the $y \in \Gamma$, $A_r(y) = A_r(x)$. A macrostate α specifies one such phase volume, call it M_α and $M_{\alpha(x)} = M(x)$. If wished we can proceed as in the previous section and speak about the distribution ρ_α, now concentrated on one phase volume M_α and random within.

Suppose that at a certain moment, along the trajectory, x is our microstate with corresponding macrostate $\alpha = \alpha(x)$. If we now apply the dynamics φ_t, a priori $\varphi_t(x)$ can fall in various different macrostates even when we know, as we do, that $x \in M_\alpha$ initially. In other words, we get a non-trivial statistics on at least some macroscopic values and it seems we fall back in the framework of section 4.4, the canonical Gibbs formalism. One can proceed along these lines but let me however deviate here and instead make an additional assumption. I assume that the macroscopic partition is so well chosen for the dynamics φ_t that (at least approximately for systems composed of many particles) in fact the dynamics is autonomous on the macroscopic variables. That means that from knowing the macrostate α at some (arbitrary) time, we also know the macrostate at a later time: almost all $y \in M_\alpha$ satisfy $\varphi_t(y) \in M_{\alpha'}$ for the same macrostate α'. Then, clearly, the image of macrostate α under φ_t is concentrated within the phase volume $M_{\alpha'}$, $\varphi_t(M_\alpha) \subset M_{\alpha'}$ or better $|\varphi_t(M_\alpha)| \leq |M_{\alpha'}|$. As a consequence, by Liouville's theorem

$$|M_\alpha| = |\varphi_t(M_\alpha)| \leq |M_{\alpha'}|$$

and thus, the entropy (4.2) is typically non-decreasing. We have now obtained something even stronger than the second law, along a microscopic trajectory

$S(\varphi_t(x)) \geq S(\varphi_s x), t \geq s$ from a typical initial microstate, i.e., one that realizes macrostates according to the autonomous equation.

Observe that the previous argument, first given in [33] and recently described in [25, 26] captures very well the intuitive meaning of a constant increase in entropy but it delegates the problem to establishing an autonomous equation for the macrovariables. That macroscopic reproducibility is well documented in experience but few are the examples where we can actually achieve it starting from microscopic models. The most famous one is the Boltzmann equation where the H−theorem was first proven for dilute gases. In contrast to what is often claimed, one does not need an extra assumption of randomization or *Stoßzahlansatz*, see [40]. In general, it remains to be seen what are the essential conditions on the dynamics that produce an H−theorem. Certainly, the chaotic nature of the microscopic dynamics can help, as adding stochasticity helps to prove it. Yet that may not be necessary.

A formulation of what an H−theorem means in a quantum context is at the end of the next section.

6   Quantum entropies

The logic and usefulness of the ideas above are not at all restricted to classical statistical mechanics. In quantum mechanics, one can go quite a bit in the same direction. Literally *a bit*, because in some sense the quantum situation is more discrete (quantized) and therefore seems more amenable to counting. Yet there are problems. One has to do with the nature of the microstates for a quantum system, the phase space. We have to make sure that we know what exactly there is to count and what we mean by a quantum history. Secondly, there is the problem of macroscopic measurements. One can always say that macroscopic variables are classical variables because they commute and we can measure them simultaneously. Yet, before taking the thermodynamic limit, they do not commute. Thirdly, there is no well-established theory of large deviations. The basic relation between entropy and fluctuations of macroscopic quantities, even in equilibrium, has not been satisfactorily settled. I will not discuss these problems here, only mention them again in context, and I restrict myself to giving some definitions which appear to be no longer very standard.

One would like to say that quantum entropy is the logarithm of the dimension of the subspace in the Hilbert space $\mathcal{H}$ of the system, that corresponds to the manifest condition. I suppose that $\mathcal{H}$ is finite dimensional (but very large). Given a microstate $\psi \in \mathcal{H}$ one then follows von Neumann [57] in writing

$$S(\psi) \equiv \sum_\alpha (\psi, P_\alpha \psi) \log d_\alpha - \sum_\alpha (\psi, P_\alpha \psi) \log(\psi, P_\alpha \psi) \tag{6.1}$$

corresponding to the decomposition

$$\mathcal{H} = \bigoplus \mathcal{H}_\alpha \tag{6.2}$$

into linear subspaces. The manifest condition (or macrostate) is represented by the projections P_α on $\mathcal{H}_\alpha$, $P_\alpha P_\beta = \delta_{\alpha,\beta} P_\alpha$, $\sum_\alpha P_\alpha = \mathrm{id}$. We write d_α for the dimension of $\mathcal{H}_\alpha$; it is the analogue of the phase space volume in the classical case.
A problem here is that if we have more than one macroscopic observable, say the magnetization in the $z-$ and in the $x-$direction for a collection of spin $1/2$-particles, in general not commuting before the thermodynamic limit is taken, then they do not have a joint eigenspace decomposition and we are in the dark as to what to write for P_α.

The sum in (6.1) reflects that the wavefunction can still correspond to different (mutually exclusive) macrostates (here labeled by the running index α). That feels like the description starting from the wavefunction as microstate is not complete. That discussion would take me way too far. Part of the problem is what we mean by statistical ensembles. As an example, the identification of probabilities on wavefunctions and density matrices does not seem one-to-one.

One can move definition (6.1) to the level of density matrices and call

$$S(\rho) \equiv \sum_\alpha \mathrm{Tr}[P_\alpha\, \rho] \log d_\alpha - \sum_\alpha \mathrm{Tr}[P_\alpha\, \rho] \log \mathrm{Tr}[P_\alpha\, \rho] \qquad (6.3)$$

the entropy of the state represented by ρ. Note that (6.3) (just like (6.1)) only depends on ρ through its projection $p(\rho)$ (a probability measure) on the macroscopic states:

$$p(\rho)(\alpha) \equiv \mathrm{Tr}[P_\alpha\, \rho] \qquad (6.4)$$

Underlying is the variational principle

$$S(\rho) = \sup_{p(\rho')=p(\rho)} - \mathrm{Tr}[\rho' \log \rho'] \qquad (6.5)$$

with the supremum reached at

$$\rho' = \sum_\alpha \frac{p(\rho)(\alpha)}{d_\alpha} P_\alpha \qquad (6.6)$$

The above mimics rather well the situation in classical statistical mechanics and the relation there between Boltzmann and Gibbs entropies and how (constrained) equilibrium is characterized by a variational principle. So we can move beyond (6.5) and construct the quantum equilibrium entropy just as one does in the Gibbs formalism, see section 4.4. The constraints are then in the form of specifying expectations like $\mathrm{Tr}[\rho'\, A_r]$, for a class of macroscopic observables A_r, and we obtain the quantum equilibrium states, generalizing (6.6) to the standard Gibbs form. The problem is however that we do not have the fluctuation formula (4.6) so important for (4.1). In fact, I even believe it is not true in general. For example, suppose we have a collection of N spin $1/2$-particles with a local Hamiltonian H at inverse temperature β. The equilibrium reference state has thus a density matrix $\exp[-\beta H]/Z$. Suppose now the additional constraint that

$$\langle M_z \rangle = \alpha_z$$

where $M_z = \sum_{i=1}^{N} \sigma_i^z/N$ is the magnetization in the z-direction. The constrained state has density $\exp[-\beta H - \lambda N M_z]/Z_\lambda$ for suitable Lagrange multiplier $\lambda = \lambda(\alpha_z)$. Its free energy is

$$\frac{1}{N} \log \mathrm{Tr}[e^{-\beta H - \lambda N M_z}] \tag{6.7}$$

On the other hand, the probability in equilibrium that the magnetization is about α_z is

$$\frac{1}{Z} \mathrm{Tr}[e^{-\beta H} P_{\alpha_z}] \tag{6.8}$$

where P_{α_z} is the projection on the space $\mathcal{H}_{\alpha_z}$ with eigenvalues of M_z around α_z. Now, in contrast with (4.6), the logarithm of (6.8) will not be given by the Legendre transform of the free energy (6.7) but rather by Legendre transforming

$$\frac{1}{N} \log \mathrm{Tr}[e^{-\beta H} e^{-\lambda N M_z}] \tag{6.9}$$

and these might give different results even in the limit $N \uparrow +\infty$ when the commutator $[H, M_z]$ is of order one.

Let me finally present the logic of any H-theorem that also works in a quantum set-up, [10].
To a density matrix ρ, we associate a macrostate $\alpha = \alpha(\rho)$ that collects the values of expectations under ρ for a selection of macroscopic observables. To a macrostate α is associated a constrained equilibrium state ρ_α which maximizes

$$-\mathrm{Tr}\,[\rho' \log \rho'], \quad \alpha(\rho') = \alpha$$

The maximum is the entropy $S(\alpha) = -\mathrm{Tr}[\rho_\alpha \log \rho_\alpha]$ of α. I write $S(\rho) = S(\alpha(\rho))$ for the entropy corresponding to the state represented by ρ.
Now dynamics comes; let $\rho(s)$ denote the density matrix at time s under a unitary evolution starting from ρ. We have, by Liouville-von Neumann,

$$\begin{aligned} S(\rho(s)) &= S(\alpha(\rho(s))) = S(\rho_{\alpha(\rho(s))}) \tag{6.10} \\ &= S(\rho_{\alpha(\rho(s))}(t-s)) \end{aligned}$$

The macroscopic autonomy-assumption of Section 5 is written as follows:

$$\text{if } \alpha(\rho(s)) = \alpha(\rho'(s)), \text{ then } \alpha(\rho(t)) = \alpha(\rho'(t)), \quad t \geq s$$

In (6.10), under that autonomy-assumption,

$$\alpha(\rho_{\alpha(\rho(s))}(t-s)) = \alpha(\rho(t))$$

and hence, by the variational characterization of the entropy

$$S(\rho_{\alpha(\rho(s))}(t-s)) \leq S(\rho(t))$$

That gives the quantum $H-$theorem

$$S(\rho(s)) \leq S(\rho(t)), \quad s \leq t$$

when combined with (6.10). As for classical dynamics, the main problem remains to understand when the autonomy is established.

In the rest of the paper, I will not come back to quantum aspects. It does not mean that there are no quantum versions of what follows. The relation between time-reversal and quantum entropy is discussed in [5]. The simplest example of relaxation to equilibrium for a unitary evolution on quantum spins, with an associated $H-$theorem can be found in [11]. An extension to some quantum versions of a fluctuation relation that connects irreversible work with free energy or to so called transient fluctuation theorems is in [38, 9, 70, 54, 55, 69, 74].

7 Time-reversal and entropy

In the present section, I give a general argument for believing in the main observation under (2.1). It is a compact upgrade of what was written in [46] in collaboration with K. Netočný. Those that are happy with examples are referred to Section 2.1 and to [48] for even more examples.

7.1 Mathematical set-up

I start with formalities that generate a quite broad class of possible scenarios. Illustrations come in the next subsection.
As we have seen before in Sections 4.3–4.4, it is not necessary for the second law that the microscopic dynamics be time-reversal invariant. From now on however, I will only consider the case that the microscopic dynamics is reversible; an extension to microscopically irreversible dynamics is in [45].

Let Γ be the phase space. There are dynamics on Γ, invertible transformations f_t, possibly depending on t, and I write

$$\varphi_t = f_t \circ f_{t-1} \circ \ldots f_1, \quad t = 1, \ldots, \tau$$

for the time-inhomogeneous updating after τ steps. If we reverse the order (for fixed τ), we get the reversed dynamics

$$\widetilde{\varphi_t} = f_{\tau-t+1} \circ \ldots \circ f_{\tau-1} \circ f_\tau$$

I skip the continuous time version.

The phase space Γ supports a measure ρ that is left invariant by φ_τ: $\rho(\varphi_\tau^{-1}B) = \rho(B)$. Γ is further equipped with an involution π that also leaves ρ invariant. I assume dynamical reversibility in the sense that for all t,

$$f_t \circ \pi = \pi \circ f_t^{-1}$$

As a consequence, $\pi\tilde{\varphi}_t^{-1}\pi = f_\tau \circ \ldots f_{\tau-t+1}$.

The statistical physics begins from considering a collection $\mathcal{A} = \{A_r\}$ of functions $A_r : \Gamma \to \mathbf{R}$ indexed via subscript r. It divides Γ in volumes M_α containing all $x \in \Gamma$ with $A_r(x) = \alpha_r$, plus/minus some tolerance that I ignore here. The label α runs the contracted description.
A statistics μ on $\mathcal{A}, \mu : \mathcal{A} \to \mathbf{R}$ assigns a real number to every function A_r.
It is also understood that the collection $\mathcal{A}$ is globally invariant under π so that we can define $\pi(A_r(x)) = A_r(\pi x) = A_{\pi r}(x)$ and $\pi\alpha$. The time-reversal of a statistics μ is written $\tilde{\mu}$ with $\tilde{\mu}(A_r) = \mu(\pi A_r)$.

To a statistics μ I associate a probability measure on Γ. That is done via a variational principle (which I assume is well posed).
The entropy of a statistics μ is the supremum

$$S(\mu) = \sup - \int_\Gamma g(x) \log g(x)\, \rho(dx) \tag{7.1}$$

over all probability densities $g \geq 0$ with

$$\int g(x)\, A_r(x)\, \rho(dx) = \mu(A_r), \quad \int g(x)\, \rho(dx) = 1$$

That entropy governs the fluctuations of ρ.
I denote by g_μ the density that reaches the supremum in (7.1) and ρ_μ is the probability measure on Γ with density g_μ with respect to ρ. The variable and μ-dependent entropy of x is

$$S_\mu(x) = -\log g_\mu(x) \tag{7.2}$$

The probability ρ_μ evolves under the dynamics and gives rise to a new statistics

$$\mu_\tau(A_r) = \int A_r(\varphi_\tau(x))\, g_\mu(x)\, \rho(dx)$$

at later time τ.

Consider now a trajectory $\omega = (\alpha(0), \alpha(1), \ldots, \alpha(\tau))$ collecting all $x \in \Gamma$ for which $A_r(x) = \alpha_r(0), \ldots, A_r(\varphi_\tau(x)) = \alpha_r(\tau)$ for all r. Its time-reversal is $\Theta\omega = (\pi\alpha(\tau), \pi\alpha(\tau-1), \ldots, \pi\alpha(0))$. The probability to see ω under the dynamics φ_t started from ρ_μ is denoted by $P_\mu(\omega)$. The probability to see $\Theta\omega$ under the reversed dynamics $\tilde{\varphi}_t$ started from $\rho_{\widetilde{\mu_\tau}}$ is denoted by $\tilde{P}_{\widetilde{\mu_\tau}}(\Theta\omega)$. Observe that it is natural to run the reversed dynamics because if for some $x \in \Gamma, y_0 = x, \ldots, y_n = \varphi_\tau x$, then $(\pi y_\tau, \pi y_{\tau-1}, \ldots, \pi y_0)$ is an orbit for the reversed dynamics.

Now comes the major claim that constitutes a part of the first observation in Section 2. Take $x \in M_0$, i.e., with the $A_r(x) = \alpha_r(0)$ and suppose that $\varphi_\tau(x) \in M_\tau$, i.e., with the $A_r(\varphi_\tau x) = \alpha_r(\tau)$. Then,

$$S_{\mu_\tau}(\varphi_\tau x) - S_\mu(x) = \log \frac{P_\mu(\omega)}{\widetilde{P}_{\widetilde{\mu_\tau}}(\Theta\omega)} \tag{7.3}$$

It is only a part of the first claim in 2 because (7.3) really talks about a transient behavior and not a steady state situation of a driven system maintained in a nonequilibrium state. Nevertheless, (7.3) is the mother of all further relations, see below in Section 7.3, but let us first see why (7.3) is true. Observe that

$$P_\mu(\omega) = e^{-S_\mu(x)} \rho[\cap_{t=0}^\tau \varphi_t^{-1} M_t]$$

while

$$\widetilde{P}_{\widetilde{\mu_\tau}}(\Theta\omega) = e^{-S_{\mu_\tau}(\varphi_\tau x)} \rho[\cap_{t=0}^\tau \tilde{\varphi}_t^{-1} \pi M_{\tau-t}]$$

But the last factor can be rewritten by using that $\rho(B) = \rho(\varphi_\tau^{-1} \pi B)$,

$$\rho[\cap_{t=0}^\tau \tilde{\varphi}_t^{-1} \pi M_{\tau-t}] = \rho[\cap_{t=0}^\tau \varphi_\tau^{-1} \pi \tilde{\varphi}_t^{-1} \pi M_{\tau-t}]$$

Finally use that $\varphi_\tau^{-1} \pi \tilde{\varphi}_t^{-1} \pi = \varphi_{\tau-t}^{-1}$ to conclude (7.3).

7.2 Illustrations

A simple illustration of the above notation is to take for Γ the phase space of constant energy for N particles in a volume V. The dynamics $f_1 = \ldots = f_\tau, \varphi_\tau = f^\tau$ is a discretization of a conservative Hamiltonian dynamics with ρ the projection of the Liouville measure on $\Gamma : \rho(B) = |B|$. One can imagine a finite partition of Γ corresponding to values of some set of macroscopic variables and each phase volume M_α collects the microstates that show the same manifest condition or macrostate α.

A statistics μ gives an initial probability measure on the macrostates, $\mu(\alpha)$ and the constrained equilibrium ρ_μ, that solves the variational principle (7.1) is

$$\rho_\mu(B) = \sum_\alpha \mu(\alpha) \frac{\rho(B \cap M_\alpha)}{|M_\alpha|}$$

Its density with respect to ρ is thus $g_\mu(x) = \mu(\alpha)/|M_\alpha|$ when $x \in M_\alpha$. The entropy (7.2) then equals $S_\mu(x) = \log |M_{\alpha(x)}| - \log \mu(\alpha(x))$. Its expectation under ρ_μ is

$$\langle S_\mu \rangle_\mu = \sum_\alpha \mu(\alpha) \log |M_\alpha| - \mu(\alpha) \log \mu(\alpha)$$

If μ is concentrated on just one macrostate α, then $\langle S_\mu \rangle_\mu = \log |M_\alpha|$ which corresponds to the Boltzmann entropy if that macrostate is selected from a microstate x as $\alpha_r = A_r(x)$.

A second illustration is obtained if we write ρ_μ in the Gibbs form. Say,

$$g_\mu(x) = \frac{1}{Z} e^{-\sum_r \lambda_r A_r(x)}$$

which solves the variational principle (7.1) for a suitable choice of λ_r made from the $\mu(A_r)$. The entropy $S(\mu)$ is the usual equilibrium Gibbs entropy in the ensemble determined by the A_r. The variable entropy $S_\mu(x)$ has ρ_μ−expectation exactly equal to $S(\mu)$. That is generally true; taking expectations of (7.3) under ρ_μ gives the change of the Gibbs entropy in terms of the expected breaking of time-reversal symmetry.

7.3   Open and driven

Just like in equilibrium statistical mechanics, it is useful to consider reservoirs that interact with the system and to derive a thermodynamics in terms of quantities that depend solely on the system's state. In other words, we want to integrate out the degrees of freedom of reservoirs, heat or particle baths, and introduce variables that describe what is going on in terms of the evolution of the system. In nonequilibrium statistical mechanics, there are plenty of effective models that describe just that. We have seen two examples of that, one a Markov diffusion process and the other a Markov jump process in Section 2.1. It is however interesting to see whether the extensions of (7.3) that were obtained there, have a more general validity. The goal is therefore to integrate out the reservoirs from the right-hand side in (7.3) and to obtain an expression which is again of the same form (as the right-hand side of (7.3)) but now in terms of probabilities for a history of the system only. That question was discussed in [46, 45] and I give here a summary of the positive result.

I consider now a system in contact with k reservoirs. The microstate x is decomposed as $x = (x_S, x_1, \ldots, x_k)$ with x_S representing the system variables and x_v stands for the microstate of the v−th reservoir. The macrostates α will now only be macroscopic in so far as the reservoirs are concerned; the system remains described by x_S. So we have macroscopic observables $A_r^v(x_S, x_v)$ whose values are determined from knowing the state x_S of the system and the state x_v of the v−th reservoir. An α is given if we know x_S and also $A_r^v(x_S, x_v)$ for all r and v. In the course of time x_S and x_v can change and hence the $A_r^v(x_S, x_v)$ are variable in time. That is how we get non-constant trajectories ω.
Yet, the intensive variables that characterize each reservoir are kept steady. Otherwise, I would not call them reservoirs. That is the steady approximation: the temperature, pressure or the electro/chemical potentials of each reservoir remain

fixed and unchanged over the time. It does not mean that there can be no time-dependent force acting on the system making the dynamics inhomogeneous as we had it before. For example, the Hamiltonian of the system can change while it is in contact with a heat bath at fixed temperature, see Section 8.

A second specification, besides the fact that the intensive variables of the reservoir remain fixed during the evolution, is that the reservoirs are spatially separated and locally coupled to the system. What I want is that from knowing the trajectory $\gamma = (\eta(0), \eta(1), \ldots, \eta(\tau))$ of the system, I know exactly what currents $J_r^v(\gamma)$ have been flowing in what reservoir. So, to the change $\eta(t) \to \eta(t+1)$ at time t corresponds a displacement $A_r^v(\eta(t+1), x_v(t+1)) - A_r^v(\eta(t), x_v(t))$ that can be calculated from the couple $(\eta(t), \eta(t+1))$ alone. In physical realizations, that is obtained from local conservation laws.

As a consequence of the previous hypothesis, trajectories ω as we had them before in (7.3) for the total system, are of the form $\omega = ((\eta(0), U(\eta(0))), \ldots, (\eta(\tau), U(\eta(0)) + J(\gamma))$ where $U(\eta(0))$ specifies the values $A_r^v(\eta(0), x_v(0)) = U_r^v(\eta(0))$ depending on $\eta(0)$ and from then on, all the $A_r^v(\eta(t), x_v(t)), t \geq 1$ are completely determined by the trajectory $\gamma = (\eta(0), \eta(1), \ldots, \eta(\tau))$ of the system only. In particular, we know

$$A_r^v(\eta(\tau), x_v(\tau)) = U_r^v(\eta(0)) + J_r^v(\gamma)$$

where $J_r^v(\gamma)$ is the time-integrated flux of type r flowing into the $v-$th reservoir.

For the reference measure ρ I take the product $dx = dx_S\, dx_1 \ldots dx_k$ over all momenta and positions of the particles. The initial statistics μ determines the intensive variables of the reservoirs and a probability distribution for the microstate x_S of the system. Denote by $h(x_S)$ the probability density of the system. That means that the initial density $g_0 = g_\mu$ can be written as

$$g_0(x) = g_\mu(x) = h(x_S) \prod_{v=1}^{k} \frac{e^{-\sum_r \lambda_r^v A_r^v(\eta, x_v)}}{Z_v(\eta)} \tag{7.4}$$

The h and the Lagrange multipliers λ_r^v sit in μ. g_μ corresponds to an equilibrium state of the environment conditioned on the state of the system, as distributed via h.

The dynamics defines the statistics $\widetilde{\mu_\tau}$ at time τ as we had it before. The only thing that changes with respect to (7.4) for the new $g_{\widetilde{\mu_\tau}}$ is that we now get a distribution h_τ instead of h for the system and, in general, new λ_r^v. As I have mentioned before however, we want to model reservoirs with fixed intensive quantities. Therefore, introduce

$$g_\tau(x) = h_\tau(x_S) \prod_{v=1}^{k} \frac{e^{-\sum_r \lambda_r^v A_r^v(\eta, x_v)}}{Z_v(\eta)} \tag{7.5}$$

where, compared with (7.4), I have changed the distribution of the system and left the λ_r^v untouched. It is that density (7.5) that I will use to start the reversed dynamics.

Let us first look at $P_\mu(\omega)$ of (7.3). According to our set-up,

$$P_\mu(\omega) = \int dx\, g_0(x) \prod_{t=0}^{\tau} \delta(x_S(t) - \eta(t)) \prod_{r,v} \delta(A_r^v(x_S, x_v) - U_r^v(\eta(0)))$$

or

$$P_\mu(\omega) = h(\eta(0)) \frac{e^{-\sum_{r,v} \lambda_r^v U_r^v(\eta(0))}}{Z_v(\eta(0))} \times$$

$$\int dx \prod_{t=0}^{\tau} \delta(x_S(t) - \eta(t)) \prod_{r,v} \delta(A_r^v(x_S, x_v) - U_r^v(\eta(0)))$$

depends only on the trajectory $\gamma = (\eta(0), \eta(1), \ldots, \eta(\tau))$ of the system. In the same way, for $\Theta\omega$ under the reversed dynamics (indicated with a tilde),

$$\tilde{P}_{\widetilde{\mu_\tau}}(\Theta\omega) = h_\tau(\eta(\tau)) \frac{e^{-\sum_{r,v} \lambda_r^v [U_r^v(\eta(0)) + J_r^v(\gamma)]}}{Z_v(\eta(\tau))} \times$$

$$\int dx \prod_{t=0}^{\tau} \delta(\tilde{x}_S(t) - \pi\eta(\tau - t)) \prod_{r,v} \delta(A_r^v(x_S, x_v) - U_r^v(\eta(0) - J_r^v(\gamma)))$$

By the same reasons that led us to (7.3), upon dividing, we get that the remaining integrals in $P_\mu(\omega)$ and $\tilde{P}_{\widetilde{\mu_\tau}}(\Theta\omega)$ cancel each other to yield

$$\log \frac{P_\mu(\omega)}{\tilde{P}_{\widetilde{\mu_\tau}}(\Theta\omega)} = \sum_v \log \frac{h(\eta(0)\, Z_v(\eta(n)))}{h_\tau(\eta(\tau)\, Z_v(\eta(0)))} + \sum_{r,v} \lambda_r^v J_r^v(\gamma) \tag{7.6}$$

The last term is the time-integrated dissipation into the reservoirs (the change of entropy in the environment), which is a function of the trajectory γ of the system. Only that term can be extensive in time, the other terms are temporal boundary terms. Of course, when the system does not get frustrated by the presence of more reservoirs that tell it opposite things, that term becomes also a total time-derivative. For example, suppose there is just one heat bath coupled to the system and no work is done: conservation of energy requires $J(\gamma) = H(\eta(0)) - H(\eta(\tau))$ where H is the Hamiltonian of the system.

The term $\log Z_v(\eta(0))/Z_v(\eta(\tau))$ is the difference in equilibrium free energy of the v-th reservoir for boundary conditions $\eta(\tau)$ and $\eta(0)$ as imposed by the system. These terms in (7.6) are not only boundary terms in time but also in the boundary of the system. One can then expect that they are typically vanishing under weak coupling assumptions, see also in Sections 8.1 and 8.2.

The identity (7.6) complements (7.3) and fulfills the first observation in Section 2. If we allow the steady state condition $h_\tau = h$ and we take the expectation of (7.6), we get the mean steady state entropy production in the world. We can however now also study its fluctuations. That brings us to the second main observation alluded at in Section 2.

8 Jarzynski relation

Thermodynamic potentials are everywhere in applications of thermodynamics. They are tabulated and predict what processes are workable, under what conditions. For example, for a system that can extract heat from an environment at constant temperature T, the energy that is available to do work is exactly the free energy $F \equiv U - TS$, that is its energy U minus the heat term TS where S is the entropy of the system. Turning it around, it suffices to measure the work done under isothermal conditions in changing the parameters of the system and it will be equal to the free energy difference. That however is only valid if the thermodynamic process involved is sufficiently slow, quasi-static, a scenario that cannot be hoped for in many cases. It was therefore very welcome that an extended relation between free energy and work was proposed and exploited in a series of papers since the pioneering work of Jarzynski in 1997, [28]. That identity reads

$$e^{-\beta \Delta F} = \langle e^{-\beta W} \rangle \tag{8.1}$$

In the left-hand side ΔF is what we want to measure, the difference in free energies between two equilibria, say with parameter values κ_τ and κ_0 in the system Hamiltonian. That parameter could for example correspond to a spring constant. The right-hand side is an average over all possible paths that take the system in equilibrium for parameter value κ_i in its initial Hamiltonian to a state where that parameter is changed into κ_f. The work done W depends on the path if the process is not adiabatic (i.e., without heat transfer) or if it is not quasi-static. The protocol, i.e., the sequence of forcing in the time-dependent Hamiltonian, is always kept fixed.

Derivations of the Jarzynski relation (8.1) have been made in various ways and in various approximations, see [8, 28, 29, 30, 31, 32, 48, 46]. From such a relation free energy differences can be measured even in situations where the process of changing the parameters is not so well controlled. That has already been experimentally realized in, e.g., molecular systems [27, 43, 65]. An important statistical problem there is to estimate how many runs one needs to estimate the right-hand side in (8.1). After all, one would like to see trajectories for which the dissipated work is negative. Assuming a Gaussian shape for the (so called second law breaking) tail of the distribution of the dissipated work (as for example done in [65]) is practically useful but theoretically, remains unmotivated.

I will not discuss the various applications and restrict myself to showing how relation (8.1) can directly be obtained from (2.1) or from (7.6).

8.1 In Markov approximation

I start by giving the derivation in the context of a Markov jump dynamics. It is originally due to Crooks, in [8], but can be directly transformed into an illustration of (2.1).

One imagines a time-dependent Hamiltonian H_t for a system in contact with a heat bath at inverse temperature β. An effective dynamics can for example be obtained from a weak coupling limit, where then the driving protocol has to vary on the same time scale as the dissipation processes. At any rate, we start here from a discrete time-inhomogeneous Markov chain on a finite state space with transition probabilities $p_t(\eta, \eta')$ for $\eta \to \eta'$. That governs the thermal transitions in exchanging heat with the reservoir. There is detailed balance with respect to H_t: for any pair of states η, η',

$$\frac{p_t(\eta, \eta')}{p_t(\eta', \eta)} = e^{-\beta[H_t(\eta') - H_t(\eta)]} \tag{8.2}$$

If we start the system in equilibrium for H_0, the probability to see a trajectory $\gamma = (\eta(0), \eta(1), \ldots, \eta(\tau))$ is

$$P_\beta(\gamma) = \frac{e^{-\beta H_0(\eta(0))}}{Z_0} \, p_1(\eta(0), \eta(1)) \ldots p_\tau(\eta(\tau - 1), \eta(\tau))$$

Expectations, needed for the right-hand side of (8.1), are denoted by $\langle G \rangle_\beta = \sum_\gamma G(\gamma) P_\beta(\gamma)$. Obviously, that process does not instruct us about the dynamics of changing the parameters in the Hamiltonian. One can think of an instantaneous change in which H_t is modified into H_{t+1} after every thermal transition.

The total change in energy is $\Delta U = H_\tau(\eta(\tau) - H_0(\eta(0))$ and the total heat that flows in the heat bath in the thermal transitions (8.2) is

$$J(\gamma) = -\sum_{t=1}^{\tau} [\, H_t(\eta(t)) - H_t(\eta(t - 1))\,] \tag{8.3}$$

The total work is therefore defined as

$$W(\gamma) = J(\gamma) + \Delta U = \sum_{t=0}^{\tau-1} [H_{t+1}(\eta(t)) - H_t(\eta(t))] \tag{8.4}$$

The claim is now that

$$\langle e^{-\beta W} \rangle_\beta = e^{-\beta \Delta F} \tag{8.5}$$

where $\Delta F = -1/\beta \log Z_\tau / Z_0$ is the difference in free energies corresponding to H_τ and H_0 respectively.

The simplest way to prove (8.5) is to use the relations (7.6) or (2.1) between entropy production and time-reversal. The reversed dynamics just reverses the protocol and starts from the equilibrium distribution for H_τ. So we let

$$\widetilde{P_\beta}(\gamma) = \frac{e^{-\beta H_\tau(\eta(0))}}{Z_\tau}\, p_\tau(\eta(0),\eta(1))\ldots p_1(\eta(\tau-1),\eta(\tau)) \tag{8.6}$$

and compute the promising

$$R(\gamma) = \log \frac{P_\beta(\gamma)}{\widetilde{P_\beta}(\Theta\gamma)} \tag{8.7}$$

A simple computation that uses (8.2) gives

$$R(\gamma) = \beta[H_\tau(\eta(\tau)) - H_0(\eta(0))] + \log \frac{Z_\tau}{Z_0} - \beta \sum_{t=1}^{\tau}[H_t(\eta(t)) - H_t(\eta(t-1))]$$

Hence, from the definitions (8.3)–(8.4), one arrives at

$$R = \beta\Delta U - \beta\Delta F + \beta J = \beta W - \beta\Delta F \tag{8.8}$$

Now comes the first time to profit from the form of R as in (2.1) or here in (8.7). The point is that we have the normalization condition $\langle \exp(-R)\rangle_\beta = 1$ or, explicitly,

$$\sum_\gamma P_\beta(\gamma) \frac{\widetilde{P_\beta}(\Theta\gamma)}{P_\beta(\gamma)} = 1$$

Inspection learns however that upon substituting (8.8) for (8.7), that normalization is exactly equivalent with (8.5) and we are done.

8.2 As application of Section 7.3

As I have illustrated in the previous lines, in our scheme, the Jarzynski relation follows essentially from a normalization condition. Let us see what that means in a more general context.

The main point is to arrive at (8.8) for the source term of time-reversal breaking. But simply look back at (7.6) and apply it for one heat bath:

$$R(\gamma) = \log \frac{h(\eta(0)\, Z(\eta(\tau))}{h_\tau(\eta(\tau))\, Z(\eta(0))} + \beta J(\gamma) \tag{8.9}$$

We choose $h(\eta) = \exp[-\beta H_0(\eta)]/Z_0$ and we choose $h_\tau(\eta) = \exp[-\beta H_\tau(\eta)]/Z_\tau$ (we are absolutely free to do that) and by the very construction we have

$$\langle e^{-R}\rangle_\beta = 1 \tag{8.10}$$

Hence, when we suppose that $Z(\eta(\tau))/Z(\eta(0)) \simeq 1$ (which is a weak coupling condition), then

$$\langle e^{-\beta\Delta U+\beta\Delta F-\beta J}\rangle_\beta = 1$$

and (8.1) follows from the first law, $\Delta U + J = W$.

There are other and mathematically even simpler derivations of (8.1). In fact, we already had it in (4.7). In the present section I wanted to relate it more closely to the relation between time-reversal and entropy production. What really happens is in (2.1). First use the observation that R is the total entropy production when the system starts and ends in equilibrium at the same temperature but for a different Hamiltonian. Since that total entropy production is the change of entropy in the system $= \beta\Delta U - \beta F$ plus the entropy production in the environment $= \beta J$, we get $R = \beta W - \beta\Delta F$. Next use that R is the logarithm of a probability density over trajectory-space and apply the normalization condition (8.10). Observe however that in the experimental realizations or verifications of the Jarzynski relation $\beta W - \beta\Delta F$ is not equal to the total entropy production since one does not wait to equilibrate the system. Therefore, the observation that for some trajectories $\gamma, \beta W(\gamma) - \beta\Delta F < 0$ does not imply transient violations of the second law.

9 Fluctuation relations

One of the ever returning themes in statistical mechanics is to find the right balance between dynamical and statistical considerations. Since the start of kinetic gas theory, there was a fruitful exchange of ideas between the theory of heat and the theory of dynamical systems. The thermodynamic formalism has become a standard chapter for studies in dynamical systems and, ever since Clausius, heat is understood as motion.

More recently, there has been a fruitful revival of connecting the two theories. In particular, programs are running for understanding the effect of nonlinearities on transport coefficients and for defining nonequilibrium ensembles in terms of Sinai-Ruelle-Bowen measures, [13, 21, 66]. A decade ago, [14] found numerically a remarkable symmetry in the fluctuations of the phase space contraction of a dynamical system. That phase space contraction played the role of entropy production within the effective set-up. The context is that of simulation via molecular dynamics and of thermostated dynamics, see [67]. Of course, even though they resemble Newtonian equations, these models are not microscopic and it is not clear how to derive them in some effective regime. They are mostly numerically interesting. There is also the inconvenience that different thermostats may give rise to different rates of phase space contraction under the same macroscopic conditions so that one needs very specific thermostats to get the phase space contraction coincide with the physical entropy production, see, e.g., [72]. Nevertheless, Gallavotti and Cohen went on to prove a fluctuation symmetry for the steady distribution of the time-averages of the phase space contraction rate in some strongly chaotic dynamics and they hypothesized that this symmetry is much more general and relevant also for the construction of nonequilibrium statistical mechanics and the fluctuations of the entropy production in particular, [18]. That was confirmed by the results in [37, 42] for physically inspired stochastic dynamics but, ideologically, it could add to the feeling that intrinsic randomness, the stochastic or chaotic character of the dynamics, is essential for obtaining a universal fluctuation behavior around a strictly positive entropy production. Moreover, the suggested identification of phase space contraction with entropy production, or in other contexts, between physical entropy and various dynamical entropies, seems to reduce the concept of entropy and of the second law to a purely dynamical context. One is reminded of the words of Maxwell where he criticizes Clausius and Boltzmann for trying at one moment to derive the second law from dynamics, *as if any pure dynamical statement would submit to such an indignity.*

In [44] it was emphasized that the fluctuation symmetry of Gallavotti-Cohen results from the Gibbs formalism and that it is not so much the chaotic nature of the dynamics that should be held responsible but rather the Gibbsian nature of the space-time distribution. That is most easy to see in stochastic models where the method of [44] as applied in [48], not only simplifies the treatments in [42, 37] but also extends the study to non-Markovian dynamics and to the much more physical local fluctuation theorems, see [51].
Also for deterministic chaotic dynamics, the Gibbsian idea extends the results of [18] to the larger class of expansive homeomorphisms with the specification property, where the technology of Markov partitioning and symbolic dynamics are not available, see [52]. It was found that phase space contraction obtains its formal analogy with the physical entropy production as source term in the potential for time-reversal breaking, as already explained under example 2.1.3.

The subject of the present section is to show how the observation in Section 2, in particular equations (2.1), and how its confirmations in (7.3) and (7.6) in a statistical mechanical context, are related to and extend previously obtained fluctuation symmetries.

9.1 General idea

I explain here what a fluctuation symmetry is and how it formally arises in great generality.

Let ω be a variable distributed according to some probability measure $P(d\omega)$. Have in mind that ω is a trajectory for reduced variables and that P is the steady state distribution. Suppose that $R(\omega)$ is a real function of ω that satisfies, formally,

$$\frac{\text{Prob}[R(\omega) = Q]}{\text{Prob}[R(\omega) = -Q]} = e^Q \tag{9.1}$$

at least for a range of Q's. The symmetry expressed by (9.1) is the so called fluctuation symmetry.

It is easy to find a distribution for which (9.1) holds exactly true. As an example take R a real variable with distribution

$$P(dR) = p(R)\,dR = g(R)\,\exp[-(R-1)^2/4]\,dR$$

where $g(R) = g(-R)$ is symmetric. Then, the probability density $p(R)$ satisfies

$$\frac{p(Q)}{p(-Q)} = e^Q$$

which means that (9.1) is satisfied for all Q. In particular, a proper rescaling of an arbitrary Gaussian random variable satisfies (9.1). The opposite is of course not true: the fluctuation symmetry does not at all imply that the (large) fluctuations of the entropy production are Gaussian.

There are various modifications of (9.1). Most interesting is to consider in (9.1) not one R but a sequence S_τ indexed by τ and to require that

$$\lim_{\tau\uparrow+\infty} \frac{1}{\tau}\log\frac{\mathrm{Prob}[S_\tau(\omega) = \tau q]}{\mathrm{Prob}[S_\tau(\omega) = -\tau q]} = q \tag{9.2}$$

Suppose for example that $S_\tau = R + b_\tau$ where $b_\tau(\omega)/\tau$ goes to zero (uniformly) with τ and R satisfies (9.1) exactly. Then, S_τ will satisfy (9.2).

One can also make versions where the probabilities are time-dependent or where the probabilities in the numerator and in the denominator of (9.1) and (9.2) are not quite the same. These have obtained names in the literature as detailed or transient versus steady, global versus local fluctuation symmetries. Let me however leave that zoology for a moment and discuss how fluctuation symmetries might arise.

I start with the exact symmetry (9.1). Suppose indeed that

$$R(\omega) = \log\frac{p(\omega)}{p(\Theta\omega)} \tag{9.3}$$

where $P(d\omega) = p(\omega)d\omega$ and where Θ is some involution, $\Theta^2 = \mathrm{id}$. That is close to what we had before, e.g., in (2.1) or in (7.3) and (7.6). Observe that $R(\Theta\omega) = -R(\omega)$. As a consequence

$$\int G(\omega)\,dP(\omega) = \int G(\Theta\omega)e^{-R(\omega)}\,dP(\omega) \tag{9.4}$$

for all functions G. In particular,

$$\int G(R(\omega))\,dP(\omega) = \int G(-R(\omega))e^{-R(\omega)}\,dP(\omega) \tag{9.5}$$

Since we can take here G to be the indicator function of the event that $R(\omega) = Q$, we recover (9.1) as a special case of (9.5). Here is another special case: take

$G(R) = \exp[-zR]$ for some complex number z,

$$\int e^{-zR(\omega)}\, dP(\omega) = \int e^{-(1-z)R(\omega)}\, dP(\omega) \qquad (9.6)$$

expressing a symmetry in the generating function for the distribution of R. In fact, in the sense of Legendre transforms, (9.1) is dual to (9.6).

Clearly, the asymptotic symmetries like in (9.2) can also be obtained from modifications of (9.3) yielding an asymptotic version of, e.g., (9.6).

The translation of the above formalities to the framework of the rest of the paper is easy. The R has already appeared in Section 2 and it has reappeared in Section 7. The physical entropy production S_τ over a time-span τ equals R, essentially, and hence, a fluctuation symmetry is immediate. In conclusion, we see that once we have understood that the entropy production is the time-reversal breaking part of the Lagrangian, then we understand that its fluctuations are governed as in (9.5).

9.2 Examples of fluctuation symmetries

9.2.1 Steady state fluctuation theorem

The story began with the numerical work of [14] and was firmly brought into the context of dynamical systems by the fluctuation theorem of Gallavotti and Cohen, see [18, 66]. It asserts that for a class of dynamical systems the fluctuations in time of the phase space contraction rate obey a general law. That means the following:

One is asked to consider a reversible smooth dynamical system $x \mapsto \varphi(x), x \in \Gamma$. The phase space Γ is in some sense bounded carrying only a finite number of degrees of freedom (a compact and connected manifold). The transformation φ is a diffeomorphism of Γ. Reversibility means that there is a diffeomorphism π on Γ with $\pi^2 = 1$ and $\pi\varphi\pi = \varphi^{-1}$. It is assumed that the dynamical system satisfies some chaotic (uniformly hyperbolic) condition: it is a transitive Anosov system. It ensures a Markov partition (and the representation via some symbolic dynamics) and the existence of a natural stationary probability measure ρ, the so called Sinai-Ruelle-Bowen measure of the dynamics, with expectations

$$\rho(G) = \lim_\tau \frac{1}{\tau} \sum_0^\tau G(\varphi_t x) \qquad (9.7)$$

corresponding to time-averages for almost every randomly chosen initial point $x \in \Gamma$.

Consider now minus the logarithm of the Jacobian determinant D which arises from the change of variables implied by the dynamics; write $J = -\log D$, $J(x)$ is the phase space contraction rate and is suggested to play here the role of entropy production, see [1]. One assumes (and sometimes proves) dissipativity: the expected contraction

$$\rho(J) > 0 \tag{9.8}$$

is strictly positive.

One is interested in the fluctuations of

$$w_\tau(x) = \frac{1}{\rho(J)\tau} \sum_0^\tau J(\varphi_t(x)), \tag{9.9}$$

for large time τ. The fluctuation theorem then states that $w_\tau(x)$ has a distribution $P_\tau(w)$ with respect to the stationary state ρ such that

$$\lim_\tau \frac{1}{\tau\rho(J)w} \log \frac{P_\tau(w)}{P_\tau(-w)} = 1 \tag{9.10}$$

always. Less precise and more clear,

$$\frac{\mathrm{Prob}[\sum_0^\tau J(\varphi_t x) = q\tau]}{\mathrm{Prob}[\sum_0^\tau J(\varphi_t x) = -q\tau]} = \exp \tau q$$

for large τ. In other words, the distribution of entropy production (read: phase space contraction) over long time intervals satisfies the symmetry (9.2).

I have already explained in Section 2.1.3 how that fluctuation symmetry can be explained from our general perspective: the phase space contraction rate is the source-term for time-reversal breaking in the action. Detailed proofs are to be found in [18, 66, 52] and [20] contains the continuous time version.

Kurchan pointed out that this fluctuation theorem also holds for certain diffusion process, finite systems undergoing Langevin dynamics, [37]. That was extended by Lebowitz and Spohn in [42] to quite general finite Markov processes. The method based on (2.1) is however simpler and more powerful. I have illustrated that already with example 2.1.2. There is a family of stationary distributions ρ_u with u specifying the density. They are Bernoulli measures with density u. The R can be easily calculated along the lines of (2.9). It is exactly equal to the physical entropy production, the variable Joule heating, and we get an exact steady fluctuation symmetry (9.1). There are many more illustrations of that scheme, see, e.g., [48]. The first analytically hard steady state fluctuation theorem was proven in a model of heat conduction by Rey-Bellet and Thomas, [63, 62] whereas the physical heuristics follows exactly what was written under example 2.1.1.

The fluctuation theorem wants to speak about the fluctuations of the entropy production. To make it observationally accessible, in particular for bulk driven systems, we better not consider the global fluctuations as they will be damped exponentially in the size of the system. We are therefore interested in spatially localized fluctuations. To explain, I give an example which is not immediately related to a dynamics.

Consider the standard two-dimensional Ising model in a lattice square V with say periodic boundary conditions. Its Hamiltonian is

$$H_V(\sigma) = -\sum_{|i-j|=1} \sigma_i \sigma_j - h \sum_i \sigma_i$$

The last term (with bulk magnetic field $h \neq 0$) breaks the spin flip symmetry $\sigma \to -\sigma$. We want to find out whether the magnetization in some subsquare Λ which is much smaller than the volume V satisfies a fluctuation symmetry. The object of study is thus

$$M_\Lambda = \sum_{i \in \Lambda} \sigma_i$$

with distribution obtained from the Gibbs measure $P_V \sim \exp[-\beta H_V]$. The idea is that we first let V be very very large and only afterwards consider growing Λ. Thus, we want to show a symmetry $q \to -q$ in the behavior of

$$p_\Lambda(q) = \lim_V P_V[M_\Lambda \simeq q|\Lambda|]$$

as Λ gets larger.

Here is how to get it in complete analogy with the suggestions of Section 2 but with spin configurations replacing space-time trajectories, spin flip replacing time-reversal, the Ising Hamiltonian replacing the space-time Lagrangian and two dimensions being thought of as 1 spatial and 1 temporal dimension. Let Θ_Λ apply a local spin flip: $(\Theta_\Lambda \sigma)_i = -\sigma_i$ if $i \in \Lambda$ and $(\Theta_\Lambda \sigma)_i = \sigma_i$ otherwise. It is immediate that

$$
\begin{aligned}
R_\Lambda(\sigma) &= \log \frac{P_V(\sigma)}{P_V(\Theta_\Lambda \sigma)} &\qquad (9.11)\\
&= \beta H_V(\Theta_\Lambda \sigma) - \beta H_V(\sigma) = 2h\beta \sum_{i \in \Lambda} \sigma_i + O(|\partial\Lambda|)
\end{aligned}
$$

where the last term is of the order of the boundary of Λ. Observe that the volume V has disappeared. As always, R satisfies an exact fluctuation symmetry (9.1), for every V, hence also in the limit:

$$\langle e^{-zR_\Lambda} \rangle_{\beta,h} = \langle e^{-(1-z)R_\Lambda} \rangle_{\beta,h}$$

From here, we substitute (9.11), and see, by easy manipulations that also M_Λ satisfies a fluctuation symmetry up to corrections of order $|\partial\Lambda|/|\Lambda|$:

$$\lim_\Lambda \frac{1}{|\Lambda|} \log \frac{p_\Lambda(q)}{p_\Lambda(-q)} = 2h\beta q$$

(The prefactor $2h\beta$ can of course be scaled away by a proper normalization of M_Λ.) That is called a local fluctuation theorem.

In a dynamical context it was obtained in [15, 19] for coupled chaotic maps and in [51] for reaction-diffusion processes. Instead of considering the full spatial volume, one looks at a finite space-time window which only afterwards is growing larger. The underlying idea remains the fact that on space-time the distribution of trajectories is Gibbsian-like and we can apply exactly the same ideology as above for the Ising model. Even though these local fluctuation relations are more physical, we are still waiting for convincing experimental realizations, see however [6].

9.2.2   Transient fluctuation theorem

The transient case is exactly similar to the steady case except that the dynamics is started from a distribution that changes with time. Within the set-up of dynamical systems, that can make a great difference. Singularities in the stationary measure present extra mathematical difficulties that are not present in the transient case, see, e.g., [7]. For statistical mechanical purposes, it makes no great difference. The idea remains the same, is easier to prove but rests again on the formula (2.1). In a way, a transient fluctuation theorem is an extension of the Jarzynski relation. Just look at (9.6) for $z = 1$ and take R not starting from the steady state but from a transient state: one sees the relation (8.10). Instead of spelling out the mathematical details, let me instead turn to a recent experiment.

9.2.3   Experimental verification

An experimental test and to some extent, verification of a transient fluctuation symmetry is contained in [73]. The authors consider a colloidal particle captured in an optical trap that is translated relative to the surrounding water. The particle is micron-sized, the force is of order of pico-Newton and about 500 particle trajectories were recorded for times up to 2 seconds after initiation. The particle Hamiltonian is time-dependent

$$H_t(p, q) = \frac{p^2}{2m} + \frac{\kappa}{2} (q - a(t))^2 \tag{9.12}$$

with $a(t)$ the time-dependent position of the trap, approximated as the position of the minimum in a harmonic potential with spring constant κ. The force exerted on the particle is $F_t(q) = -\kappa(q - a_t)$. The motion of $a(t)$ is rectilinear. The total work W done on the system over a time τ is

$$W(\gamma) = \int_0^\tau dt \, \dot{a}_t \, F_t(q(t)) \tag{9.13}$$

depending on the trajectory $\gamma = (q(t))$ of the particle and the protocol (a_t). It is useful to check that (9.13) satisfies the decomposition in (8.4), here in continuous time,

$$W(\gamma) = \kappa \frac{(a_\tau - q(\tau))^2 - (a_0 - q(0))^2}{2} - \kappa \int_0^\tau dt \, v(t) \, (q(t) - a_t) \tag{9.14}$$

with $v(t) = \dot{q}(t)$. The total work need of course not equal the entropy production. As can be seen in (8.8) or as I have written already under Section 8.2, the entropy produced in the world is (β times) the total work minus the mechanical work.

Let us see what fluctuation symmetry we can expect from our main observation (2.1), or more specifically for our purpose here, from (8.7):

$$R(\gamma) = \log \frac{P_\beta(\gamma)}{\widetilde{P}_\beta(\Theta\gamma)} \tag{9.15}$$

where P_β gives the probability of the trajectory γ over a time τ, when starting from the particle in equilibrium with H_0 and for the protocol (a_t); $\widetilde{P}_\beta$ gives the probabilities when starting from the particle in equilibrium with H_τ and for the reversed protocol $(a_{\tau-t})$; Θ reverses the particle-trajectory. For the reversed protocol I write

$$\widetilde{R}(\gamma) = \log \frac{\widetilde{P}_\beta(\gamma)}{P_\beta(\Theta\gamma)}$$

so that

$$R(\Theta\gamma) = -\widetilde{R}$$

I repeat the standard computation in much detail: for a function G of $R(\gamma)$,

$$
\begin{aligned}
\langle G(R) \rangle_\beta &= \sum_\gamma G(R(\gamma)) \frac{P_\beta(\gamma)}{\widetilde{P}_\beta(\Theta\gamma)} \widetilde{P}_\beta(\Theta\gamma) \\
&= \sum_\gamma G(R(\gamma)) \, e^{R(\gamma)} \, \widetilde{P}_\beta(\Theta\gamma) \\
&= \sum_\gamma G(R(\Theta\gamma)) \, e^{R(\Theta\gamma)} \, \widetilde{P}_\beta(\gamma) \\
&= \sum_\gamma G(-\widetilde{R}(\gamma)) \, e^{-\widetilde{R}(\gamma)} \, \widetilde{P}_\beta(\gamma)
\end{aligned}
\tag{9.16}
$$

Taking in (9.16) the function $G(R)$ as 1 or 0 depending on whether $R = Q$ or $R \neq Q$, we get

$$\frac{P_\beta[R = Q]}{\widetilde{P}_\beta[\widetilde{R} = -Q]} = e^Q$$

Obviously, the $\widetilde{R}$ for the reversed protocol has under $\widetilde{P}_\beta$ the same distribution as has our original R under P_β:

$$\widetilde{P}_\beta[\widetilde{R} = -Q] = P_\beta[R = -Q]$$

and hence,

$$\frac{P_\beta[R = Q]}{P_\beta[R = -Q]} = e^Q \tag{9.17}$$

Observe that that is an exact fluctuation symmetry valid for all times τ no matter how small.

On the other hand, we know that $R = \beta(W - \Delta F)$, see, e.g., (8.8). For the time-dependence in the Hamiltonian (9.12), the difference in Helmholtz free energies $\Delta F = 0$ and therefore $R = \beta(W - \Delta F) = \beta W$. We conclude

$$\frac{P_\beta[\beta W = Q]}{P_\beta[\beta W = -Q]} = e^Q \tag{9.18}$$

again, valid for all times τ, which is what Wang et al have verified experimentally in [73].

The problem remains that R of (9.15) or, what is the same, βW of (9.13) are not exactly equal to the physical entropy production. It would be that when τ is so large that after starting the particle in equilibrium for H_0, the particle gets in equilibrium with H_τ, after pulling it through the medium with the optical trap. Then indeed, the entropy production is $\beta(W - \Delta F) = \beta W$. That would however require a relaxation of the particle in the trap after it has stopped moving and that takes time.

The real entropy production over a trajectory γ is given by the second term in (9.14). Indeed, as nearly always, it differs from R and hence, here, from βW by a temporal boundary term. It remains true therefore that the (true) entropy production, the dissipated work

$$S(\gamma) = -\beta\kappa \int_0^\tau dt\, v(t)\, (q(t) - a_t)$$

satisfies a fluctuation symmetry but only its asymptotic version (9.2). The fluctuation symmetry of $S(\gamma)$ is therefore only valid for large enough τ but, here as in other cases, remains unobservable since the fluctuation symmetry is exactly implying that the occurrence of negative entropy production over time τ is exponentially damped in τ! Quite to the contrary of what Wang et al claim, they do not observe second law violations through the fluctuation symmetry; for small times τ there is no fluctuation symmetry for the entropy production and for large times τ, it is precisely the fluctuation symmetry that tells us how unlikely (and unobservable) are second law violations.

9.2.4   Integrated fluctuation symmetries

For completeness I introduce here yet some other forms of the basic fluctuation symmetry (9.1) or of its asymptotic forms. Most interesting is perhaps to see how conditioning on a negative entropy production makes the time run backward. I start from the exact expression (9.4) to make it simpler and take $G(\omega) = F(\omega)\delta(R(\omega) - Q)$:

$$\int_{R=Q} F(\omega)\, dP(\omega) = e^Q \int_{R=-Q} F(\Theta\omega)\, dP(\omega) \tag{9.19}$$

Divide that by the exact fluctuation symmetry for $F = 1$ to get conditional expectations:

$$\langle F | R = Q \rangle = \langle F \circ \Theta | R = -Q \rangle$$

Conditioning on the opposite entropy production is cancelled by applying the time-reversal Θ.

We can also start from (9.5): take $G(R) = F(R)$ if $R > 0$ and $G = 0$ otherwise:

$$\int_{R>0} F(R(\omega))\, dP(\omega) = \int_{R<0} F(-R(\omega)) e^{-R(\omega)}\, dP(\omega)$$

Combining the choice $F(R) = 1$ with the choice $F(R) = \exp -R$ leads directly to the integrated fluctuation symmetry

$$\frac{\text{Prob}[R < 0]}{\text{Prob}[R > 0]} = \langle e^{-R} | R > 0 \rangle$$

which is sometimes easier to examine. Of course, the asymptotic fluctuation symmetries (only valid in some limit $\tau \uparrow +\infty$, as in (9.2)) have similar integrated versions.

Less known but useful are a number of inequalities for R and hence (be it in asymptotic form) for the entropy production. We can divide (9.4) by $\langle G \rangle$ for some $G > 0$ and obtain, via the Jensen inequality,

$$\langle R G \rangle \geq \langle G \rangle \log \frac{\langle G \rangle}{\langle G \circ \Theta \rangle}$$

which shows that $\langle R^n \rangle \geq 0$ for all powers n. In the same direction, since $\langle \exp -R \rangle = 1$, by applying a Chebyshev inequality,

$$\langle R \rangle \geq (e^{-\delta} - 1 + \delta)\text{Prob}[|R| \geq \delta]$$

for all $\delta \geq 0$. If we take here $\delta = q\tau > 0$ with $\tau \uparrow +\infty$, we get

$$\lim_{\tau} \text{Prob}[|R|/\tau \geq q] \leq \frac{1}{q} \lim_{\tau} \langle \frac{R}{\tau} \rangle$$

where the right-hand side gives the mean entropy production rate.

The above two inequalities imply that breaking of the detailed balance condition (of time-reversal invariance) implies a strictly positive entropy production. In most physically relevant cases that will imply that currents will flow in the direction as expected from the second law of thermodynamics. The opposite statement, whether one can have a maintained current even when time-reversal invariance is not broken, is referred to as spontaneous breaking of time-reversal invariance. An example of that is heat conduction in harmonic chains where the heat current does not vanish when taking the thermodynamic limit (no Fourier law), see [64, 56].

Such superconductors are in general excluded in classical particle systems with realistic interactions, see, e.g., [49, 50], but in quantum mechanics, where the quantum statistics introduces a global and hence nonlocal effective interaction, they make some of the most interesting phenomena in nonequilibrium statistical mechanics.

10 Response relations

The relations (9.4) are a form of Ward identities (but for a discrete symmetry) since they generate correlation function identities by differentiation with respect to z and with respect to parameters present in the distribution $P(d\omega)$. In that sense, (9.1) or its dual (9.5) are speaking about fluctuation-dissipation relations.

10.1 In linear order

Let us take again (9.4) for an antisymmetric $G, G(\Theta\omega) = -G(\omega)$. An important example is $G(\omega) = J_r(\omega)$ a current of type r. By its antisymmetry under time-reversal

$$\int G(\omega)\, dP(\omega) = \frac{1}{2} \int G(\omega)(1 - e^{-R(\omega)})\, dP(\omega) \tag{10.1}$$

The integral over dP can be realized as a steady state expectation in a nonequilibrium state driven by thermodynamic fields $E = (E_r)$. Equilibrium corresponds to $E = 0 = R$. Close to equilibrium, when the E_r are very small, R is also small and we can expand to first order in the E_r:

$$\frac{\partial}{\partial E_{r'}}[\int G(\omega)\, dP(\omega)]_{E=0} = \frac{1}{2} \int G(\omega)[\frac{\partial R(\omega)}{\partial E_{r'}}]_{E=0}\, dP_{E=0}(\omega) \tag{10.2}$$

That is a generalized Green-Kubo relation. If we have

$$R = \sum_{r'} E_{r'} J_{r'}$$

and take $G = J_r$, then (10.2) becomes

$$\frac{\partial}{\partial E_{r'}}[\int J_r(\omega)\, dP(\omega)]_{E=0} = \frac{1}{2} \int J_r(\omega) J_{r'}(\omega)\, dP_{E=0}(\omega) \tag{10.3}$$

giving the usual expression for the linear transport coefficients with the Onsager reciprocity.

That structure is not rigorously valid for R equal to the entropy production because of the temporal boundary terms but asymptotic forms are obtained when dividing by τ and letting $\tau \uparrow +\infty$. I refer to [44] for a rigorous version of the argument in a context where everything is under control.

Whether the linear response theory of irreversible thermodynamics can be rigorously derived from statistical mechanical models along the general formalism above is a technical question. The heuristics is clear and Green-Kubo relations follow from expanding to first order the fluctuation symmetry, see also in [16, 17, 42]. The question whether the space-time correlation functions present in the linear transport coefficients (10.3) are really integrable and under what conditions is an important physical question (about equilibrium dynamics) but the fluctuation symmetry is silent about that. Before worrying about rigorously deriving linear response theory, let us see what message the fluctuation symmetry perhaps holds for higher order response.

10.2 Second order

For that we look again at (9.4) but now we need it for a symmetric $G, G(\Theta\omega) = G(\omega)$, in particular for $G(\omega) = J_r(\omega)J_{r'}(\omega)$. Inspection of the resulting formula is disappointing: we cannot move beyond linear order. The fluctuation symmetry for the entropy production in all its versions is just a way of rewriting the basic observation (2.1) and it tells us nothing about the symmetric part under time-reversal of the Lagrangian $\mathcal{L}$. That part also is possibly non-trivially modified by the nonequilibrium driving. Let us take the example of heat conduction, see Section 2.1.1.

Consider the reversible reference process P_β^κ corresponding to the dynamics

$$
\begin{aligned}
dq_i &= p_i\, dt, \quad i \in V && (10.4)\\
dp_i &= -\frac{\partial U}{\partial q_i}(q)dt, \quad i \in V \setminus \partial V \\
dp_i &= -\frac{\partial U}{\partial q_i}(q)dt - \gamma\kappa_i p_i dt + \sqrt{\frac{2\gamma}{\beta_i}}dW_i(t), \quad i \in \partial V
\end{aligned}
$$

Taking $\kappa_i\beta_i = \beta$, $\forall i \in \partial V$, makes the process (10.4) reversible, as may be easily checked. As a consequence, for that choice, $P_\beta^\kappa = P_\beta^\kappa\Theta$ where $(\Theta\omega)_t = \pi\omega_{\tau-t}$ with $\omega = ((p_t, q_t), t \in [0, \tau])$ a trajectory and π reverses the sign of the momenta. The stationary density is the Gibbs measure $\rho^\beta \sim \exp{-\beta H}$ with respect to (2.4).
Let P_ρ denote the steady state path-space measure obtained from the dynamics (2.6), with stationary measure ρ. We compute the density of the process P_ρ with respect to P_β^κ. This makes (2.2) more precise. Writing the Radon-Nikodym derivative in the form

$$
dP_\rho(\omega) = e^{-A_\rho(\omega)}\, dP_\beta^\kappa(\omega)
$$

the action functional A_ρ is simply found by application of a Girsanov formula, see [47]:

$$
\begin{aligned}
-A_\rho(\omega) = \sum_{i\in\partial V} \frac{1}{2}\Big[&\int_0^\tau (\beta - \beta_i)p_i(t)dp_i(t) + \int_0^\tau (\beta - \beta_i)\frac{\partial U}{\partial q_i}(q(t))p_i(t)dt \\
&+ \int_0^\tau \gamma(\beta\kappa_i - \beta_i)p_i^2(t)dt\Big] \\
&+ \log\rho(\omega_0) - \log\rho^\beta(\omega_0)
\end{aligned}
$$

The source of time-reversal breaking is $R_\rho(\omega) = A_{\rho\pi}(\Theta\omega) - A_\rho(\omega)$ and equals (2.7) (with here in the steady state, $\rho_0 = \rho_\tau = \rho$). That is the entropy production. The symmetric part is $Y_\rho(\omega) = A_{\rho\pi}(\Theta\omega) + A_\rho(\omega)$, here equal to

$$Y_\rho(\omega) = \log \frac{\rho^\beta(\omega_\tau)\,\rho^\beta(\omega_0)}{\rho(\omega_\tau)\,\rho(\omega_0)} - \gamma\beta \sum_{i\in\partial V} \frac{\varepsilon_i(\varepsilon_i + 2)}{\varepsilon_i + 1} \int_0^\tau p_i^2(t)dt$$

where I have written $\beta/\beta_i = \kappa_i = 1 + \varepsilon_i$. We conclude that there is a term, extensive in time, which also depends on the different driving $\varepsilon_i \neq 0$. That will need to be taken into account when computing higher order response functions.

Acknowledgment. I am very grateful to Karel Netočný for many useful discussions and correspondence, in particular in connection with Section 7. Part of this text was written while visiting the Isaac Newton Institute (Cambridge) in the programme *Interaction and Growth in Complex Stochastic Systems*. Its hospitality and unique atmosphere are gratefully acknowledged.

References

[1] L. Andrey, The rate of entropy change in non-Hamiltonian systems, *Phys. Lett.* **11A**, 45–46 (1985).

[2] R. Balian, From Microphysics to Macrophysics: methods and applications of statistical physics, Vol. II. Springer-Verlag (Berlin Heidelberg) 1991.

[3] L. Bertini, A. De Sole, D. Gabrielli, G. Jona-Lasinio and C. Landim, Macroscopic Fluctuation Theory for Stationary Non-Equilibrium States, *J. Stat. Phys.* **107**, 635–675 (2002).

[4] J. Bricmont, *Bayes, Boltzmann and Bohm: Probability in Physics.* In: Chance in Physics, Foundations and Perspectives. Eds. J. Bricmont, D. Dürr, M.C. Galavotti, G. Ghirardi, F. Petruccione, and N. Zanghi, (Springer-Verlag, 2002).

[5] I. Callens, W. De Roeck, T. Jacobs, C.Maes and K. Netočný, Quantum entropy production as a measure for irreversibility, *Physica D: Nonlinear phenomena* **187**, 383–391 (2004).

[6] S. Ciliberto and C. Laroche, An experimental verification of the Gallavotti-Cohen fluctuation theorem, *J. Phys. IV (France)* **8**, 215–222 (1998).

[7] E.G.D. Cohen and G. Gallavotti, Note on Two Theorems in Nonequilibirum Statistical Mechanics, *J. Stat. Phys.* **96**, 1343–1349 (1999).

[8] G.E. Crooks, Nonequilibrium measurements of free energy differences for microscopically reversible Markovian systems, *J. Stat. Phys.* **90**, 1481 (1998).

[9] W. De Roeck and C. Maes, A quantum version of free energy – irreversible work relations, to appear in *Phys. Rev. E* (2004).

[10] W. De Roeck, C. Maes and K. Netočný, *Mathematical formulation and example of the conditions that lead to a quantum H-theorem*, private communication.

[11] W. De Roeck, T. Jacobs, C. Maes and K. Netočný, An Extension of the Kac Ring Model, *J. Phys. A: Math. Gen.* **36**, 11547–11559 (2003).

[12] Jean et Nicole Dhombres: Lazare Carnot. Fayard (Paris) 1997.

[13] J.R. Dorfman, An Introduction to chaos in nonequilibrium statistical mechanics. Cambridge University Press (Cambridge) 1999.

[14] D.J. Evans, E.G.D. Cohen and G.P. Morriss, Probability of second law violations in steady flows, *Phys. Rev. Lett.* **71**, 2401–2404 (1993).

[15] G. Gallavotti, A local fluctuation theorem, *Physica A* **263**, 39–50 (1999).

[16] G. Gallavotti, Chaotic hypothesis: Onsager reciprocity and the fluctuation dissipation theorem, *J. Stat. Phys.* **84**, 899–926 (1996).

[17] G. Gallavotti, Extension of Onsager's reciprocity to large fields and the chaotic hypothesis, *Phys. Rev. Lett.* **77**, 4334–4337 (1996).

[18] G. Gallavotti and E.G.D. Cohen, Dynamical ensembles in nonequilibrium Statistical Mechanics, *Phys. Rev. Letters* **74**, 2694–2697 (1995). Dynamical ensembles in stationary states, *J. Stat. Phys.* **80**, 931–970 (1995).

[19] G. Gallavotti and F. Perroni, *An experimental test of the local fluctuation theorem in chains of weakly interacting Anosov systems*, mp_arc # 99–320, chao-dyn/9909007.

[20] G. Gentile, Large deviation rule for Anosov flows, *Forum Math.* **10**, 89–118 (1998).

[21] P. Gaspard, Chaos, Scattering and Statistical Mechanics. Cambridge University Press (Cambridge) 1998.

[22] H.-O. Georgii, *Probabilistic Aspects of Entropy*, In: Entropy, Princeton University Press, Princeton and Oxford. Eds. A. Greven, G. Keller and G. Warnecke, 2003.

[23] C.C. Gillispie, Lazare Carnot, Savant. Princeton University Press (Princeton) 1971.

[24] S. Goldstein, *Boltzmann's Approach to Statistical Mechanics*. In: Chance in Physics, Foundations and Perspectives. Eds. J. Bricmont, D. Dürr, M.C. Galavotti, G. Ghirardi, F. Petruccione, and N. Zanghi, (Springer-Verlag, 2002).

[25] S. Goldstein and J.L. Lebowitz: On the (Boltzmann) Entropy of Nonequilibrium Systems, to appear in *Physica D* cond-mat/0304251.

[26] P.L. Garrido, S. Goldstein and J.L. Lebowitz, The Boltzmann Entropy for dense fluids not in local equilibirum, cond-mat/0310575.

[27] G. Hummer and A. Szabo, Free energy reconstruction from nonequilibrium single-molecule pulling experiments, *PNAS* **98**, 3658–3661 (2001).

[28] C. Jarzynski, Nonequilibrium Equality for Free Energy Differences, *Phys. Rev. Lett.* **78**, 2690–2693 (1997).

[29] C. Jarzynski, Equilibrium free-energy from nonequilibrium measurements: a Master-equation approach, *Phys. Rev. E* **56**, 5018–5035 (1997).

[30] C. Jarzynski, Equilibrium free energies from nonequilibrium processes, *Act. Phys. Pol. B* **6**, 1609–1622 (1998).

[31] C. Jarzynski, Microscopic analysis of Clausius-Duhem processes, *J. Stat. Phys.* **96**, 415–427 (1999).

[32] C. Jarzynski, Hamiltonian derivation of a detailed fluctuation theorem, *J. Stat. Phys.* **98**, 77 (2000).

[33] E.T. Jaynes, Gibbs vs Boltzmann Entropies, *Am. J. Phys.* **33**, 391–398 (1965). *Papers on Probability, Statistics and Statistical Physics*, Ed. R. D. Rosencrantz (Reidel, Dordrecht 1983).

[34] E.T. Jaynes, *The Evolution of Carnot's Principle*, In: Maximum-Entropy and Bayesian Methods in Science and Engineering (Vol.I), Eds. G.J. Erickson and C.R. Smith, Kluwer, 267–281 (1988).

[35] M. K-H. Kiessling, *How to implement Boltzmann's probabilistic ideas in a relativistic world?*, In: Chance in Physics, Foundations and Perspectives. Eds. J. Bricmont, D. Dürr, M.C. Galavotti, G. Ghirardi, F. Petruccione, and N. Zanghi (Springer-Verlag, 2002).

[36] R. Kubo, K. Matsuo and K. Kitahara, Fluctuation and Relaxation of Macrovariables, *J. Stat. Phys.* **9**, 51–95 (1973).

[37] J. Kurchan, Fluctuation theorem for stochastic dynamics, *J. Phys. A: Math. Gen.* **31**, 3719–3729 (1998).

[38] J. Kurchan, *A Quantum Fluctuation Theorem*, cond-mat/0007360.

[39] O.E. Lanford III, *Entropy and equilibrium states in classical statistical mechanics*, In *Statistical Mechanics and Mathematical Problems (Batelle Seattle Rencontres 1971)*, Lecture Notes in Physics No. 20 (Springer-Verlag, Berlin), 1–113 (1973).

[40] O. Lanford, *Time Evolution of Large Classical Systems*, In: LND **38**, Ed. J. Moder, Springer (1975); *Physica A* **106**, 70 (1981).

[41] J.L. Lebowitz, Microscopic Origins of Irreversible Macroscopic Behavior, *Physica A* **263**, 516–527 (1999). Round Table on Irreversibility at STATPHYS20, Paris, July 22, 1998.

[42] J.L. Lebowitz and H. Spohn, A Gallavotti-Cohen type symmetry in the large deviations functional of stochastic dynamics, *J. Stat. Phys.* **95**, 333–365 (1999).

[43] J. Liphardt, S. Dumont, S.B. Smith, I. Tinoco and C. Bustamante, Equilibrium information from nonequilibrium measurements in an experimental test of Jarzynski's equality, *Science* **296**, 1832–1835 (2002).

[44] C. Maes, The Fluctuation Theorem as a Gibbs Property, *J. Stat. Phys.* **95**, 367–392 (1999).

[45] C. Maes, Fluctuation relations and positivity of the entropy production in irreversible dynamical systems, preprint (2003).

[46] C. Maes and K. Netočný, Time-reversal and Entropy, *J. Stat. Phys.* **110**, 269–310 (2003).

[47] C. Maes, K. Netočný and M. Verschuere, Heat Conduction Networks, *J. Stat. Phys.* **111**, 1219–1244 (2003).

[48] C. Maes, F. Redig and A. Van Moffaert, On the definition of entropy production via examples, *J. Math. Phys.* **41**, 1528–1554 (2000).

[49] C. Maes, F. Redig and M. Verschuere, Entropy Production for Interacting Particle Systems, *Markov Proc. Rel. Fields* **7**, 119–134 (2001).

[50] C. Maes, F. Redig and M. Verschuere, No current without heat, *J. Stat. Phys.* **106**, 569–587 (2002).

[51] C. Maes, F. Redig and M. Verschuere, From Global to Local Fluctuation Theorems, *Moscow Mathematical Journal* **1**, 421–438 (2001).

[52] C. Maes and E. Verbitskiy, Large Deviations and a Fluctuation Symmetry for Chaotic Homeomorphisms, *Commun. Math. Phys.* **233**, 137–151 (2003).

[53] R.A. Minlos, S. Roelly and H. Zessin, Gibbs states on spacetime, *J. Potential Analysis* **13**, Issue 4 (2001).

[54] T. Monnai and S. Tasaki, *Quantum Correction of Fluctuation Theorem*, cond-mat/0308337.

[55] S. Mukamel, Quantum Extension of the Jarzynski Relation; Analogy with Stochastic Dephasing, *Phys. Rev. Lett.* **90**, 170604 (2003).

[56] H. Nakazawa, On the lattice thermal conduction, *Suppl. Prog. Theor. Phys.* **45**, 231–262 (1970).

[57] J. von Neumann, *Mathematical Foundations of Quantum Mechanics*. Princeton University Press (Princeton) 1955.

[58] L. Onsager and S. Machlup, Fluctuations and Irreversible Processes, *Phys. Rev.* **91**, 1505–1512 (1953).

[59] R. Penrose, *The Emperor's New Mind*, Oxford University Press (Oxford), 1989.

[60] M. P. Qian, M. Qian and C. Qian, Circulations of markov chains with continuous time and probability interpretation of some determinants, *Sci. Sinica* **27**, 470–481 (1984).

[61] M. P. Qian and M. Qian, *The entropy production and reversibility of Markov processes* Proceedings of the first world congress Bernoulli soc. 1988, 307–316.

[62] L. Rey-Bellet, Statistical mechanics of anharmonic lattices, mp_arc 03/97.

[63] L. Rey-Bellet and L.E. Thomas, Fluctuations of the Entropy Production in Anharmonic Chains, *Ann. Henri Poincaré* **3**, 483–502 (2002).

[64] Z. Rieder, J.L. Lebowitz and E. Lieb, Properties of a harmonic crystal in a stationary nonequilibrium state, *J. Math. Phys.* **8**, 1073–1078 (1967).

[65] F. Ritort, C. Bustamante and I. Tinoco,Jr., A two-state kinetic model for the unfolding of single molecules by mechanical force, *PNAS* **99**, 13544–13538 (2002).

[66] D. Ruelle, Smooth Dynamics and New Theoretical Ideas in Nonequilibrium Statistical Mechanics, *J. Stat. Phys.* **95**, 393–468 (1999).

[67] S. Sarman, D.J. Evans and P.T. Cummings, Recent developments in non-Newtonian molecular dynamics, *Physics Reports*, Elsevier (Ed. M.J. Klein), **305**, 1–92 (1998).

[68] J. Schnakenberg, Network theory of behavior of master equation systems, *Rev. Mod. Phys.* **48**, 4, 571–585 (1976).

[69] H. Tasaki, *Jarzynski Relations for Quantum Systems and Some Applications*, technical note, cond-mat/0009244.

[70] S. Tasaki and T. Matsui, *Fluctuation Theorem, Nonequilibrium Steady States and MacLennan-Zubarev Ensembles of L^1-Asymptotic Abelian C^* Dynamical Systems*, mp_arc 02-533.

[71] S.R.S. Varadhan, *Large Deviations and Entropy*, In: Entropy, Princeton University Press, Princeton and Oxford. Eds. A. Greven, G. Keller and G. Warnecke, 2003.

[72] C. Wagner, R. Klages and G. Nicolis, Thermostating by deterministic scattering: Heat and shear flow, *Phys. Rev. E* **60**, 1401–1411 (1999).

[73] G.M. Wang, E.M. Sevick, E. Mittag, D.J. Searles and D.J. Evans, Experimental Demonstration of Violations of the Second Law of Thermodynamics for Small Systems and Short Time Scales, *Phys. Rev. Lett.* **89**, 050601 (2002).

[74] S. Yukawa, A Quantum Analogue of the Jarzynski Equality, *J. Phys. Soc. Jpn.* **69**, 2367 (2000).

Christian Maes
Instituut voor Theoretische Fysica
K.U. Leuven, Belgium
email: Christian.Maes@fys.kuleuven.ac.be

Poincaré Seminar 2003, 193 – 226
© Birkhäuser Verlag, Basel, 2004

Work Fluctuations, Transient Violations of the Second Law and Free-Energy Recovery Methods: Perspectives in Theory and Experiments

Félix Ritort

Abstract. In this report I discuss fluctuation theorems and transient violations of the second law of thermodynamics in small systems. Special emphasis is placed on free-energy recovery methods in the framework of non-equilibrium single molecule pulling experiments. The treatment is done from a unified theoretical-experimental perspective and emphasizes how these experiments contribute to our understanding of the thermodynamic behavior of small systems.

1 Biophysics and statistical physics

Living systems are the most notable example of how matter can organize into states of extremely high complexity. The investigation of the structural organization of biological matter was boosted since the discoveries of the double helix structure of the DNA by Watson and Crick and the ensuing discovery of the structure of various proteins half century ago. Microscopic and spectroscopic techniques have greatly developed since then and current research is revealing an unprecedented richness of details about the functional behavior of living systems at the molecular and cellular level.

Biophysics is an area of science at the interface of physics, chemistry and biology. It is probably the most important interdisciplinary area of research whose knowledge requires a good understanding of how matter functions at the physical, chemical or biological level. While physics during the past has traditionally avoided the study of complex systems as imperfect, unapproachable or even uninteresting, the fact is that complexity is becoming a more and more common abode for physicists [1]. Among all possible disciplines in physics, statistical physics occupies a privileged position as the natural framework to understand the behavior of biological systems at the molecular level. Stochasticity, fluctuations, metastability and thermal activation are concepts that are commonly used in statistical physics, yet they are also relevant to understanding the great variety of tasks carried out by biomolecules.

Thermodynamics is the discipline that describes the exchange processes of energy and matter that occur at the molecular and cellular level. However, thermodynamics, a science inherited in the 18th century from the times of the industrial revolution, has been inspired by motors and steam engines that proved to be indispensable during that time. It is fair then to question the relevance and applicability of all this knowledge when scientists immerse into the realm of the very small, far from the initial context that inspired Carnot and others. There has been a recent interest in the study of the so called work fluctuations and transient violations of the second law in systems driven to a non-equilibrium state. Fluctuation theorems quantify the probability of those non-equilibrium trajectories that, taken individually, violate some of the inequalities of thermodynamics. For macroscopic systems these trajectories are known to be irrelevant and unobservable, however at the level of the small, when the energies interested are of order of several times $k_B T$, these rare trajectories might become important. Although thermodynamic inequalities are known to describe the behavior of average values, it is important to explore the implications and relevance of these deviations in our understanding of energy transformation processes at the molecular level. A quantitative experimental observation and measurement of these trajectories has only recently become possible. This report describes these experiments from a unified theoretical-experimental perspective and emphasizes how these experiments contribute to our understanding of the thermodynamic behavior of small systems.

Sec. 2 is a short reminder about the second law of thermodynamics. It serves to explain the importance of fluctuations in small systems and short times. Section 3 describes work fluctuations in the framework of stochastic systems. Particular emphasis is put in the case where the system, initially in an equilibrium state, is perturbed arbitrarily far from equilibrium. We then discuss the non-equilibrium work relation originally derived by Jarzynski. Sec. 4 describes the current state of the art regarding experimental measurements of work fluctuations. Sec. 5 presents a digression on single molecule experiments as an excellent framework to investigate work fluctuations and free-energy recovery methods applied to biomolecules. Sec. 6 describes some of the experiments conducted in the unfolding of small RNA molecules under the action of an external force and the test of the Jarzynski equality. Sec. 7 illustrates a model where work fluctuations can be analytically computed as well as a comparison with the experiments reported in the preceding section. Finally, Sec. 8 presents some conclusions and perspectives.

2 Few facts about the second law of thermodynamics

To put in perspective the content of the present article we start by recalling few facts about the second law of thermodynamics [2, 3]. Let us consider a gas consisting of N molecules enclosed in a given vessel of volume V. The vessel is in contact with a thermal bath at temperature T and the gas inside is kept in equilibrium (i.e., its macroscopic properties remain stationary), see Fig. 1. Particles in the gas collide with the walls of the container exerting a pressure P that is function of the volume V and the temperature T. Their relation defines the equation of state of the gas. In these conditions heat is continuously exchanged between the gas and the bath through the walls of the vessel. The state of the gas can be modified by changing (e.g., expanding) the volume of the container from an initial volume V_i to a final volume V_f. If the transformation is done by keeping the system always in contact with the bath at temperature T the process is called isothermal. If the transformation is done slow enough then the gas goes through a sequence of equilibrium states and the process is called reversible. In general the transformation will not be reversible and the gas will be driven to a non-equilibrium state after the volume has been expanded. For the transformation $V_i \rightarrow V_f$ the first law of thermodynamics states that energy is conserved,

$$\Delta E = \Delta Q + W = \Delta Q + P\Delta V \tag{1}$$

with $\Delta V = V_f - V_i$. From (1) we see that the variation of the energy of the gas ΔE is the sum of the work exerted upon the system W plus the net heat supplied from the bath to the system ΔQ. The difference between heat and the other state variables E and V is important. If the volume or energy characterize the thermodynamic state of the system, the amount of heat contained does not as it is fully interchangeable with work depending on the path followed during the transformation. In a general transformation, part of the total work exerted upon the system is lost and dissipated in the form of heat to the surroundings. This is the content of the second law as stated by Clausius,

$$\Delta Q \leq T\Delta S \tag{2}$$

where $S(V,T)$ is a state function called entropy. The amount of heat lost during the process is called dissipated work W_{dis}. It is given by the difference between the maximum amount of heat that can be supplied to the system and the actual heat supplied (i.e., the right- and left-hand sides of (2)),

$$W_{\mathrm{dis}} = T\Delta S - \Delta Q \quad \text{with} \quad W_{\mathrm{dis}} \geq 0 \quad . \tag{3}$$

Another way to state the content of the second law is in terms of the Helmholtz free energy $F = E - TS$. Using (1,2,3) we have,

$$\Delta F = \Delta E - T\Delta S = \Delta Q + W - T\Delta S = W - W_{\mathrm{dis}} = W_{\mathrm{rev}} \tag{4}$$

where W_{rev} is the so-called reversible work, identical to the free energy change ΔF associated to the initial and final equilibrium states. Only in a reversible transformation the equality (2) is satisfied and $W = W_{\mathrm{rev}}$ or $W_{\mathrm{dis}} = 0$.

It has been known since the early days of statistical mechanics that the work can fluctuate [1]. To better clarify what we mean by this let us go back to the previous example and consider the gas of molecules enclosed in the vessel depicted in Fig. 1. Let us imagine that we repeat many times the experiment of the expansion of the container from V_i to V_f by following always the same variation

---

[1]The existence of rare fluctuations have given rise to several paradoxes, see [1] for an historical perspective

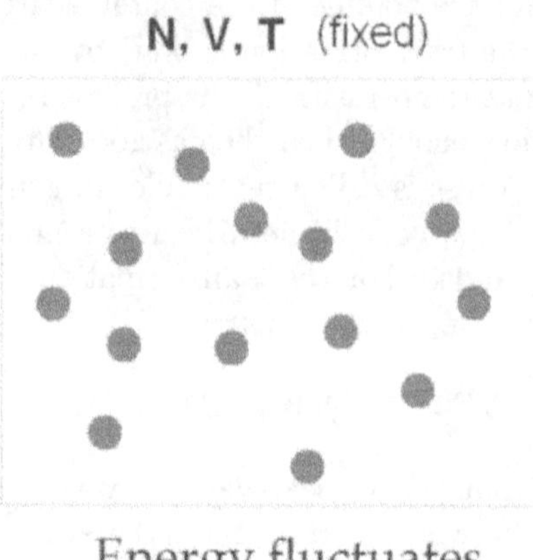

Figure 1: Schematic representation of a vessel containing molecules that exchange energy and momentum with the surrounding bath through different mechanisms such as collisions with the walls. If the number of particles is small pressure fluctuations (due to the fluctuations in the collision rate of the molecules against the walls) could be observable with highly sensitive instruments.

protocol $V(t)$ where t is the time and the whole expansion lasts for a time t_0 [2]. Then for each experiment a different work value $W = \int P dV$ would be obtained as the pressure itself is a fluctuating variable. For a macroscopic system pressure fluctuations are unobservable due to the large number of molecules (of the order of the Avogadro number $N_A \sim 10^{23}$). However, for small systems, fluctuations could be observable. The result that intensive quantities display thermally induced fluctuations was put forward by Niquist for the case of voltage fluctuations across a resistance [5] and generalized later on by Welton and Callen [6]. Let us focus on the case where the gas contains about 100 molecules and is kept in equilibrium inside a volume V at the room temperature $T = 298K$. Let us imagine that at every interval of time (e.g., $\tau = 1$ millisecond) we count the number of collisions of the molecules against the wall, $N_c(\tau)$, as well as the average momentum $<p_c>$ transferred by the colliding molecules to the walls along each time interval. Then, the pressure exerted by the molecules on the walls P will be proportional to $N_c(\tau)$ and $<p_c>$. An histogram of the values of P thereby collected would show a Gaussian distribution centered around a mean value P_{mean} (which would also coincide with the most probable value) and a variance that decreases like $1/N$. This is the content of the law of large numbers. Sometimes the measured pressure will be large, sometimes it will be small: the fluctuations will become

---

[2]The simplest protocol would be an expansion of the volume at a constant rate r, $\dot{V} = r = \frac{\Delta V}{t_0}$ or $V(t) = V_i + rt$

observable as the number of molecules is reduced. Initially these fluctuations will show a characteristic Gaussian profile. However, upon reduction of the number of molecules, deviations from the Gaussian behavior will be observed. The same happens in the non-equilibrium experiment where the volume is expanded. If the time t_0 is short enough or the number of molecules N is small enough, the work $W = \int P dV$ exerted upon the system will fluctuate from one non-equilibrium trajectory to another, and the fluctuations will become more noticeable as t_0 and N decrease. If the gas is initially in an equilibrium state, then (2,3) holds in average. However, for some trajectories the work will be such that $W_{\text{dis}} < 0$ and the inequality (2,3) will be reversed. These particular set of trajectories are called violating trajectories or transient violations of the second law.

The theory describing thermal fluctuations in equilibrium systems was put forward long ago by Einstein. However, this theory describes fluctuations of extensive quantities (such as the energy) in different ensembles in the macroscopic regime where fluctuations are subleading of order $\mathcal{O}(\sqrt{V})$, yet large. The relative magnitude of these fluctuations compared to the actual value of the energy content is of order $\mathcal{O}(1/\sqrt{V})$. In macroscopic samples relative fluctuations are of order $\mathcal{O}(1/\sqrt{N_A})$. This gives estimates for the magnitude of relative fluctuations of the order of 10^{-11}. A theory of fluctuations for small systems (such as those relevant to biophysics) where relative fluctuations are much larger (e.g., of order one) should be explored if the theory of fluctuations, as has been developed for macroscopic systems, reveals inadequate to explain the behavior observed in future experiments.

3   Fluctuation theorems in non-equilibrium systems

As we discussed in the previous section the work is quantity that fluctuates among different repetitions of the same experiment. Moreover, as compared to other quantities such as the internal energy, the entropy or the heat transferred, the amount of work exerted upon the system is a directly measurable quantity. A relevant question is then to ask what can we learn by measuring work fluctuations.

To answer this question we consider the dynamical evolution of a system in contact with a large bath or reservoir where heat and/or matter can be continuously exchanged. Both system and bath may appear inextricably linked and no partial description of the system can be achieved without considering the behavior of the bath and its interaction with the system. The mathematical treatment of the combined system plus bath complex represents a formidable theoretical challenge and some coarse-graining strategies must be considered to tackle this question. A common strategy is to consider a reduced level in the description of the dynamics of the system where many details of the bath have been eliminated in favor of a few number of parameters (such as its temperature, its pressure or its chemical potential). These parameters are thought to be sufficient to characterize the heat and/or matter exchange between system and bath. After this reduction is adopted, few requirements have to be imposed on the dynamics in order to reproduce many of the observable properties. In particular, dynamics must be microscopically reversible (or satisfying detailed balance) and ergodic (all configurations must be accessible) [3] to guarantee that the system reaches thermal equilibrium after long times so the net heat exchange between system and bath asymptotically decays to zero.

3.1 Work fluctuations in stochastic systems

It is in the framework of such coarse-grained dynamics that we want to focus our discussion and investigate the origin of work fluctuations. A prominent example of such reduced dynamics are stochastic Markov processes. The content of this section will be useful to establish the notation that we will use throughout the paper. We now deepen the mathematical level of our discussion and consider a general system described by an energy function $E(\mathcal{C})$ where $\mathcal{C}$ is a generic configuration (in the example in Sec. 2 of a gas of N molecules, $\mathcal{C}$ would stand for the positions and momenta of all molecules inside the vessel) in contact with a bath at temperature T. The dynamics are assumed to be discrete in time with elementary step Δt. A trajectory of the system is characterized by the sequence of configurations $\mathcal{T} \equiv \{\mathcal{C}_k; 0 \leq k \leq N_s\}$ where k is the index for the discrete time step and N_s is the total number of time steps. The time corresponding to step k is then given by $t = k\Delta t$ with $t = 0(k = 0)$ and $t_f(k = N_s/\Delta t)$ denoting the initial and final times respectively. The continuous-time limit is recovered if $\Delta t \rightarrow 0, N_s \rightarrow \infty$ with t_f finite. Dynamics are then defined by the set of probabilities $P_k(\mathcal{C})$ for the system to be found at configuration $\mathcal{C}$ at time-step k. The $P_k(\mathcal{C})$ satisfy a master equation. For a Markov process the time evolution of these probabilities depends upon the form of the rates $W_k(\mathcal{C}'|\mathcal{C})$, defined as the transition probability per unit time to go from configuration $\mathcal{C}$ to $\mathcal{C}'$ at time-step k. These rates are assumed to lead to an ergodic dynamics (where any pair of configurations are always connected by at least one trajectory) and satisfy the detailed balance condition,

$$\frac{W_k(\mathcal{C}'|\mathcal{C})}{W_k(\mathcal{C}|\mathcal{C}')} = \exp\Big(-\beta(E(\mathcal{C}') - E(\mathcal{C}))\Big) \qquad (5)$$

where $\beta = 1/k_B T$, k_B being the Boltzmann constant. Under very general conditions this dynamics guarantees that the system reaches a stationary state where configurations are populated according to the Boltzmann weight. The solution to the master equation gives the time evolution for the system.

Now we will treat the case where the system is perturbed in a prescribed way and consider the ensemble of all possible non-equilibrium trajectories that start from an initial state characterized by the distribution $P_0(\mathcal{C})$. Because dynamics is

---

[3]Ergodicity is not an essential property if one considers equilibrium restricted to a given region of phase space, the sole condition is that all configurations contained in that region of phase space must be accessible during the dynamical process.

stochastic, it will generate an ensemble of non-equilibrium trajectories by repeating the same experiment many times [4]. In addition to the configuration C, and in order to characterize the perturbation protocol, we introduce a parameter λ that specifies the value of the control parameter that is changed throughout the non-equilibrium process [5]. An important remark is now in place. The control parameter is a variable that can change in time but does not fluctuate. The temporal sequence of values $\{\lambda_k; 0 \leq k \leq N_s\}$ defines the perturbation protocol and this sequence of values never changes from experiment to experiment. Somehow, the control parameter plays a role akin to the temperature of the bath. In particular, for a fixed value of λ, we require that dynamics is such that the system asymptotically reaches the thermodynamic state corresponding to that value of λ. To understand how rates depend on the value of λ we reason as follows. The control parameter usually shifts the energy levels of the system according to the relation,

$$E_\lambda(C) = E(C) - \lambda A(C) \tag{6}$$

where $A(C)$ is the observable coupled to the parameter λ [6]. The simplest assumption is then to enforce detailed balance for the perturbed rates,

$$\frac{W_\lambda(C'|C)}{W_\lambda(C|C')} = \exp\left(-\beta(E_\lambda(C') - E_\lambda(C))\right) = \frac{W(C'|C)}{W(C|C')} \exp\left(-\beta\lambda\Delta A\right) \tag{7}$$

where we used (5) in the r.h.s. and the definition $\Delta A = A(C') - A(C)$. We now consider the variation of energy along a given trajectory $\Delta E(T) = E_{\lambda_f}(C_f) - E_{\lambda_0}(C_0)$ where C_I, C_f are the initial and final configurations for that trajectory and λ_0, λ_f are the initial and final values of the control parameter as defined by the protocol (trajectory independent). From (6) this is given by,

$$\Delta E(T) = \left[\sum_{k=0}^{N_s-1} (E_{\lambda_k}(C_{k+1}) - E_{\lambda_k}(C_k)) \right] - \left[\sum_{k=0}^{N_s-1} A(C_k)\Delta\lambda_k \right] = \Delta Q(T) + W(T) \tag{8}$$

with $\Delta\lambda_k = \lambda_{k+1} - \lambda_k$. This decomposition was proposed originally by Crooks [9] to identify work and heat by using the first law of thermodynamics (1). The first

---

[4]The same result holds for deterministic (e.g., Hamiltonian) dynamics. In this case the ensemble of non-equilibrium trajectories is determined by the ensemble of initial configurations sampled with probability $P_0(C)$. The set of phase space points then behaves as an incompressible fluid, a consequence of the Liouville theorem. Hamiltonian dynamics can be seen as a particular limit of stochastic dynamics, where rates $W_k(C'|C)$ vanish except along the constant energy surface $E(C) = E(C')$ and are deterministic, i.e., rates are different from zero only for pairs of configurations C, C' connected by the equations of motion. Dynamics is reversible and corresponds to (5) with $W_k(C'|C) = W_k(C|C')$. The case of Hamiltonian dynamics was originally addressed by Jarzynski in his original derivation of the non-equilibrium work relation [7]. The stochastic case has been analyzed also for general Markov processes by Crooks and Jarzynski [8, 9, 10] and for Langevin dynamics by Kurchan [11]. For a discussion of the similarities and differences between deterministic and stochastic dynamics see [12].

[5]For simplicity we only consider the case of one control parameter. For many control parameters the generalization is straightforward.

[6]For instance, if λ is a magnetic or gravitational field then A stands for the magnetization and the height of the center of mass respectively

term in (8) is identified as the heat transferred from the bath to the system and
the second with the work exerted upon the system. We concentrate our attention
on the the work exerted upon the system along a given trajectory $\mathcal{T}$,

$$W(\mathcal{T}) = \sum_{k=0}^{N_s-1} \left(\frac{\partial E_\lambda(\mathcal{C}_k)}{\partial \lambda}\right)_{\lambda=\lambda_k} \Delta\lambda_k = -\sum_{k=0}^{N_s-1} A(\mathcal{C}_k)\Delta\lambda_k \equiv -\int_0^t ds\dot{\lambda}(s)A(\mathcal{C}(s))ds \tag{9}$$

where we have applied the continuous-time limit [7] in the last term in the r.h.s.
of (9). As the trajectory is stochastic the work is a fluctuating quantity that can
be characterized by its probability distribution $\mathcal{P}(W)$ defined as,

$$\mathcal{P}(W) = \sum_{\mathcal{T}} P(\mathcal{T})\delta(W - W(\mathcal{T})) \tag{10}$$

where $\mathcal{T}$ stands for the trajectory and was already defined. The importance of
$\mathcal{P}(W)$ relies upon the fact that it is a quantity that is experimentally measurable
and therefore is suitable to quantitatively characterize work fluctuations along
non-equilibrium trajectories.

3.2 The fluctuation theorem (FT)

Fluctuation theorems (FTs) provide specific relations for the quantity $\mathcal{P}(W)$ in
(10) for general non-equilibrium processes. In fact, until now nothing was said
about the type of non-equilibrium process and the treatment given in the previous
section was general. We defined concepts such as the initial and final state, the
perturbation protocol $\lambda(t)$, the trajectory $\mathcal{T}$ and the work and heat along a given
trajectory. The main difference between a general non-equilibrium process and a
reversible one is the enormous and various type of situations one can encounter in
the first case. General physically meaningful statements about the properties of the
distribution (10) quite probably do not exist and a specific type of non-equilibrium
process has to be adopted to come up with specific results. Several fluctuation theorems have appeared in the literature depending on the particular non-equilibrium
context. Many fall into the category of entropy production FTs. The first example in this class was proposed by Evans, Cohen and Morriss [13] for systems in
steady states. The entropy production there defined bears some resemblance with
the work (9) that is exerted by the external non-conservative forces that act upon
the system. Several related theoretical results have followed [14, 15, 16] as well as
experiments [17, 18]. A comprehensive review can be found in [19]. Other more
complex scenarios can be envisaged, for example in the case where the system is in
a non-stationary aging state[8]. In this case, no work is performed upon the system

---

[7]In the continuous-time limit, both $\{\mathcal{C}_k, \lambda_k\}$ become $\mathcal{C}(t), \lambda(t)$ defining the real-time trajectory
and perturbation protocol respectively.

[8]Non-equilibrium aging states are widespread in condensed matter physics. The most common
example are structural glasses quenched below their glass transition temperature. The aging state
is characterized by strong violations of the fluctuation-dissipation theorem [21].

and the relevant quantity turns out to be the released heat from the system to the bath [22, 23].

The content of this article deals particularly with systems initially in equilibrium with the bath that are driven to a non-equilibrium state by the action of an applied perturbation [20]. Therefore,

$$P_0(\mathcal{C}) = P_{\mathrm{eq}}(\mathcal{C}) = \frac{\exp(-\beta E_{\lambda_0}(\mathcal{C})}{Z} \tag{11}$$

where $Z = \sum_{\mathcal{C}} \exp(-\beta E_{\lambda_0}(\mathcal{C})$. This case has been studied by Jarzynski [7, 8] and Crooks [9, 10]. We omit details of the derivation as these can be found in the references. A particularly interesting identity has been derived by Crooks [10] who considered the forward and reverse paths in a non-equilibrium process. The forward process (F) is characterized by the protocol function $\lambda_F(t)$ with the initial state in equilibrium at the value $\lambda_F(0)$. The reverse process (R) is characterized by the inverted protocol function $\lambda_R(t) = \lambda_F(t_0 - t)$ with t_0 being the total time for the forward process and the initial state for the reverse process being in equilibrium at the value $\lambda_R(0)$ (which is equal to $\lambda_F(t_0)$). The following result is obtained [10],

$$\frac{P_F(W)}{P_R(-W)} = \exp\left(\frac{W - \Delta F}{k_B T}\right) = \exp\left(\frac{W_{\mathrm{dis}}}{k_B T}\right) \quad . \tag{12}$$

Simple manipulation of this ratio and integration of one side of the relation from $-\infty$ to ∞ gives,

$$\int_{-\infty}^{\infty} P_F(W) \exp\left(-\frac{W}{k_B T}\right) = \exp\left(-\frac{\Delta F}{k_B T}\right) \quad . \tag{13}$$

This is the content of the Jarzynski equality originally derived in [7] for Hamiltonian dynamics (see the footnote 4). In what follows, if not stated otherwise, we will only consider the forward process in a non-equilibrium experiment and drop the subscript F for the work probability distribution P_F. We will use the symbol $\overline{(\ldots)}$ to denote the average over all non-equilibrium trajectories generated by a given protocol. The Jarzynski equality (JE) reads,

$$\overline{\exp\left(-\frac{W}{k_B T}\right)} = \int_{-\infty}^{\infty} P(W) \exp\left(-\frac{W}{k_B T}\right) = \exp\left(-\frac{\Delta F}{k_B T}\right) \quad . \tag{14}$$

This expression can be written in a more compact form,

$$\overline{\exp\left(-\frac{W_{\mathrm{dis}}}{k_B T}\right)} = 1 \quad . \tag{15}$$

The Jarzynski equality provides a simple way to derive the second law. Using Jensen's inequality [24] $\langle \exp(x) \rangle \geq \exp(\langle x \rangle)$ in (15) we obtain,

$$\overline{W_{\mathrm{dis}}} \geq 0 \quad . \tag{16}$$

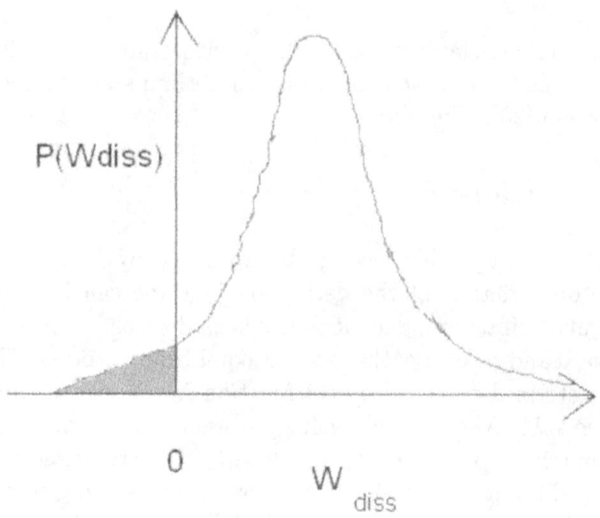

Figure 2: A typical work probability distribution for a small system is useful to characterize work fluctuations. Transient violations of the second law are a particular class of trajectories with work values characterized by the fact that the the Clausius inequality (2,3) is reversed.

An important aspect of the JE (15) is that it introduces a way to quantitatively estimate transient violations of the second law, i.e., the fraction of trajectories whose dissipated work is negative. The reason is easy to understand by inspection of (15). The average of the exponential in the l.h.s of (15) equals 1 only if trajectories with $\overline{W_{\rm diss}} < 0$ exist. An analysis of the JE in the near-equilibrium regime is useful. Close to equilibrium the distribution $P(W)$ can be approximated by a Gaussian [9],

$$P(W) = \frac{1}{\sqrt{2\pi\sigma^2}} \exp\left(-\frac{(W - W_{\rm mean})^2}{2\sigma^2}\right) \tag{17}$$

where $W_{\rm mean}$ is the value of the work at the center of the Gaussian and $\sigma^2 =$

---

[9]The exact form of the $P(W)$ can be very complicated, however the Gaussian approximation is expected to hold in the non-equilibrium regime where work trajectories do not deviate much from the reversible one. In this way the Gaussian approximation appears tightly related to linear response theory. In the latter, quantities deviate from their equilibrium values proportionally to the intensity of the perturbation. In a similar way, the deviation of the average work value (the first moment of $P(W)$) from its reversible value ΔF is expected to be linear with the perturbing speed $\dot{\lambda}$, see also [25] for a quantitative estimate of this statement. In other special cases, however, the Gaussian result can be exact even for arbitrarily strong perturbations. This is the example of a bead dragged through water [26].

$\overline{W^2} - \overline{W}^2$ its variance. Substitution of (17) into (14) gives,

$$R = \frac{\sigma^2}{2\overline{W_{\text{dis}}}k_B T} = 1 \quad .\tag{18}$$

Throughout the paper we will refer to R as the *fluctuation-dissipation ratio* as it involves a ratio between work fluctuations σ_2 and dissipation $\overline{W_{\text{dis}}}$ [10]. This result is a particular form of the fluctuation-dissipation theorem in the linear response regime. In general, $R \neq 1$ far from equilibrium. More about the dependence of R with the value of the dissipated work will be said later in Sec. 7. We infer from (18) that in the near-equilibrium regime the variance of the work is of the same order of the average dissipated work. In the reversible limit $W_{\text{dis}} \to 0$ the JE trivially holds as $W_{\text{dis}} = 0$ (or $W = \Delta F$). In this limit the number of trajectories with $W_{\text{dis}} > 0$ equals the number with $W_{\text{dis}} < 0$, therefore the reversible limit is the case where the fraction of violating trajectories is maximal. This may look rather unexpected as transient violations might be thought to be a characteristic of non-equilibrium processes.

Transient violations of the second law are expected to decrease fast as the average value of the dissipated work $\overline{W_{\text{dis}}}$ increases (for instance, if the size of the system increases) becoming unobservable in the thermodynamic limit. How transient violations are suppressed as the system size increases can be understood also from the JE. We rewrite (15) as,

$$1 = \overline{\exp\left(-\frac{W_{\text{dis}}}{k_B T}\right)} = P_+ \left[\exp\left(-\frac{W_{\text{dis}}}{k_B T}\right)\right]_+ + P_- \left[\exp\left(-\frac{W_{\text{dis}}}{k_B T}\right)\right]_-\tag{19}$$

where P_+, P_- are the probabilities to generate a trajectory with $W_{\text{diss}} > 0, W_{\text{dis}} < 0$ respectively, i.e., $P_+ + P_- = 1$. In analogous way, the square brackets $[..]_+, [..]_-$ denote averages $\overline{(..)}$ but restricted over the subsets of trajectories with $W_{\text{dis}} > 0, W_{\text{dis}} < 0$ respectively. In Fig. 2 P_+, P_- would correspond to the green and white areas under the distribution. Using this decomposition it is easy to understand how work trajectory values contribute to the r.h.s. of (19) enforcing the validity of the JE. The value of W_{dis} increases proportionally to the size N of the system (because the work is an extensive quantity). Because all four quantities appearing in the r.h.s. of (19) are positive, to impose a sum of both terms of 1, the factor p_- has to be exponentially small with the average value of the dissipated work (i.e., the size of the system) $p_- \sim \exp(-\mathcal{O}(N))$ to compensate the divergence of the corresponding average $[..]_-$. This implies $p_+ = 1 - \exp(-\mathcal{O}(N))$ so transient violations are exponentially suppressed with the system size, yet they have to be weighed for the JE to hold.

---

[10]The use of the term *fluctuation-dissipation ratio* for the quantity R has not to be confused with that adopted in glassy systems and usually denoted by x [21]. For glassy systems x describes the ratio between a time derivative of the correlation function and the response function. Albeit similar, the two-quantities are not identical as they refer to different non-equilibrium scenarios.

3.3 Free energy recovery from non-equilibrium experiments

An important consequence of the JE (14) is that non-equilibrium experiments can be used to recover equilibrium free-energy differences [7],

$$\Delta F = -k_B T \log \left(\overline{\exp\left(-\frac{W}{k_B T}\right)} \right) \quad . \tag{20}$$

The non-equilibrium work relation (20) is useful to find the equilibrium free-energy change along a given reaction when it is not possible to carry it out reversibly. The idea is to repeat non-equilibrium experiments many times and evaluate the exponential average in the r.h.s of (20) to derive the corresponding work in a reversible process. This formula has been used to recover the free-energy change in the folding-unfolding reaction for small RNA molecules, see Sec. 6 for a detailed exposition. However, there are practical difficulties in the applicability of (20) as the number of trajectories included in the exponential average must be actually infinite. This is unrealizable in practice as non-equilibrium experiments can be performed only a finite number of times and the finiteness of the number of trajectories introduces a bias. In what follows we will use ΔF_{JE} to denote the estimate for the equilibrium free energy ΔF obtained by using the JE given in (20) with a finite number of trajectories. As we remarked in the previous paragraph, the number of trajectories required to evaluate the JE grows exponentially with the average value of the dissipated work. The dependence of the bias and error with the number of pulls has been estimated in some cases [27, 28]. In general this dependence can be quite complicated as it depends on the behavior of the left tails of the distribution $P(W)$ which are difficult to analyze in general.

If the non-equilibrium experiment is done in the near-equilibrium regime then it is better to use the fluctuation-dissipation (FD) estimate (18),

$$\Delta F_{FD} = \overline{W} - \frac{\sigma^2}{2 k_B T} \quad . \tag{21}$$

In most cases it is difficult to determine whether the non-equilibrium process is done in the near equilibrium regime, so this estimate has to be taken with caution. A description of the bias and error for the estimates (20,21) in the near-equilibrium case has been recently given. Both are exact in the limit of infinite number of non-equilibrium trajectories. Interestingly, when the number of repeated experiments is small the JE estimate (20) provides a better estimate than the FD (21) does [28].

4 Experimental observation of work fluctuations

For sake of clarity we discuss now some examples where work fluctuations are experimentally measurable. Some of them have been already measured, others might be in the near future.

We start this tour discussing recent experiments on simple systems. This is the case of a micron-sized polystyrene bead confined in an optical trap and dragged through a solvent (e.g., water) of viscosity η at constant speed v. In average the viscous drag on the bead exerts a force γv that counteracts the force inside the trap $f(t) = -kx(t)$ where t denotes the time, k is the stiffness of the trap and $x(t)$ is the distance of the bead to the center of the trap. The control parameter is the position of the center of the trap $x_0(t)$ that moves at a constant speed $\dot{x}_0 = v$ and the fluctuating variable is the position $x(t)$ of the bead inside the trap (or equivalently the force $f(t)$ acting on the bead). The work along a trajectory of duration t_f is given by $W = \int_0^{t_f} ds f(s)v ds$. For such case work fluctuations were theoretically predicted [26] and recently measured [29].

Moving to more complex systems, recent experiments have studied the response of biomolecules to mechanical force [30, 31, 32]. The advent of nanotechnologies has opened the possibility to exert very small forces on nanosized systems (from piconewtons $1pN = 10^{-12}N$ using optical or magnetic tweezers, to nanonewtons $1nN = 10^9 N$ using AFMs). These techniques allow researchers to manipulate and study individual biomolecules one by one. Work fluctuations have been already observed in the unfolding of small RNA molecules (around 100 pair bases) under the action of mechanical force. More will be said below in Sec. 5. In these experiments the RNA molecule is held through linker polymers to two micron-sized beads. One is held by suction on the tip of a micropipette, the other is confined in the optical trap. The molecule is pulled as the distance $x(t)$ between the center of the optical trap and the tip of the micropipette increases at a given rate. $x(t)$ therefore defines the control parameter. The fluctuating variable in this case is the force $f(t)$ exerted on the whole system that is measured through the deflection of the bead in the trap. The work along a given trajectory is again $W = \int_0^{t_f} f(s)\dot{x}(s)ds$ [11]. Work fluctuations are observed during the unfolding process due to the stochastic behavior of the breakage force at which the molecule unfolds. Similar experiments are expected to be conducted also for proteins [34], albeit the large molecular weight of such molecules might render the quantitative evaluation of work fluctuations difficult.

One might speculate also on the importance of work fluctuations on the behavior of molecular motors and their efficiency. In this case, work fluctuations are observable through single molecule experiments by measuring the mechanical force exerted upon the motor as it translocates along the template. Examples of these motors are DNA [38, 37] or RNA [35, 36, 39] polymerases during the replication and transcription process, helicases and topoisomerases that unwind the DNA [41] or the ribosome during the transcription process. Other cases include the condensation process of DNA inside the viral capside during the infection cycle [40], and gene regulatory mechanisms (such as transcription factors) ruled

---

[11]In these experiments usually $x(t)$ is the distance measured from the center of the bead in the trap (rather than the center of the trap) to the center of the bead in the micropipette. This introduces a correction to the work that is negligible in most cases [33].

by protein-DNA interactions that expose large segments of condensed DNA to the replication machinery [42]. Work fluctuations are predicted to be observable in all these systems. Quantitative investigations will be surely conducted in the future.

Despite of their inherent interest, the measurement of work fluctuations in biomolecules has two important drawbacks: accuracy and reproducibility. Indeed, few single molecule experiments are fully reproducible due to the complexity of conditions and external factors required. Reproducibility at the single molecule level is specially serious in biomolecular processes requiring protein activity as many external factors strongly affect the outcome of the experiment. Accuracy is also an issue specially for measurements with nanometer resolution where stability and drift of the machines (e.g., optical tweezers) still impede high accuracy results.

Accurate and reproducible measurements of work fluctuations might be easier in systems with reduced complexity within the traditional domain of physics. One example is the already mentioned experiment of the bead in an optical trap moved through a solvent. High accuracy recent experimental measurements of the work between non-equilibrium steady-states confirm that such measurements are indeed possible [43]. Another example that has called our attention recently is the case of magnetic nanoparticle systems in a magnetic field [44]. Magnetic measurements in microsquids in Grenoble (France) have shown how it is possible to observe magnetization reversal of single magnetic nanoparticles of magnetic moment $\mu \sim 1000\mu_B$ (μ_B is the Bohr magneton) at low temperatures [45]. For magnetic nanoparticle systems the control parameter is the external magnetic field that can be switched at a constant speed (ramping experiments). The fluctuating or stochastic variable is the value of the field at which the magnetization reverses. The work along a given trajectory is then given by $W = -\mu \int_0^{t_f} M(t)\dot{H}(t)dt$. The aspect that makes these systems specially interesting is the possibility to use SQUID quantum (i.e., high-precision) technology to measure the magnetic moment in favor of a higher accuracy. Reproducibility is also easier to achieve as many physical properties of nanoparticles can be externally tuned, for example the height of the activation barrier and consequently the relaxation time of the nanoparticle as well. Other specific properties of magnetic nanoparticle systems makes them specially suitable to measure work fluctuations [44]. We may see these experiments done in the near future.

4.1   Measurement of heat fluctuations

Up to now we discussed about work fluctuations but nothing was said about heat fluctuations. The reason is simple. Work is much easier to measure than heat. Although transferred heat can be measured by using a small thermometer probe (by recording its change of temperature) heat fluctuations are another matter. The easiest procedure to measure heat along a trajectory is to use the first law of thermodynamics where $\Delta Q = \Delta E - W$. Knowledge of both the work and the energy change along a trajectory immediately gives the heat exchanged between system and the bath. Measuring the energy content of the system can be hopeless in many cases. Only in some special cases this is possible. Here I discuss two possible situations.

The first one corresponds to the case where no energy change occurs between the initial and final configuration for all non-equilibrium trajectories. This situation is realized in the magnetic example [44] discussed in the previous section where the reversal symmetry of the system under a field induces a zero energy change $E_f = E_i = -\mu H_0$ if the field is changed from $-H_0$ to H_0 in a ramping experiment (here we assume H_0 to be large enough for the initial and final magnetization to align in the direction of the field). In general, $\Delta E = 0$ can be accomplished in any non-equilibrium cycle assuming that the initial and final states are identical. In the case of the unfolding of the RNA molecule under applied force this can be achieved by considering non-equilibrium trajectories where the molecule first unfolds and then refolds along a given cycle.

The second situation corresponds to the case where, due to the inherent simplicity of the system, the energy is known. A relevant example is the particle confined in an optical trap. In that case the energy of the bead in the trap is well approximated by $E = (1/2)kx^2$ and therefore the value of ΔE is known for each trajectory. The distribution of exchanged heat shows interesting features as compared to the work that have been recently discussed by Zohn and Cohen [46].

5 Single molecule experiments

The advent of nanotechnologies has provided instruments and tools for scientists to manipulate individual molecules and follow their dynamical trajectories as they carry out specialized molecular tasks [47]. The research of molecular reactions performed by individual molecules offers new insight on the importance of fluctuations and stochasticity in small systems. Mechanical force has been recognized as essential to understand the fate of many chemical reactions [48, 49]. Several force-microscopies are currently available to investigate the individual behavior of biomolecular complexes. Atomic force microscopy (AFM), optical and magnetic tweezers tweezers have become common tools that allow scientists to measure the response of these systems to applied external force. These techniques cover different but overlapping ranges of forces: AFM covers a range of forces spanning from several tens of pN up to hundreds of pN, optical tweezers span the intermediate region between 1pN and 100pN and magnetic tweezers are sensitive to tenths of pN.

Thermal fluctuations are important whenever the energies involved in molecular processes are of the order of several $k_B T$. A quick estimate of the forces participating in this regime can be obtained as follows. The typical distance d involving conformational changes at the biomolecular level is of the order of 1nm [12].

---

[12] This is only a rough estimate, for instance the base pair distance in DNA is around one third of a nanometer. This is the minimal distance that polymerases have to cover to elongate one base pair the newly synthesized strand.

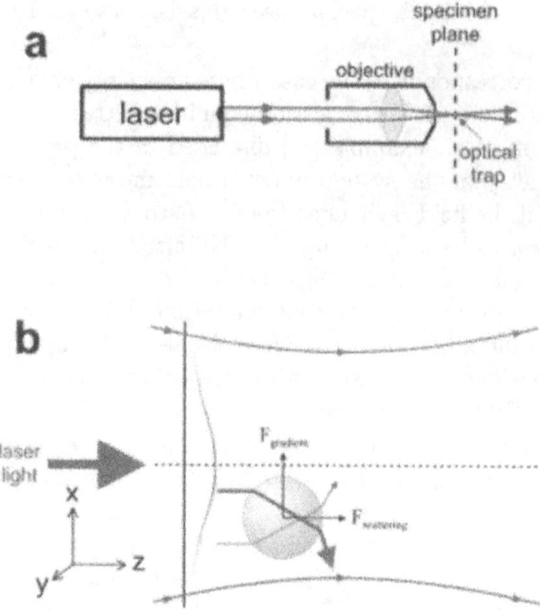

Figure 3: A single laser tweezers setup. (a) The laser light is focused into a spot by using an objective. (b) A Gaussian profile of light intensity generates a confining potential due to conservation of light momentum. The trapping force is induced by the difference in the index of refraction between the polystyrene bead and the surrounding water.

At room temperature $T = 298K$, $Fd = k_B T$ this gives a force $F \simeq 4pN$. Optical tweezers are ideal to investigate a large region of intermediate forces around this value [51]. Nowadays, optical tweezers are used to investigate many processes operated by biomolecules, ranging from the elastic deformation of nucleic acids or proteins to the specific action of enzymes acting on molecular substrates [50]. A typical experimental setup is shown in Fig. 3. Optical tweezers use light momentum conservation to generate a force gradient on polystyrene beads (of a diameter between 1 and 3 microns) that are immersed in water. Light deflection inside the beads arises from the difference in the index of refraction between the beads and water. In this way a confining potential can be generated by focusing a beam of light inside the chamber. To a high degree the confining potential can be considered as harmonic. Single beam tweezers can generate confining forces of the order of several tens of pN [13]. Dual tweezers use two counter propagating beams to gen-

---

[13]The confining force depends on wavelength of the light. Typical wavelength values are in the range 700-1000nm, lower frequencies are inadvisable as they can lead to light absorption and subsequent heat convection effects around the bead.

erate higher forces (up to 150 pN) and have the advantage (by measuring the total amount of deflected light) that recurrent calibration is not required to measure forces. A fluid chamber is fixed in a movable stage or frame that is controlled by a piezo actuator. The chamber is made out of two parallel glass plates separated by a thin layer of parafilm. Inside the chamber there is a glass micropipette that can trap beads of the size of the micron by air suction. The two counter propagating laser beams can confine another bead in the optical trap. To measure forces on molecules a tether is attached to the two beads (one in the micropipette, the other in the trap). Attachments are designed by chemical treatment of the surface of the beads and chemical modification of the ends of the molecule (called labeling). As the stage is moved the force on the bead in the trap (and therefore, on the tether) can be measured. The distance between beads is then measured by using a light lever and a force-extension curve (FEC) can be recorded. Optical tweezers have been used in different fields ranging from physics to biology. A survey of their applications can be found in [50].

DNA plays a central role in biophysics [52, 53]. Accordingly its mechanical properties have been extensively investigated during the past 10 years [54, 55]. Initial investigations on the elastic response of double-stranded DNA under tension [56] have revealed that DNA behaves like an entropic spring as predicted by the worm-like chain model of polymer theory [57]. However, at difference with other polymers DNA shows structural transitions at modest forces (around or below 100pN) depending on how the molecule is pulled. For example, torsionally unconstrained double-stranded DNA shows a highly cooperative overstretching transition around 65pN [58, 59]. At the origin of this behavior there is the double-helix structure of DNA and the associated uncoiling of the two strands. However, if both strands are pulled from the same end of the DNA molecule, then DNA sequentially unzips at constant force following a curve that depends on the particular nucleotide sequence [60, 61]. The elastic response of DNA has produced many experimental [62, 63, 64, 65, 66] as well as theoretical investigations [67, 68, 69, 70, 71, 72] to characterize its structural transitions. A particular force-extension curve (FEC) showing the characteristic overstretching transition of double-stranded DNA is shown in Fig. 4.

6 Pulling experiments on RNA

RNA is an essential molecule in biochemistry. It plays an intermediate role between DNA (which encodes the genetic information and represents the "software" in living organisms) and proteins (which perform specialized tasks inside the cells and represent the "hardware"). Such intermediate role has been emphasized after the discovery that certain RNA molecules (called ribozymes) have catalytic activities that are essential in many regulational processes [73]. The relevance of RNA has motivated many single molecule studies. Compared to DNA, optical tweezers measurements in RNA present additional difficulties to the experimentalist. Not

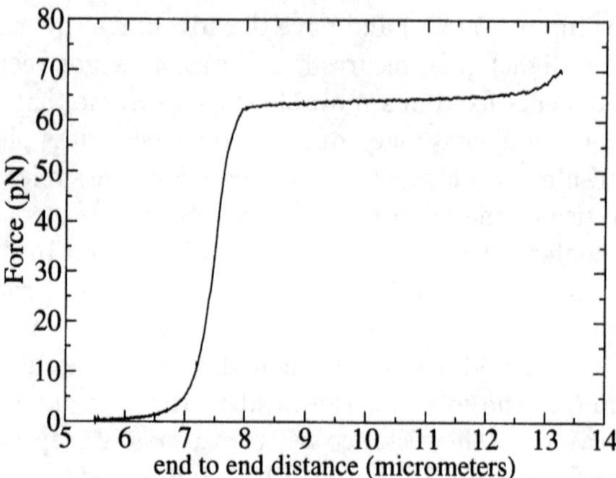

Figure 4: Force extension curve for a torsionally unconstrained DNA molecule of the λ bacteriophage, 24000 base pairs long in a water buffer at 100mM NaCl concentration and 7pH. The molecule has a contour length of approximately $8\mu m$ and shows the characteristic overstretching transition around 65 pN.

only RNA requires more elaboration in the synthesis of the molecular constructs, it is also a molecule very sensitive to the surrounding environment and degrades easier. Moreover, RNA domains have extensions of few tens of nanometers after unfolding, thus requiring more careful and precise measurements.

Liphardt et al. [74] have pulled RNA molecules and studied their unfolding by applying external force using optical tweezers. The molecular construct consists of two hybrid DNA-RNA handles that are annealed to the ends of a small RNA molecule, Fig 5. As the molecular construct is pulled the force-extension curve (FEC) reflects the elastic behavior of the handles (well described by a worm-like chain model [57]) until a force is reached where the molecule unfolds and a jump in the force and distance is observed.

Very interesting dynamical effects were later observed in small RNA hairpins depending on the pulling rate. For slow pulling rates the molecule was seen to follow always the same trajectory and unfold at a reproducible value of the critical force [14]. At this force coexistence and hopping between the folded and unfolded conformations has been observed characteristic of cooperative unfolding [74] [15]. More interesting, as the pulling rate increases larger hysteresis and stochastic fluctuations in the value of the breakage force were observed. Typically the average value of the breakage force tends to increase with the pulling rate. This dependence has been investigated by Evans and Ritchie [76, 77] who have applied

---

[14]Reproducibility of trajectories has always limitations imposed by the unavoidable drift of the optical tweezers machine, see the remark at the end of Sec. 4. The accuracy in the value of the breakage force can be well controlled.

[15]The dependence of the value of the transition force and the hopping frequency on the sequence of the RNA molecule has been studied in [75]

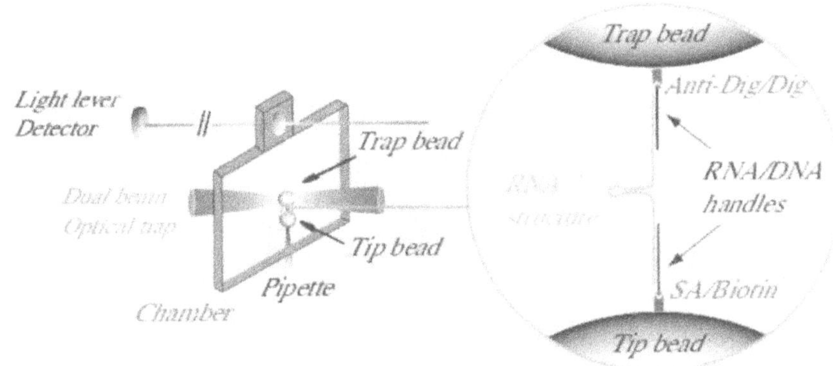

Figure 5: Typical experimental setup when pulling RNA molecules. The molecular construct consists of an RNA molecule attached by its ends to two RNA/DNA hybrid handles (to avoid formation of secondary structures in the handles). As compared to DNA, RNA single molecule experiments present additional difficulties as RNA quickly degrades and the resolution required to observe the unravelling of the molecule is much higher and of the order of the nanometer.

Kramers theory [78] to describe the activated dynamics of a particle jumping over a force-dependent barrier as described by Bell [79]. The study of the loading rate dependence of the breakage force in this type of systems has led to new developments in what is now commonly referred as single-molecule force spectroscopy [80, 81], a technique that is useful to investigate the energy landscape of molecular interactions. Typical unfolding curves showing the pulling rate dependence of the breakage force and the resulting hysteresis effects are shown in Fig. 6.

Hummer and Szabo have realized [82] that the non-equilibrium work relation (20) can be used in single molecule experiments to reconstruct the free energy landscape along the force coordinate. The JE (20) has been experimentally tested in [83] for the P5abc hairpin by repeated measurements of the work done along the unfolding trajectory at different pulling speeds. For P5abc in EDTA buffer the unfolding free energy change is well known from its secondary structure and therefore is a useful example to test the validity of the JE. Typical work histograms are shown in Fig. 7 for three pulling speeds. As expected, as the pulling rate increases the average value of the dissipated work increases reaching values of the order of $4k_BT$ at the fastest pulling speeds. The main result in [83] is that the JE can be used to predict the free energy change for the folding-unfolding transition in the P5abc hairpin with a precision within $1k_BT$ using a modest number of pulls (around 100). Moreover, the JE provides a better estimate for the equilibrium free-energy change than the FD estimate does (21). The advantage of the former as compared to the later has been verified in the near-equilibrium regime (where (18) holds), when the number of repeated pulls is not too large [28].

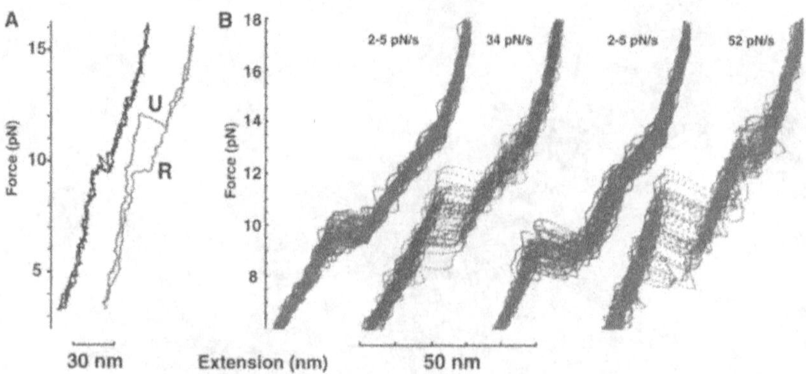

Figure 6: Non-equilibrium pulls in the RNA molecule P5abc at different pulling speeds in a buffer in the absence of magnesium. Panel A shows a reversible (left blue, pulling speed equal to 3-4pN/s) trajectory and an irreversible trajectory (right red, pulling speed 52pN/s). Panel B shows unfolding trajectories for two pulling speeds, 34pN/s (green) and 52pN/s(red) compared to near-equilibrium pulls at 2-5pN/s (blue). Figure taken from [74].

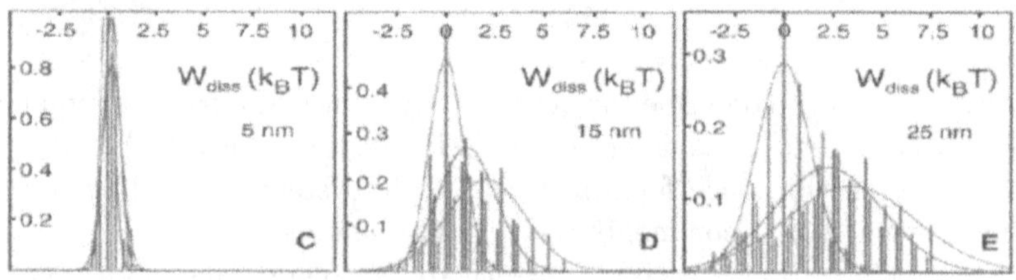

Figure 7: Work distributions for the P5abc molecule at different pulling speeds (3-5pN/s blue, 34pN/s green, 52pN/s red) measured at different distances along the pulling process. Figure taken from [83].

For the case of the P5abc in EDTA buffer the value of the dissipated work is small [16]. In general, for larger molecules the JE is expected to give less reliable estimates for the equilibrium free-energy as the value of the average dissipated work increases. An example is shown in Fig. 8. Typical unfolding curves for a three way RNA junction are shown in Fig. 8. The validity of the JE for such cases is currently investigated [85]. Other more complex cases imply the unfolding of even larger RNA molecules consisting of many domains such as the recently investigated L21 RNA ribozyme [84].

---

[16]In this buffer conditions kinetic barriers are low. High kinetic barriers and strong irreversibility are obtained either by going to faster pulling speeds (however, this is not easy to accomplish due to limited experimental capabilities) or in different buffer conditions. For the latter, high kinetic barriers and many intermediate states are obtained in the presence of divalent cations such as magnesium that establish specific tertiary contacts between some bases.

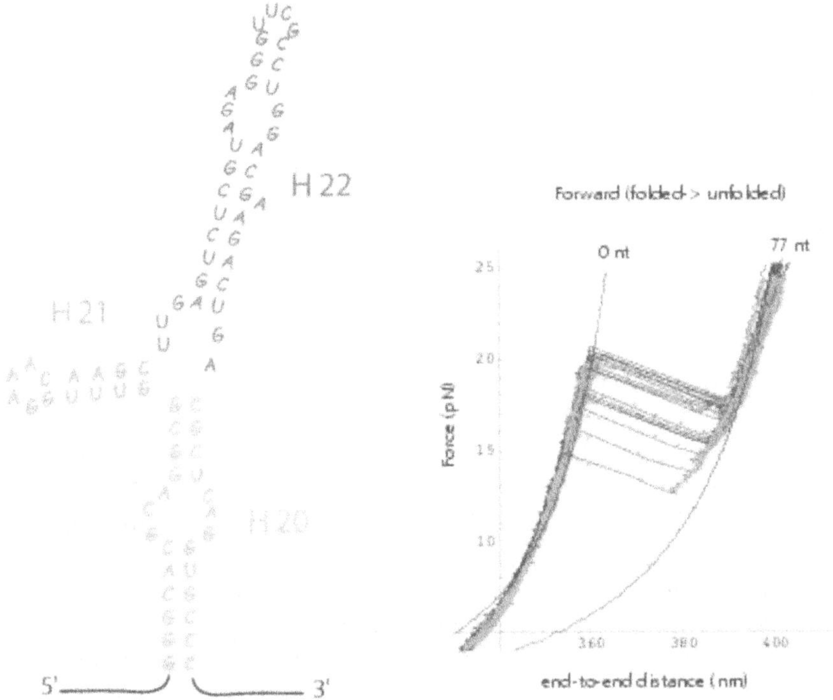

Figure 8: The left figure shows the structure of the junction in the 16S domain of the 30S ribosomal RNA subunit. The right figure shows some unfolding curves at 3.5pN/s. Courtesy of Delphine Collin.

7   Modeling the experiment

In [25] a two-state model has been studied to justify the non-equilibrium experiments in [83]. The main goal was to confirm that indeed it is possible to obtain the equilibrium free energy (within an error equal to $1k_BT$) by using the JE with a limited number of pulls done in that experiment. Interestingly, with this model it is possible to go quite far and find out several results regarding the kinetics of the unfolding process. This allows to make also specific predictions about the kinetic dependence of the dissipated work that can be experimentally tested as well as quantitative statements about the validity of the JE for two-state systems. Moreover, it is possible to do explicit calculations for the work distribution $P(W)$ and, if desired, go beyond the Gaussian case (17). Two-state models provide phenomenological descriptions of systems that can exist in two different forms, therefore the following considerations are expected to be applicable to many systems beyond the folding-unfolding dynamics of RNA molecules. In fact, the two-state model has been shown to provide a good description of the folding-unfolding dynamics of small DNA or RNA hairpins that display strong cooperativity [86, 87] as well as structural transitions in polymers [88]. The model is represented in Fig. 9 where the two conformations (folded and unfolded) are separated by an intermediate barrier located at a distance $\Delta x_{f \to u}$ from the folded state and $\Delta x_{u \to f}$ from the unfolded state, the value $x_m = \Delta x_{f \to u} + \Delta x_{u \to f}$ being the total distance between the folded and the unfolded states. The free energy difference between the two states is denoted as ΔF_0 and the height of the barrier is indicated as B. Transition rates between the folded and the unfolded state are thermally activated and force dependent [79],

$$k_{f \to u}(f) = k_m k_0 \exp(-\beta(B - f\Delta x_{f \to u}))$$
$$k_{u \to f}(f) = k_m k_0 \exp(-\beta(B - \Delta F_0 + f\Delta x_{u \to f})) \tag{22}$$

where f is the external force, $\beta = 1/k_B T$, k_0 is a microscopic attempt frequency and k_m is a contribution arising from the handles, the bead in the trap and the machine [17]. The rates (22) satisfy detailed balance, a necessary condition for the equilibrium regime to be characterized by Boltzmann populations of the folded and unfolded states. The dynamics of the two-state model under the action of an external force has been analyzed in detail for the case of no-refolding process along the unfolding curve [76, 77], also called a first-order Markov process. This particular case is analytically tractable and specific predictions about the form of the work distribution can be made [89].

In the theoretical treatment of a non-equilibrium pulling experiment the force can be taken as the control parameter [18] and increased at an approximately constant rate $r = \dot{f}$. The work exerted along a given trajectory is taken as $W = \int x df$. The probability distribution cannot be exactly evaluated in closed form and only the moments of the distribution can be computed in a perturbative scheme where the average dissipated work is assumed to be several times $k_B T$. The results have been given in [25] for the first two moments. These give the average dissipated work and its variance, from which the value of the fluctuation-dissipation ratio R can be inferred. The first two moments are the most relevant quantities as they can be directly compared with the experimental results. A general result for the average dissipated work can be derived in the linear-response regime where the pulling speed is slow compared to the hopping frequency at the transition force f_t (i.e., the value of the force at which the folded and unfolded populations of the

---

[17]In (22) we continue to use the term F for the Gibbs free energy. To be precise we should use instead G as for the experimental conditions the temperature and pressure of the bath are held constant.

[18]Strictly speaking this is not true. As remarked in Sec. 4 the control parameter in pulling experiments using optical tweezers is not the force but the distance between the center of the optical trap and the tip of the micropipette. The pulling speed r is always an average value of the force-dependent speed along the unfolding curve. Under this approximation (which typically introduces a small correction), using the force or the distance as the control parameter turns out to be equivalent as $d(fx) = f dx + x df$, see footnote (**) in [25] for a related remark.

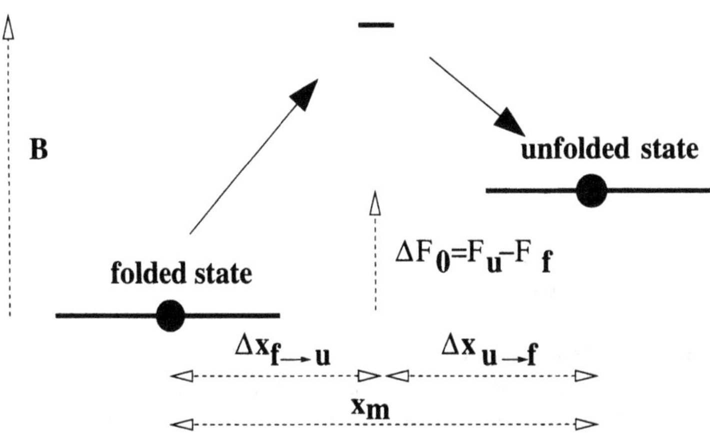

Figure 9: The two-state model with an intermediate barrier. The parameters are the free-energy gap ΔF_0, the unfolding distance x_m, the height B of the intermediate barrier and the distance of the intermediate barrier to the folded state $\Delta x_{f \to u} = x_m - \Delta x_{u \to f}$. Figure taken from [25].

RNA molecules are equal in equilibrium). The linear-response regime is therefore characterized by the dimensionless parameter ρ defined as,

$$\rho = \frac{r}{f_t k_{\text{total}}(f_t)} \tag{23}$$

where $k_{\text{total}}(f_t) = k_{u \to f}(f_t) + k_{f \to u}(f_t)$ is the total rate at the transition force. When $\rho < 1$ the average dissipated work is given by,

$$\overline{W_{\text{dis}}} \sim \rho \Delta F_0 + \mathcal{O}(\rho^2) \tag{24}$$

The linear dependence of (24) can be used to derive estimates for the relaxation time of the molecule that might complement other type of kinetic measurements (such as the measurement the folding-unfolding hopping frequency right at the transition force). In Fig. 10 we show the results for these quantities as a function of the pulling rate using some kinetic parameters in (22) to fit the experimental data. The dashed line in the left panel of Fig. 10 shows the linear response prediction (24). Note that the experimental points fall off the linear response curve, showing that the pulling rates investigated in [84] explore the far from equilibrium regime. This is an important result because it shows that, despite of the smallness of the value of the average dissipated work (in the range $2 - 4k_BT$) the experiments were carried out far from equilibrium reinforcing the validity of the Jarzynski re-

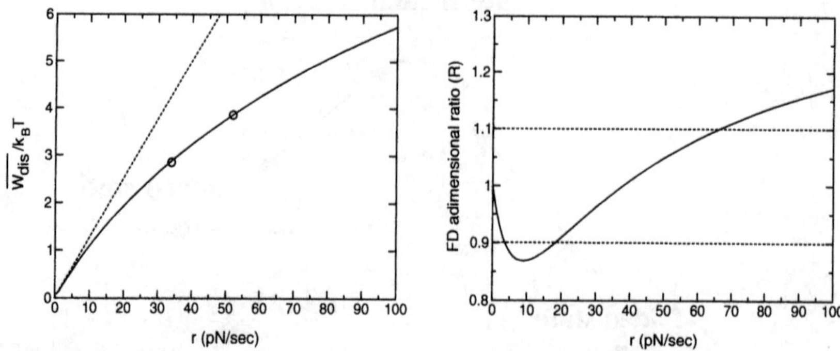

Figure 10: Average dissipated work (left) and fluctuation-dissipation ratio R (right) as a function of the pulling speed. The circles in the left figure are the experimental values. The values of the kinetic parameters characterizing the rates (22) have been chosen to fit the experimental data. The dashed line in the left figure shows the linear-response formula (24). The two horizontal dashed lines in the right figure limit a region of pulling speeds where the FD estimate (21) is expected to approximate well the free-energy change during the folding-unfolding reaction. Figure taken from [25].

lation in such regime [19]. This conclusion is substantiated by the dependence of the fluctuation-dissipation ratio (right panel in Fig. 10) which shows a strong non-monotonic behavior for pulling speeds above 20pN/s. Further evidence endorsing the fact that experiments were carried out far from the equilibrium regime is inferred from the shape of the work probability distributions $P(W)$ [20]. The results are shown in the left panel of Fig. 11 and were obtained from numerical simulations of the model using the kinetic values of the fitting parameters as derived from Fig. 10. Gaussian behavior is a fingerprint of the near-equilibrium regime, see the discussion in the paragraph containing footnote 9 in Sec. 3.2. As a comparison we show in the right panel of Fig. 11 the histograms obtained from the experiments. Both theory and experiments bear a close resemblance. Fig. 11 reveals the existence of long tails at both sides of the work distribution that strongly deviate from the Gaussian behavior.

Finally we provide an answer to the original question with which we started this section. Can we support the main result of the experiment [83] where a small

---

[19]Indeed, had the experiments been carried out in the near-equilibrium regime, then the recovery of the equilibrium free-energy change using the JE would be expected due to the smallness of the average values of the dissipated work. The fact that the value of the dissipated work is small, yet the system is far from equilibrium, is consequence of the smallness of the RNA molecule.

[20]It must be emphasized though that this statement would not be valid if, by some reason, the parameters used to fit the kinetic data were completely off from the actual values. Although far-fetched, this possibility cannot be ruled out as the two-state model here considered is probably a crude approximation to the real description of the unfolding process (see the discussion in Ref. [49]). New experiments in other RNA hairpins are required to reach a better understanding.

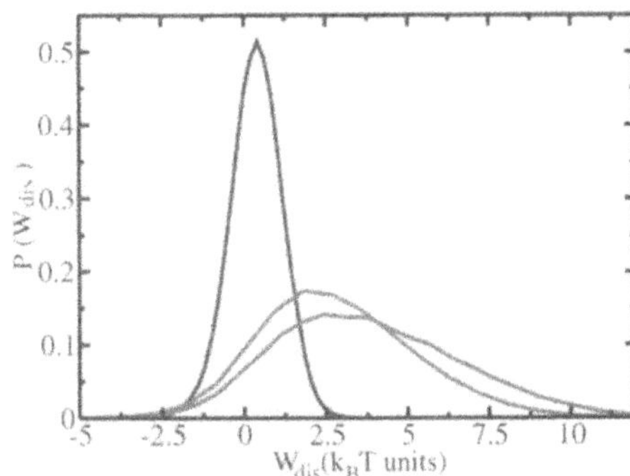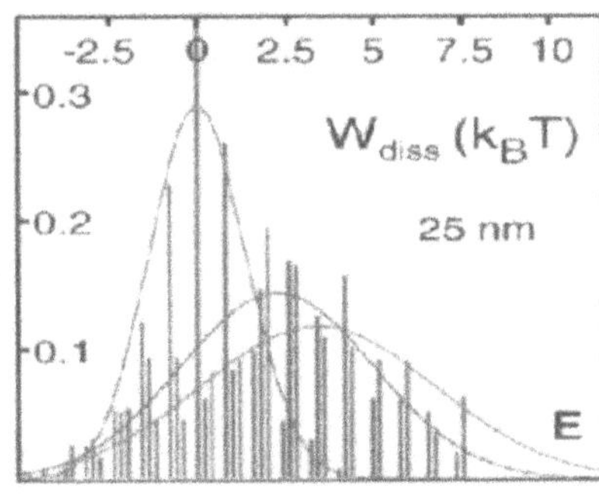

Figure 11: Work distributions obtained for the model (left) compared to the experimental results. Strong deviations from a Gaussian behavior are predicted, specially in the left tails of the distribution. Figures taken from [25, 83].

number of pulls (around 60) was enough to obtain the equilibrium free energy (within an error of $1k_BT$) by using the JE? In Fig. 12 we compare the bias error obtained from the two estimates (20,21) as well as from the average dissipated work $\overline{W}_{\rm dis}$ along the force coordinate. The different bias errors are defined as,

$$B^{\rm dis} = \overline{W} - \Delta F = \overline{W}_{\rm dis} \tag{25}$$

$$B^{FD} = \Delta F_{FD} - \Delta F = \overline{W}_{\rm dis}(1 - R) \tag{26}$$

$$B^{JE} = \Delta F_{JE} - \Delta F = -\log\left(\overline{\exp(-\frac{W_{\rm dis}}{k_BT})}\right) \tag{27}$$

where (18) and (20,21) have been used. The bias error depends on the number of pulls $N_{\rm pulls}$. To obtain these bias values we have averaged (25,26,27) over a large number of sets of experiments, each set characterized by $N_{\rm pulls}$ repeated pulls. Full convergence to the correct free-energy, as the number of pulls increases, corresponds to a vanishing bias throughout the force axis. From Fig. 12 we can see how the values obtained from the average dissipated work (25) and the FD estimate (26) quickly converge to limiting curves characterized by a finite bias. However, the bias obtained from the JE (27) slowly converges to zero and practically vanishes only for $N_{\rm pulls} \sim 10^6$. Also, from the JE bias (27) shown in Fig. 12 we learn that 100 pulls are enough to get an estimate of the free-energy within $1k_BT$ of error for the folding-unfolding reaction, by using non-equilibrium work values at the two largest pulling speeds. We mention also that the FD estimate works well in the large force region of the force axis (around 20pN) but not in the intermediate force region (around 14pN) where it develops a bump. The reason why the FD works so well at high forces has it root in the behavior of the fluctuation-dissipation ratio R shown in the right panel in Fig. 10. There R has been evaluated from a pulling protocol where the force is ramped from 0 to a

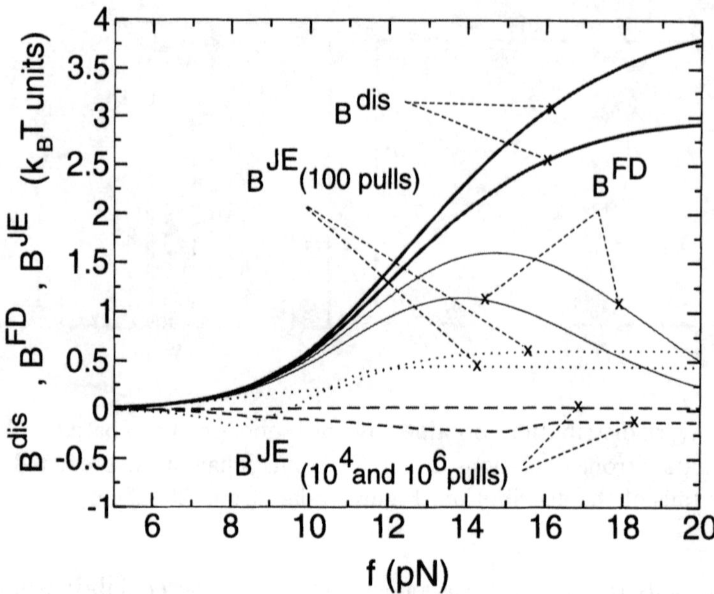

Figure 12: Behavior of the different bias defined in (25,26,27) along the force coordinate. Figure taken from [25].

value around 20pN. In that case R is close to 1 for a large region of pulling speeds (delimited by the two horizontal dashed lines). Whenever $R \sim 1$ the FD estimate is expected to work well if the number of pulls is not too small.

The number of pulls required to obtain the equilibrium free energy with an error within $1k_BT$ can be estimated by measuring B^{JE} averaged over many sets, each one containing N_{pulls} repeated pulls. The dependence of the bias B^{JE} with N_{pulls} is shown in Fig. 13. The decay of the bias with N_{pulls} can be very well approximated by a power law $(N_{\mathrm{pulls}})^{-\alpha(r)}$ where the exponent $\alpha(r)$ depends on the pulling rate (or the average dissipated work as they are related each other). The bias B^{JE} shows as a crossover to a $1/N_{\mathrm{pulls}}$ behavior for $N_{\mathrm{pulls}} > 1000$ in agreement with the prediction by Wood [90]. For a Gaussian process in the near-equilibrium regime the value of the exponent $\alpha(r)$ has been estimated numerically [28] and is relatively close to the values found in this case. From Fig. 13 we see the number of pulls required for the bias B^{JE} to be equal to $1k_BT$ (indicated as the horizontal dashed line). This number of pulls is then shown in the inset of Fig. 13 as a function of the average dissipated work (also the pulling speeds are indicated). Under certain assumptions (see [91]), and only for modest values of the average dissipated work [91], this number of pulls can be shown to approximately grow as $\exp(R_- \frac{\overline{W_{\mathrm{dis}}}}{k_BT})$ with R_- a constant of order unity that characterizes the left side tail of the work distribution. For the experimental values of the pulling speed considered in the experiment (the region limited by the square box shown in the

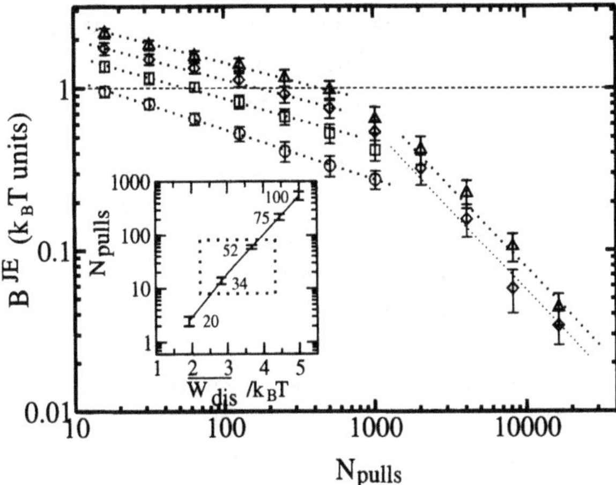

Figure 13: Main panel: Bias error (in units of k_BT) for the Jarzynski average (27) as function of the number of pulls for different pulling rates (from bottom to top: 34 (green),52 (red),75 (violet),100 (black) pN/s). Data have been averaged over 1000 sets and error bars correspond to 100 sets. Inset: Number of pulls necessary to obtain an estimate for the equilibrium free energy within k_BT and fit to the estimate $N_{\text{pulls}} \sim \exp(R_- \frac{\overline{W_{\text{dis}}}}{k_BT})$ which yields $R_- \simeq 1.5$.

figure in the Inset) the required number of pulls to get the desired accuracy is of the order of several tens as done in the experiment.

All in all, the two-state model reproduces quantitatively many aspects of the non-equilibrium behavior observed in the experiment [84] and justifies the test of the validity of the JE there claimed.

8   Conclusions

Thermodynamics represented a great step in the development of science. It provided a general framework to understand all natural processes that involve the transformation of different sorts of energy (mechanical, chemical, electromagnetic) into work and heat. While work can be viewed as useful energy, heat represents energy that is not useful. The second law of thermodynamics limits the amount of useful work that can be extracted from heat. As heat abounds in nature it seems plausible that the level of organization that we see today in the form of biological matter originates from certain properties that characterize heat exchange processes.

Statistical mechanics provided a mechanistic picture of the abstract concepts of thermodynamics in terms of the average behavior of a large number of atoms or molecules and their interactions. According to this picture, thermodynamic quantities are not strictly constant but fluctuate around their average values. However, the amount of these fluctuations is small relative to the value of the thermodynamic quantities themselves. Much larger fluctuations are hardly observable and become irrelevant as the macroscopic level is approached.

Fluctuation theorems go beyond this statistical level of description by quantifying fluctuations arbitrarily large whose magnitude can be of the same order of the average value. This is the content of the non-equilibrium work relation originally derived by Jarzynski [21]. In that case, work trajectories quite far from the average or most probable trajectory, have to be properly weighed for the equality to be satisfied. As the system size increases (or as the time increases for steady state systems) the probability to observe these rare trajectories quickly decreases. Were we repeat many times the dynamical experiment, the time we should wait until finding a trajectory that notably reduces the bias error associated with the equality, increases exponentially with the size of the system, ultimately reaching values that are of the order of the Poincare recurrence time. We then considered the suggestive fact that most of the non-equilibrium trajectories that enforce the validity of the Jarzynski equality, are also those that inspired many of the paradoxes underlying the statistical interpretation of heat and that were proposed in the early days of statistical mechanics.

What is the fundamental value of these rare trajectories described by fluctuation-theorems? If fluctuation theorems were just theorems, the value would be predominantly academic. There is of course interest in using the non-equilibrium work relation to obtain free energies for transformations that cannot be carried out reversibly. However, it might be possible that fluctuation theorems have an added fundamental value. They could provide physicists with a tool to explore the validity of the principles underlying some energy transformation processes. In the same way that classical mechanics proved inadequate to describe energy exchange between radiation and matter at the atomistic level, one could imagine that current theories describing thermal exchange processes occurring at very small length-scales or short times should be accordingly revised. In fact, all fluctuation theorems use in one way or another the concept of microscopic reversibility. This condition ensures that systems thermalize if left to evolve for a long time. However, it might be possible that microscopic reversibility holds only in average, that transitions at the microscopic level have unexpected properties with important consequences for biology and life [22]. If this were the case, while the average behavior would be well described by current dynamical theories, rare fluctuations might display a more refined pattern beyond our current expectation.

Biological matter tends to organize reaching fantastic levels of complexity. Although it is often tacitly assumed that our current understanding of physics

---

[21]We did not mention in this feature extensions of the classical non-equilibrium work relation to the quantum regime. Although several papers have recently appeared in the literature [92, 93, 94, 95], the concept of a quantum trajectory and quantum work are to be clarified and the first experimental attempt to test the corresponding quantum relations is still to be done.

[22]Ideas such as *purposiveness* of changes have appeared recurrently in the context of natural selection in biology, see for instance [96].

will provide clues to fill into the many "details" that surround the organization of biological matter, the truth is that bold ideas will be probably needed to go beyond the present state of the art. Biological matter will become a common laboratory for physicists in order to test and understand many of the questions that transcend the behavior of ordinary matter. Single molecule experiments have opened a vein of research for physicists, that require the combination of a general knowledge of physics, chemistry and biology to grasp the most relevant aspects required to unravel the behavior of living matter at the most fundamental level.

Acknowledgments. I acknowledge the warm hospitality of the Bustamante, Tinoco and Liphardt labs at UC Berkeley where this work has been done. I thank C. Bustamante, D. Collin, C. Jarzynski, S. Smith, I. Tinoco and E. Trepagnier for useful discussions. I wish to thank also J. Liphardt for discussions and a critical reading of the manuscript. This work is supported by the David and Lucile Packard Foundation, the European community (STIPCO network), the Spanish research council (Grant BFM2001-3525) and the Catalan government.

References

[1] R. B. Laughlin, D. Pines, J. Schmalian, B. P. Stojkovic and P. Wolynes, *Proc. Nat. Acad. Sci. USA* **97**, 32 (2000).

[2] An excellent book is that by H. B. Callen, *Thermodynamics and Introduction to thermostatistics* (Wiley, New York, 1985).

[3] A beautiful book in a narrative style has been written by P. W. Atkins *The 2nd Law* (W. H. Freeman and Company, New York, 1984).

[4] S. G. Brush, *Kinetic theory* (Vol.1 Pergamon 1965;Vol2 Pergamon 1966).

[5] H. Nyquist, *Phys. Rev.*,**32**, 110 (1928)

[6] H. B. Callen and T. A. Welton, *Phys. Rev.* **83**, 34 (1951).

[7] C. Jarzynski, *Phys. Rev. Lett.* **78**, 2690 (1997).

[8] C. Jarzynski, *Phys. Rev. E* **56**, 5018 (1997).

[9] G. E. Crooks, *J. Stat. Phys.* **90**, 1481 (1998).

[10] G. E. Crooks, *Phys. Rev. E* **61**, 2361 (2000).

[11] J. Kurchan, *J. Phys. A (Math. Gen.)* **31**, 3719 (1998).

[12] C. Jarzynski, in *Dynamics of Dissipation*, P. Garbaczewski, R. Olkiewicz, Eds., (Springer, Berlin 2002).

[13] D.J. Evans, E. G. D. Cohen and G. P. Morriss, *Phys. Rev. Lett.* **71**, 2401 (1993).

[14] G. Gallavotti and E.G.D. Cohen, *Phys. Rev. Lett.* **74**, 2694 (1995).

[15] G. Gallavotti and E.G.D. Cohen, *J. Stat. Phys.* **80**, 931 (1995).

[16] C. Maes, *J. Stat. Phys.* **95**, 367 (1999).

[17] S. Ciliberto and C. Laroche, J. Phys. IV (France) **8**, 215 (1998)

[18] S. Ciliberto, N. Garnier, S. Hernandez, C. Lacpatia, J.-F. Plinton and G. Ruiz Chavarria, *Preprint arXiv:nlin.CD/0311037.*

[19] Subsequent work has been reviewed in D. Evans and D. Searles, *Adv. Phys.* **51**, 1529 (2002).

[20] D.J. Evans and D.J. Searles, *Phys. Rev. E* **50**, 1645 (1994).

[21] For a review see A. Crisanti and F. Ritort, *J. Phys. A (Math. Gen.)* **36**, R181 (2003).

[22] A. Crisanti and F. Ritort, *Preprint arXiv:condmat/0307554.*

[23] F. Ritort, *Preprint arXiv:condmat/0311370.*

[24] D. Chandler, *Introduction to Modern Statistical Mechanics*, Oxford University Press, (1987).

[25] F. Ritort, C. Bustamante and I. Tinoco Jr., *Proc. Nat. Acad. Sci. USA* **99**, 13544 (2002).

[26] O. Mazonka and C. Jarzynski, *Preprint arXiv:cond-mat/9912121.*

[27] D.M. Zuckerman and T.B. Woolf, *Chem. Phys. Lett.* **351**, 445 (2002); *Phys. Rev. Lett.* **89**, 180602 (2002).

[28] J. Gore, F. Ritort and C. Bustamante, *Proc. Nat. Acad. Sci. USA* **100**, 12564 (2003).

[29] G.M. Wang, E.M. Sevick, E. Mittag, D.J. Searles, and D.J. Evans, *Phys. Rev. Lett.* **89**, 050601 (2002).

[30] C. Bustamante, S.B. Smith, J. Liphardt and D. Smith, *Curr. Opin. Struct. Biol.* **10**, 279 (2000).

[31] C. Bustamante, J.C. Macosko and G.J.L. Wuite, *Nature Rev. Mol. Cell. Bio.* **1**, 130 (2000).

[32] G. Bao, *J. Mech. Phys. Sol.* **50**, 2237 (2002).

[33] J.M. Schurr, private communication.

[34] M. Carrion-Vazquez, A.F. Oberhauser, S.B. Fpwler, P.E. Marszalek, S. E. Broedel, J. Clarke and J.M. Fernandez, *Proc. Nat. Acad. Sci. USA* **96**, 3694 (1999).

[35] J. Gelles and R. Landick, *Cell* **93**, 13 (1998).

[36] M.D. Wang, M.J. Schnitzer, H. Yin, R. Landick, J. Gelles and S.M. Block, *Science* **282**, 902 (1998).

[37] A. Goel, R.D. Astumian and D. Herschbach, *Proc. Nat. Acad. Sci. USA* **100**, 9699 (2003).

[38] G.J.L. Wuite, S.B. Smith, M. Young, D. Keller and C. Bustamante, *Science* **404**, 103 (2000).

[39] N.R. Forde, D. Izhaky, G.R. Woodcock, G.J. L. Wuite abd C. Bustamante, *Proc. Nat. Acad. Sci. USA* **99**, 11682 (2002).

[40] D.E. Smith, S.J. Tans, S.B. Smith, S. Grimes, D.L. Anderson and C. Bustamante, *Nature* **413**, 748 (2001).

[41] T. Ha, I. Rasnick, W. Cheng, H.P. Babcock, G.H. Gauss, T.M. Lohman and S. Chu, *Nature* **419**, 638 (2002).

[42] J.F. Leger, J. Robert, L. Bourdieu, D. Chatenay and J. Marko, *Proc. Nat. Acad. Sci. USA* **95**, 12295 (1998).

[43] E. Trepagnier, C. Jarzynski, F. Ritort, G. Crooks, C. Bustamante and J. Liphardt, Verification of a generalized second law of thermodynamics, submitted.

[44] F. Ritort, Work fluctuations and non-equilibrium temperatures in two-state systems, submitted.

[45] W. Wernsdorfer, E. Bonet-Orozco, K. Hasselbach, A. Benoit, B. Barbara, N. Demoncy, A. Loiseau, H. Pascard and D. Mailly, *Phys. Rev. Lett.* **78**, 1791 (1997).

[46] R. Van Zon and E.G.D. Cohen, *Phys. Rev. Lett.* **91**, 110601 (2003).

[47] C. Bai, C. Wang, X.S. Xie and P.G. Wolyness, *Proc Nat. Acad. Sci. USA* **96**, 11075 (1999).

[48] I. Tinoco Jr. and C. Bustamante, *Biophys. Chcm.* **101–102**, 513 (2002).

[49] C. Bustamante, Y.R. Chemla, N.R. Forde and D. Izhaky, submitted.

[50] A resource letter can be found in M.J. Lang and S.M. Block, *Am. J. Phys.* **71**, 201 (2003).

[51] S.B. Smith, Y. Cui and C. Bustamante, *Methods. Enzymol.* **361**, 134 (2002).

[52] M.D. Frank-Kamenetskii, *Phys. Rep.* **288**, 13 (1997).

[53] For an excellent book see C.R. Calladine and H. Drew, *Understanding DNA* (Acaddemic London 1997).

[54] T. Strick, J-F. Allemand, V. Croquette and D. Bensimon, *Prog. Biophys. Mol. Biol.* **74**, 115 (2000).

[55] C. Bustamante, Z. Bryant and S.B. Smith, *Nature* **421**, 423 (2003).

[56] S.B. Smith, L. Finzi and C. Bustamante, *Science* **258**, 1122 (1992).

[57] C. Bustamante, J.F. Marko, E.D. Siggia and S. Smith, *Science* **265**, 1599 (1994).

[58] P. Cluzel, A. Lebrun, C. Heller, R. Lavery, J.-L. Viovy, D. Chatenay and F. Caron, *Science* **271**, 792 (1996).

[59] S.B. Smith, Y. Cui, C. Bustamante, *Science* **271**, 795 (1996).

[60] B. Essevaz-Roulet, U. Bockelmann and F. Heslot, *Proc. Nat. Acad. Sci. USA* **94**, 11935 (1997).

[61] C. Danilowicz, V. W. Coljee, C. Bouzigues, D. K. Lubensky, D. R. Nelson and M. Prentiss, *Proc. Nat. Acad. Sci. USA* **100**, 1694 (2003)

[62] M.D. Wang, H. Yin, R. Landick, J. Gelles and S.M. Block, *Biophys. J.* **72**, 1335 (1997).

[63] J.F. Leger, G. Romano, A. Sarkar, J. Robert, L. Bourdieu, D. Chatenay and J.F. Marko, *Phys. Rev. Lett.* **83**, 1066 (1999).

[64] M. Rief, H. Clausen-Schaumann and H.E. Gaub, *Nature* **6**, 346 (1999).

[65] M. C. Williams, I. Rouzina and V.A. Bloomfield, *Acc. Chem. Res.* **35**, 159 (2002).

[66] Z. Bryant, M.D. Stone, J. Gore, S.B. Smith, N. Cozzarelli and C. Bustamante, *Nature* **424**, 338 (2003).

[67] A. Lebrun and R. Lavery, *Nucleic Acids Res.* **24**, 2260 (1996).

[68] C. Bouchiat and M. Mezard, *Phys. Rev. Lett.* **80**, 1556 (1998).

[69] J. Marko, *Phys. Rev. E* **57**, 2134 (1998).

[70] A. Ahsan, J. Rudnick and R. Bruinsma, *Biophys. J.* **74**, 132 (1998).

[71] S. Cocco and R. Monasson, *Phys. Rev. Lett.* **83**, 5178 (1999).

[72] H. Zhou, Y. Zhang and Z. Ou-Yang, *Phys. Rev. E* **62**, 1045 (2000).

[73] R.F. Gesteland, T.R. Cech and J.F. Atkins, Editors of *The RNA world*, 2nd edit., Cold Spring Harbor Laboratory Press, Cold Spring Harbor, NY.

[74] J. Liphardt, B. Onoa, S.B. Smith, I. Tinoco Jr. and C. Bustamante, *Science* **292**, 733 (2001).

[75] S. Cocco, R. Monasson and J. Marko, *Eur. Phys. J. E* **10**, 153 (2003).

[76] E. Evans and K. Ritchie, *Biophys. J.* **72**, 1541 (1997).

[77] E. Evans and K. Ritchie, *Biophys. J.* **76**, 2439 (1999).

[78] P. Hanggi, P. Talkner and M. Borkovec, *Rev. Mod. Phys.* **62**, 251 (1990).

[79] G.I. Bell, *Science* **200**, 618 (1978).

[80] E. Evans, *Annu. Rev. Biomol. Struct.* **30**, 105 (2001).

[81] C. Friedsam, A.K. Wehle, F. Kühner and H.E. Gaub, *J. Phys. Cond. Matt.* **15**, S1709 (2003).

[82] G. Hummer and A. Szabo, *Proc. Natl. Acad. Sci. USA* **98** 3658 (2001).

[83] J. Liphardt, S. Dumont, S.B. Smith, I. Tinoco Jr. and C. Bustamante, *Science* **296**, 1832 (2002).

[84] B. Onoa, S. Dumont, J. Liphardt, S.B. Smith, I. Tinoco Jr. and C. Bustamante, *Science* **299**, 1892 (2003).

[85] F. Ritort, D. Collin, C. Jarzynski, S. B. Smith, I. Tinoco Jr. and C. Bustamante, unpublished.

[86] G. Bonnet, O. Krichevsky, A. Libchaber, *Proc. Natl. Acad. Sci. USA* **95** 8602 (1998).

[87] S.J. Chen and K.A. Dill, *Proc. Natl. Acad. Sci. USA* **97** 646 (1997).

[88] M. Rief, J.M. Fernandez and H.E. Gaub, *Phys. Rev. Lett.* **81**, 4764 (1998).

[89] M. Mañosas and F. Ritort, unpublished.

[90] R. H. Wood, *J. Phys. Chem.* **95**, 4838 (1991).

[91] F. Ritort, unpublished.

[92] J. Kurchan, *Preprint arXiv:condmat/0007360*.

[93] H. Tasaki, *Preprint arXiv:condmat/0009244*.

[94] S. Yukawa, *J. Phys. Soc. Jpn.* **69**, 2367 (2000).

[95] S. Mukamel, *Phys. Rev. Lett.* **80**, 170604 (2003).

[96] L. Margulis and D. Sagan, *What is life?*, (University of California Press, Berkeley, 1995).

Félix Ritort
Departament of Physics
Faculty of Physics
University of Barcelona
Diagonal 647
E-08028 Barcelona, Spain

and

Department of Physics
University of California
Berkeley CA 94720, USA
email: ritort@ffn.ub.es

Poincaré Seminar 2003, 227 – 264
© Birkhäuser Verlag, Basel, 2004

The Entropy of Black Holes: A Primer

Thibault Damour

Abstract. After recalling the definition of black holes, and reviewing their energetics and their classical thermodynamics, one expounds the conjecture of Bekenstein, attributing an entropy to black holes, and the calculation by Hawking of the semi-classical radiation spectrum of a black hole, involving a thermal (Planckian) factor. One then discusses the attempts to interpret the black-hole entropy as the logarithm of the number of quantum micro-states of a macroscopic black hole, with particular emphasis on results obtained within string theory. After mentioning the (technically cleaner, but conceptually more intricate) case of supersymmetric (BPS) black holes and the corresponding counting of the degeneracy of Dirichlet-brane systems, one discusses in some detail the "correspondence" between massive string states and non-supersymmetric Schwarzschild black holes.

1 Black holes

Let us start by briefly reviewing the concept of black hole. Within a couple of months after Einstein's discovery of the final form of the field equations of General Relativity, Karl Schwarzschild succeeded in writing down the exact form of the general relativistic analog of the "gravitational field of a mass point" [1], namely (in the coordinates introduced by Johannes Droste) the general spherically-symmetric solution of Einstein's vacuum field equations ($R_{\mu\nu} = 0$):

$$ds^2_{\text{SCHW}} = g^{\text{SCHW}}_{\mu\nu}(x^\lambda)\, dx^\mu\, dx^\nu =$$
$$-\left(1 - \frac{2GM}{c^2 r}\right) c^2\, dt^2 + \frac{dr^2}{1 - \frac{2GM}{c^2 r}} + r^2(d\theta^2 + \sin^2\theta\, d\varphi^2). \quad (1.1)$$

Here, G denotes Newton's constant and M denotes the (total) mass of the considered object, as it can be measured from infinity (*e.g.*, by comparing the motion of far-away test particles in the metric (1.1) to that of test particles in the Newtonian potential $U(\mathbf{x}) = GM/r$).

It was soon noticed that the metric (1.1) has an apparently singular behaviour at the "Schwarzschild radius" $r_S \equiv 2GM/c^2$. For instance, a clock at rest in the metric (1.1) and located at a radius r ($r > r_S$) exhibits, when its ticks are "read" from infinity via electromagnetic signals, a redshift equal to

$$\frac{[ds]^{\text{read at}}_{\text{infinity}}}{[ds]^{\text{measured}}_{\text{locally}}} = \frac{\sqrt{-g_{00}(r = \infty)}}{\sqrt{-g_{00}(r)}} = \frac{1}{\sqrt{1 - \frac{2GM}{c^2 r}}}. \quad (1.2)$$

The redshift (1.2) goes to infinity as $r \to r_S$. Therefore the "Schwarzschild sphere" $r = r_S = 2GM/c^2$ has the observable characteristic of being an infinite-redshift surface. Similarly, one finds that the force needed to keep a particle at rest at a radius $r > r_S$ goes to infinity as $r \to r_S$.

It took many years, and the work of many scientists (notably Oppenheimer and Snyder, Kruskal and Penrose), to decipher the physical meaning of these infinities occurring at the "Schwarzschild sphere". This understanding is summarized in the spacetime diagram of Fig. 1.

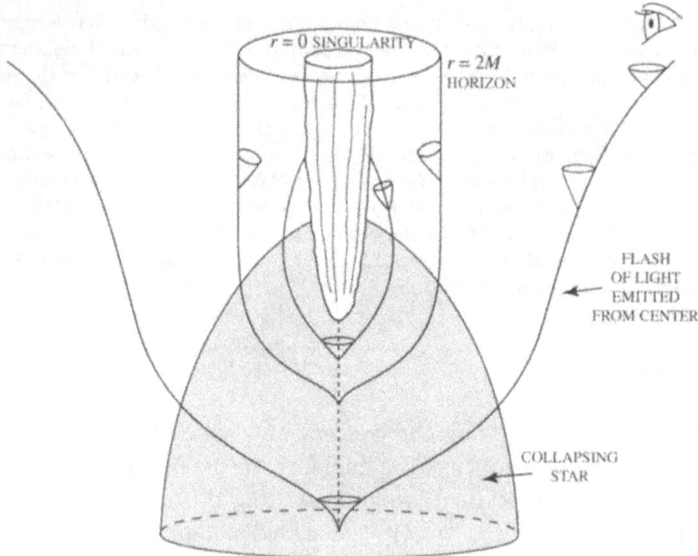

Figure 1: Spacetime representation of the formation of a black hole by the collapse of a star. The horizon is the spacetime history of a bubble of light (*i.e.*, a null hypersurface) which stabilizes itself under the strong pull of relativistic gravity

The 2-dimensional surface $r = r_S$, becomes, when adding the time dimension, a 3-dimensional hypersurface $\mathcal{H}$ in spacetime. The hypersurface $\mathcal{H}$ is a fully regular submanifold of a locally regular spacetime. To see the regularity of $\mathcal{H}$, one needs to change coordinates near $r = r_S$. For instance, one can use the ingoing Eddington-Finkelstein coordinates (v, r, θ, φ), where $v = t + r_*$, with r_* defined as

$$r_* = \int \frac{dr}{1 - \frac{2GM}{c^2 r}} = r + \frac{2GM}{c^2} \ln\left(\frac{c^2 r}{2GM} - 1 \right). \tag{1.3}$$

In these coordinates the Schwarzschild metric reads

$$ds^2_{\text{SCHW}} = -\left(1 - \frac{2GM}{c^2 r} \right) dv^2 + 2dvdr + r^2(d\theta^2 + \sin^2 \theta \, d\varphi^2). \tag{1.4}$$

In these coordinates the horizon is located at $r = r_S$, the other coordinates (v, θ, φ) taking arbitrary, but finite values. [Note that a finite value of the new "time variable" v corresponds to an infinite (and positive) value of the original Schwarzschild time coordinate t.] One easily checks that the geometry (1.4) is regular near $r = r_S$.

One can then see that the hypersurface $\mathcal{H}$ $(r = r_S)$ is a quite special submanifold in that it is a *null hypersurface*, *i.e.*, a co-dimension-1 surface which is locally tangent to the light cone. In other words, the (co-)vector normal to the hypersurface, say ℓ_μ (such that $\ell_\mu \, dx^\mu = 0$ for all directions dx^μ tangent to the hypersurface) is a null vector: $0 = g^{\mu\nu} \ell_\mu \ell_\nu = g_{\mu\nu} \ell^\mu \ell^\nu$. As a consequence ℓ^μ is both normal and tangent to the hypersurface (because $\ell_\mu \ell^\mu = 0$). The local light cone, $g_{\mu\nu}(x) \, dx^\mu \, dx^\nu = 0$, is tangent to $\mathcal{H}$ along the special direction ℓ^μ. [This gives a special, fibered structure to $\mathcal{H}$, generated by the lines tangent to ℓ^μ, called the "generators" of $\mathcal{H}$.] Physically, the tangency between $\mathcal{H}$ and the local light cone (and the fact that the spatial sections of $\mathcal{H}$ are compact) means that $\mathcal{H}$ is the boundary between the part of spacetime from which light can escape to infinity, and the part out of which light cannot so escape. This boundary is called the (future) *Horizon* (hence the notation $\mathcal{H}$).

Fig. 1 illustrates the fact that the Horizon $\mathcal{H}$ is a dynamical, time-evolving structure. In the simple, spherically-symmetric situation assumed in Fig. 1, the horizon $\mathcal{H}$ is the spacetime history of a special bubble of light that was emitted from the center of a collapsing star, and which, after expanding out from the center, stabilized itself, under the strong pull of relativistic gravity, into an asymptotically stationary configuration, which is precisely the (time-independent) Schwarzschild sphere $r = r_S \equiv 2GM/c^2$. The structure illustrated by Fig. 1 is called a *black hole*[1]. The "interior" of $\mathcal{H}$ is the interior of the black hole (spacetime region where light is "trapped", *i.e.*, cannot escape to infinity), while $\mathcal{H}$ (which marks the great "divide" between trapped light, and light escaping at infinity) is the horizon, or the *surface of the black hole*. Note the presence, inside the black hole, of an infinite spacetime volume which ends, in its future, in a spacelike singularity at $r = 0$ (where the curvature blows up as r^{-3}). This singularity is a "big crunch" singularity: it is of a cosmological type (the coordinate r of Eq. (1.1) being "time-like" when $r < r_S$), and a time-reverse analog of the familiar Friedmann big-bang singularity.

A lot of work (for references, see, *e.g.*, [2, 3, 4, 5, 6]) went into understanding how general is the picture of Fig. 1. The stability (near and outside $\mathcal{H}$) of the spherical collapse situation of Fig. 1 under small, non-spherically-symmetric perturbations, the existence of multi-parameter generalizations of the Schwarzschild metric (due to Reissner, Nordstrøm, Kerr, Newman *et al.*) featuring regular horizon structures, and the proof of several "no-hair" theorems (by Israel, Carter, etc.), led to the following (conjectural) picture. The gravitational collapse of any type of matter configuration will generically lead to the formation of a black hole, whose

---

[1]Note the interesting historical coincidence that the name of the discoverer of the solution Eq.(1.1) means *black shield*.

surface is a null hypersurface $\mathcal{H}$, and will asymptotically settle into a stationary state, which is completely described (if the black hole is isolated) by a small number of parameters. When considering, as long range fields, only gravity $(g_{\mu\nu})$ and electromagnetism (A_μ), the final, stationary configuration of a black hole is described by three parameters, its total mass M, its total angular momentum J, and its total electric charge Q, and is given by the following Kerr-Newman solution

$$ds^2_{\mathrm{KN}} = -\frac{\Delta}{\Sigma}\,\omega_t^2 + \frac{\Sigma}{\Delta}\,dr^2 + \Sigma d\theta^2 + \frac{\sin^2\theta}{\Sigma}\,\omega_\varphi^2\,, \qquad (1.5a)$$

$$A^{\mathrm{KN}}_\mu\,dx^\mu = -\frac{Qr}{\Sigma}\omega_t\,, \qquad (1.5b)$$

$$\frac{1}{2}\,F^{\mathrm{KN}}_{\mu\nu}\,dx^\mu \wedge dx^\nu = \frac{Q}{\Sigma^2}\,(r^2 - a^2\cos^2\theta)\,dr \wedge \omega_t$$

$$+\frac{2Q}{\Sigma^2}\,ar\cos\theta\sin\theta\,d\theta \wedge \omega_\varphi\,, \qquad (1.5c)$$

where (setting for simplicity $G = c = 1$)

$$a \equiv \frac{J}{M}\,, \quad \Delta = r^2 - 2Mr + a^2 + Q^2\,, \quad \Sigma = r^2 + a^2\cos^2\theta\,,$$

$$\omega_t = dt - a\sin^2\theta\,d\varphi\,, \quad \omega_\varphi = (r^2 + a^2)\,d\varphi - a\,dt\,. \qquad (1.6)$$

The Kerr-Newman field configuration (1.5) is a black hole (with a regular horizon $\mathcal{H}$) if and only if the three parameters M, J, Q satisfy the inequality (recall the notation $a \equiv J/M$, and the choice of units $c = G = 1$ that will often be used below)

$$a^2 + Q^2 \le M^2\,. \qquad (1.7)$$

The black holes that saturate the inequality (1.7), *i.e.*, those that satisfy $a^2 + Q^2 = M^2$, are called *extremal*. Note that a Schwarzschild black hole $(a = Q = 0)$ can never be extremal, while a Reissner-Nordstrøm black hole $(a = 0)$ is extremal when $|Q| = M$ (*i.e.*, $|Q| = G^{\frac{1}{2}}M$), and a Kerr black hole $(Q = 0)$ is extremal when $|a| = M$ (*i.e.*, $J = GM^2$). By analyzing the spacetime geometry (1.5a) one finds that the horizon of a Kerr-Newman black hole is located (when viewed at "constant time")[2] on the surface $r = r_+$ with

$$r_+ \equiv M + \sqrt{M^2 - a^2 - Q^2}\,. \qquad (1.8)$$

---

2   Energetics and thermodynamics of black holes

It is very fruitful to consider a black hole as a kind of *gravitational soliton, i.e.*, as a physical object, localized "within the horizon $\mathcal{H}$", generating the "external fields" (1.5), and possessing a total mass M, a total angular momentum J and a total electric charge Q. By dropping massive, charged test particles, moving on generic non radial orbits, one then expects that it is possible to change the values of M, J and Q, *i.e.*, to evolve from an initial Kerr-Newman black hole state (M, J, Q) to a final one $(M + \delta M, \ J + \delta J, \ Q + \delta Q)$ where $\delta M = E$, $\delta J = p_\varphi$ and $\delta Q = e$. Here $E = -p_t = -\partial S/\partial t$ is the conserved energy of the test particle freely moving (geodesic motion) in the background Kerr-Newman configuration (M, J, Q), $p_\varphi = \partial S/\partial\varphi$ its conserved (z-component of the) angular momentum, and e its electric charge. [S denotes here the action of the test particle, satisfying the Hamilton-Jacobi equation $g^{\mu\nu}(p_\mu - eA_\mu)(p_\nu - eA_\mu) = -\mu^2$, where $p_\mu = \partial S/\partial x^\mu$ is its 4-momentum and μ its mass.] It was first realized by Penrose [7] that such a process can lead, in certain (special) circumstances, to a *decrease* of the total mass of the black hole: $\delta M < 0$. A detailed analysis of the changes of M, J and Q during the absorption of test particles, then led Christodoulou [8] and Christodoulou and Ruffini [9] to introduce the concepts of *reversible transformation* and of *irreversible transformation* of a black hole. More precisely, starting from $\delta M = -p_t$, $\delta J = p_\varphi$, $\delta Q = e$ and from the mass-shell constraint $g^{\mu\nu}(p_\mu - eA_\mu)(p_\nu - eA_\nu) = -\mu^2$ (which gives a quadratic equation to determine the energy $-p_t$ in terms of p_φ, e and the other components of p_μ, namely p_r and p_θ), they derived, when considering a particle crossing the horizon, *i.e.*, being absorbed by the black hole, the following equality

$$\delta M - \frac{a\,\delta J + r_+\,Q\,\delta Q}{r_+^2 + a^2} = \frac{r_+^2 + a^2\cos^2\theta}{r_+^2 + a^2}\,|p^r|\,, \qquad (2.1)$$

the right-hand side of which is proportional to the *absolute value* of the radial momentum of the particle as it enters the black hole. [The latter absolute value comes from taking a positive square root, similarly to the usual special relativistic result $E = +\sqrt{\mu^2 + \mathbf{p}^2}$.] The positivity of the latter quantity then leads to deriving the *inequality*

$$\delta M \geq \frac{a\,\delta J + r_+\,Q\,\delta Q}{r_+^2 + a^2}\,. \qquad (2.2)$$

A *reversible* transformation is then defined as a transformation which saturates the inequality (2.2), *i.e.*, where one replaces $\geq$ by an equality sign. It is called "reversible" because, after having (algebraically) added δJ and δQ (and the corresponding $\delta M = $ R.H.S. of (2.2)) to a black hole, the subsequent reversible "addition" of $\delta' J = -\delta J$, $\delta' Q = -\delta Q$ (and the corresponding, saturated $\delta' M = -\delta M$) will lead to a final state $(M + \delta M + \delta' M, J + \delta J + \delta' J, Q + \delta Q + \delta' Q) \equiv (M, J, Q)$, *i.e.*, identical to the initial state. By contrast, a sequence of infinitesimal transformations where at least one of the elementary processes (2.2) contains the strict inequality sign $>$ cannot be reversed.

By integrating the differential equation obtained by saturating (2.2) one obtains the Christodoulou-Ruffini *mass formula* of black holes,

$$M^2 = \left(M_{\text{irr}} + \frac{Q^2}{4M_{\text{irr}}} \right)^2 + \frac{J^2}{4M_{\text{irr}}^2} , \tag{2.3}$$

where the *irreducible mass* M_{irr} is an (integration) *constant* under reversible processes, and varies in the following irreversible manner,

$$\delta M_{\text{irr}} \geq 0 , \tag{2.4}$$

under general, possibly irreversible, processes. One also finds that the irreducible mass can be explicitly (rather than implicitly as in Eq. (2.3)) expressed in terms of M, J and Q as

$$M_{\text{irr}} = \frac{1}{2} \sqrt{r_+^2 + a^2} . \tag{2.5}$$

Note that Eq. (2.3) says that the "free energy" of a black hole, *i.e.*, the maximal energy which can be extracted by depleting (in a reversible) way J and Q is $M - M_{\text{irr}}$. This free energy has both Coulomb ($\propto Q^2$) and rotational ($\propto J^2$) contributions. It vanishes in the case of a Schwarzschild black hole ($J = 0 = Q$).

For our present purpose, a crucial aspect of the global energetics of black holes, given by Eq. (2.3), is the striking similarity of the irreversible increase (2.4) of the irreducible mass with the second law of thermodynamics, *i.e.*, the irreversible increase of the total entropy S of an isolated system. A general theorem of Hawking [10] led to a deeper understanding of the irreversible behaviour (2.4). Indeed, starting from the basic geometrical fact that a horizon is a null hypersurface, and analyzing the global properties of the generators of the horizon, Hawking proved that the area A of successive time sections[3] of the horizon of a black hole cannot decrease,

$$\delta A \geq 0 . \tag{2.6}$$

He also proved that, if one considers a system made of several, separate black holes, the sum of the areas of all the horizons cannot decrease

$$\delta \left(\sum_a A_a \right) \geq 0 . \tag{2.7}$$

In the particular case of a Kerr-Newman hole the result (2.6) yields back (2.4). Indeed, one easily finds from Eq. (1.5a), when using $r = r_+$ (*i.e.*, $\Delta = 0$ and $dr = 0$) that the inner geometry of the horizon is $d\sigma^2 = \gamma_{AB}(x^C) dx^A dx^B = (r_+^2 + a^2 \cos^2 \theta) d\theta^2 + \sin^2 \theta (r_+^2 + a^2)^2 d\varphi^2 / (r_+^2 + a^2 \cos^2 \theta)$ (with $A, B = 1, 2 = \theta, \varphi$ in the example at hand), so that the area of (a time section) of the horizon is

$$A_{\text{KN}} = \int \int (r_+^2 + a^2) \sin \theta \, d\theta \, d\varphi = 4\pi (r_+^2 + a^2) = 16\pi M_{\text{irr}}^2 . \tag{2.8}$$

---

[3]By "time sections" of the horizon we mean some slices $v = cst$ where v is some generalization of the (regular) Eddington-Finkelstein time coordinate.

In view of Eqs. (2.4), (2.6), (2.7) it is tempting to attribute to a black hole, considered as a physical object, not only a total mass M, a total angular momentum, and a total electric charge, but also a total *entropy*, proportional to the area of the horizon, say

$$S_{\mathrm{BH}} = \alpha A \,, \tag{2.9}$$

where α is a constant with dimension of inverse length squared. By varying the mass formula (2.3) one then finds the "first law of the thermodynamics of black holes",

$$dM = \Omega \, dJ + \Phi \, dQ + T_{\mathrm{BH}} \, dS_{\mathrm{BH}} \,, \tag{2.10}$$

where (see Eq. (2.2)) $\Omega = a/(r_+^2 + a^2)$ can be interpreted as the angular velocity of the black hole, $\Phi = Q \, r_+/(r_+^2 + a^2)$ as its electric potential (see [2] for discussions of Ω and Φ), and where

$$T_{\mathrm{BH}} \equiv \frac{1}{\alpha} \frac{\partial M}{\partial A} = \frac{\kappa}{8\pi\alpha} \tag{2.11}$$

is expected to represent the "temperature" of the black hole. The quantity κ in Eq. (2.11) is the *surface gravity* of a black hole, generally defined as the coefficient relating the covariant directional derivative of the horizon normal vector ℓ^μ along itself to ℓ^μ : $\ell^\nu \nabla_\nu \ell^\mu = \kappa \, \ell^\mu$. [Here, ℓ^μ is normalized so that, on the horizon, $\ell^\mu \partial_\mu = \partial_t + \Omega \partial_\varphi$ where ∂_t is the usual Killing vector of time translations.] The surface gravity may be thought of as the *redshifted* acceleration of a particle staying "still" on the horizon. [As we said above, the proper-time acceleration of a particle sitting on the horizon is actually infinite, but the infinite redshift factor associated to the difference between the proper time and the "time" associated to the generator ℓ^μ compensates for this infinity.]

In the case of a Kerr-Newman hole, the value of the surface gravity reads $(r_\pm \equiv M \pm \sqrt{M^2 - a^2 - Q^2})$

$$\kappa = \frac{1}{2} \frac{r_+ - r_-}{r_+^2 + a^2} = \frac{\sqrt{M^2 - a^2 - Q^2}}{r_+^2 + a^2} \,. \tag{2.12}$$

Note that the surface gravity of a Schwarzschild black hole is given by the usual formula for the surface gravity of a massive star, *i.e.*, $\kappa = GM/r_S^2 = M/(2M)^2 = 1/(4M)$. Note also that the surface gravity (and therefore the expected "temperature") of extremal black holes *vanishes*.

The "first law of black-hole thermodynamics" (2.10) has been derived above in the particular (but particularly important!) case of Kerr-Newman black holes. A general derivation, for generic (non isolated) black-hole equilibrium states, has been given by Carter [11] and by Bardeen, Carter and Hawking [12]. For a recent review of black-hole thermodynamics, see [13]. See also these references for a discussion of the "zeroth law of black-hole thermodynamics", namely the constancy of the "temperature" (2.11), *i.e.*, of the surface gravity κ, all over the horizon. [In the "membrane" approach to black-hole physics [16], the uniformity of κ for stationary states is viewed as a consequence of the Navier-Stokes equation because the "surface pressure" of a black hole happens to be equal to $p \equiv \kappa/8\pi$.]

Before going on to considering *quantum* aspects of the irreversibility of black hole evolution, let us complete our brief description of the *classical* dynamics and thermodynamics of black holes by mentioning that the irreversibility (2.4) or (2.6) affecting a *global* characteristic of a black hole has its roots in the irreversible behaviour of a *local* characteristic of the horizon. It proved fruitful to view a black hole as a kind of *membrane* [14, 15, 16], fibered by the generators (see above), and endowed with familiar physical properties: pressure $p = \kappa/8\pi$, shear viscosity η, bulk viscosity ζ, resistivity ρ. For instance, one can write an Ohm's law relating the electromotive force tangent to the horizon to the current flowing in the horizon, and a Navier-Stokes equation relating the time-evolution of the black-hole surface momentum density to the (surface) gradient of the pressure, to the effects of the shear and of the expansion of the "2-dimensional" fluid of generators and to a possible external flux of momentum. Starting from Einstein's equations one finds, *e.g.*, that the (surface) resistivity of black holes is $\rho = 377$ ohm (*i.e.*, 4π), and that their (surface) shear viscosity is $\eta = (16\pi)^{-1}$. Of particular interest for our present purpose is the existence of a *local* (*i.e.*, at each point of the surface of the black hole) version of the "second law of thermodynamics". Let us "localize" the global attribution (2.9) by attributing an *entropy*

$$dS_{\mathrm{BH}} = \alpha \, dA \qquad (2.13)$$

to each area element dA of the horizon. This local entropy is then found to evolve, as a consequence of Einstein's equations, in the following irreversible manner [17, 16]

$$\frac{d}{dt}\left(dS_{\mathrm{BH}}\right) - \tau \frac{d^2}{dt^2}\left(dS_{\mathrm{BH}}\right) = \frac{dA}{T_{\mathrm{BH}}}\left[2\eta\,\sigma_{\mathrm{AB}}\,\sigma^{\mathrm{AB}} + \zeta\,\theta^2 + \rho\left(\mathbf{J}_{\mathrm{BH}} - \sigma_{\mathrm{BH}}\,\mathbf{v}\right)^2\right].$$

$$(2.14)$$

Here, d/dt is a convective derivative (along the generators), $\tau \equiv \kappa^{-1}$ a characteristic time scale, η the shear viscosity mentioned above, σ_{AB} the shear tensor of the 2-dimensional fluid motion, $\zeta = -\eta$ the bulk viscosity, θ the local rate of dilatation of the fluid motion, ρ the (surface) resistivity, $\mathbf{J}_{\mathrm{BH}} - \sigma_{\mathrm{BH}}\mathbf{v}$ the horizon's conduction current (total current $\mathbf{J}_{\mathrm{BH}}$ minus convection current $\sigma_{\mathrm{BH}}\mathbf{v}$, where σ_{BH} is the horizon's surface charge density), and where T_{BH} can be interpreted as the *local* temperature of the horizon

$$T_{\mathrm{BH}} \equiv \frac{\kappa}{8\pi\alpha}, \qquad (2.15)$$

where κ (defined as above by $\ell^\nu \nabla_\nu \ell^\mu \equiv \kappa\,\ell^\mu$) is the local value of the surface gravity. [The surface gravity is uniform, on the horizon, for stationary black holes, but is generically non-uniform for evolving black holes.]

In the limit of an adiabatically slow evolution of the black-hole state, Eq. (2.14) reduces to the usual thermodynamical law giving the local increase of the entropy of a fluid element heated by the dissipations associated to viscosity and the Joule's law. [For non adiabatic evolutions, the L.H.S. of Eq. (2.14) can be interpreted, similarly to the famous a-causal Lorentz-Dirac radiation-reaction equation, in terms of an anticipated response of the black hole to external sollicitations [18, 16].]

3 Bekenstein's entropy and Hawking's temperature

The striking similarity between the irreversible increase of the area of the horizon of a black hole, reviewed in the last Section, and the second law of thermodynamics, motivated Bekenstein [19, 20, 21] to introduce the concept of black-hole entropy, say S_{BH}. [See the Appendix for John Wheeler's account [22] of how Jacob Bekenstein started to think about the concept of black-hole entropy and see Bekenstein's own account in [23].] He gave several arguments suggesting that S_{BH} is (as anticipated in the previous Section) proportional to the area A of the black hole, and of the form

$$S_{\mathrm{BH}} = \hat{\alpha} \, \frac{c^3}{\hbar G} \, A \equiv \hat{\alpha} \, \frac{A}{\ell_P^2} \,, \tag{3.1}$$

where $\hat{\alpha}$ is a universal dimensionless number of order unity, and where $\ell_P \equiv \sqrt{\hbar G/c^3}$ denotes the Planck length. [We set Boltzmann's constant to unity.]

One of the arguments used by Bekenstein was an analysis of the *quantum limitations* on the existence of reversible transformations, in the sense of Christodoulou [8]. Indeed, from Eq. (2.1) above the inequality (2.2) can be saturated only in the limit where the particle captured by the black hole has *exactly* zero radial momentum *on the horizon*. He considered that, in order of magnitude, quantum effects impose a minimal (proper) size, for a particle of mass μ, of order its Compton wavelength $\lambda_c = \hbar/\mu c$, so that a particle can never be located exactly on the horizon, but only within an uncertainty $\sim \lambda_c$ in proper distance. Bekenstein then found that this uncertainty implied the existence of a lower bound on the difference between the L.H.S. and the R.H.S. of Eq. (2.2). When estimating the corresponding lower bound on the increase of the area of a black hole as it captures a particle, he found a value of order $\hbar G/c^3 = \ell_P^2$, seemingly independent of all the characteristics of both the black hole and the particle.

Considering that, as a particle goes down a black hole, one looses (because of the no-hair theorems) an information of the order of one bit, and following Brillouin in identifying information with negative entropy [24], he estimated that the dimensionless constant $\hat{\alpha}$ in Eq. (3.1) was probably numerically $\simeq (\ln 2)/(8\pi)$. In other words, Bekenstein's black-hole entropy was viewed by him as a measure of the information about the interior of a black hole which is inaccessible to an external observer. He gave several other arguments leading to a result of the form (3.1), including an argument based on the analysis of quantum limitations on the efficiency of Carnot cycles using a black hole as heat sink. The latter argument suggested the need to attribute to a black hole a temperature of order of that corresponding (in the sense of Eqs. (2.9)–(2.11) above, with $\alpha \equiv \hat{\alpha} \, c^3/(\hbar G)$) to the entropy (3.1), *i.e.*,

$$T_{\mathrm{BH}} = \frac{1}{8\pi\hat{\alpha}} \frac{\hbar}{c} \kappa \,, \tag{3.2}$$

where κ denotes the surface gravity of the black hole. [A factor G/c^2 disappeared between (3.1) and (3.2) because we measure here κ in the usual way surface gravities are measured: $\kappa \sim GM/r^2 = c^2/\mathrm{length}$.] Finally, Bekenstein conjectured the validity of a *generalized version of the second law of thermodynamics*, stating that the sum of the black-hole entropy (3.1) and of the ordinary entropy in the exterior of the black hole never decreases.

Bekenstein's arguments could not yield a precise determination of the dimensionless coefficient $\hat{\alpha}$ in Eq. (3.1) or (3.2) (nor could they really establish the universality of this coefficient). Soon after Bekenstein's original suggestions, Hawking (who, initially, did not believe that it made sense to attribute a nonzero temperature (3.2) to a black hole) discovered the universal phenomenon of quantum radiance of black holes [25, 26]. This work gave a precise meaning to a black-hole temperature of the form (3.2). More precisely, Hawking found

$$T_{\mathrm{BH}} = \frac{1}{2\pi} \frac{\hbar}{c} \kappa \,, \tag{3.3}$$

which exactly corresponds to Eq. (3.2) if (and only if) the dimensionless coefficient $\hat{\alpha}$ there is equal to

$$\hat{\alpha} = \frac{1}{4} \,. \tag{3.4}$$

In view of the importance of Hawking's result [25, 26], let us (following [27]) give a simplified derivation of the phenomenon of quantum radiance. Let us consider (for simplicity) a massless scalar field $\varphi(x)$ propagating in a Schwarzschild background. The quantum excitations of $\varphi(x)$ can be decomposed into classical mode functions, say

$$\varphi_{\omega\ell m}(t,r,\theta,\varphi) = \frac{e^{-i\omega t}}{\sqrt{2\pi|\omega|}} \frac{u_{\omega\ell m}(r)}{r} Y_{\ell m}(\theta,\varphi) \,, \tag{3.5}$$

which must satisfy

$$0 = \Box_g \, \varphi = \frac{1}{\sqrt{g}} \, \partial_\mu(\sqrt{g} \, g^{\mu\nu} \, \partial_\nu \, \varphi) \,. \tag{3.6}$$

Before tackling the application of Eq. (3.6) to a Schwarzschild background, let us recall the essentials of the theory of particle creation by a background field. The quantum operator $\hat{\varphi}(x)$ describing (real, massless) scalar particles can be decomposed both with respect to some "in" basis of modes (describing free, incoming particles), say

$$\hat{\varphi}(x) = \sum_i \hat{a}_i^{\mathrm{in}} \, p_i^{\mathrm{in}}(x) + (\hat{a}_i^{\mathrm{in}})^+ \, n_i^{\mathrm{in}}(x) \,, \tag{3.7}$$

and with respect to an "out" basis of modes (describing outgoing particles), say

$$\hat{\varphi}(x) = \sum_i \hat{a}_i^{\mathrm{out}} \, p_i^{\mathrm{out}}(x) + (\hat{a}_i^{\mathrm{out}})^+ \, n_i^{\mathrm{out}}(x) \,. \tag{3.8}$$

Here the $\hat{a}^{\text{in}}$, $(\hat{a}^{\text{in}})^{+}$, $\hat{a}^{\text{out}}$, $(\hat{a}^{\text{out}})^{+}$ are two sets of annihilation and creation operators (with, $e.g.$, $[\hat{a}_i^{\text{in}}, (\hat{a}_j^{\text{in}})^{+}] = \delta_{ij}$) corresponding to a decomposition in modes which can physically be considered as incoming positive-frequency ones $(p_i^{\text{in}}(x))$, incoming negative-frequency ones $(n_i^{\text{in}}(x)$; which can be taken to be the complex conjugate of $p_i^{\text{in}}(x)$ in our case), outgoing positive-frequency ones $(p_i^{\text{out}}(x))$ and outgoing negative-frequency ones $(n_i^{\text{out}}(x))$. These mode functions are normalized so that, say, $(p_i^{\text{in}}, p_j^{\text{in}}) = \delta_{ij}$, $(p_i^{\text{in}}, n_j^{\text{in}}) = 0$, $(n_i^{\text{in}}, n_j^{\text{in}}) = -\delta_{ij}$, where (φ_1, φ_2) denotes the standard (conserved) Klein-Gordon scalar product.

The "in" vacuum is defined by $\hat{a}_i^{\text{in}}|\text{in}\rangle = 0$, and the mean number of i-type "out" particles present in the in-vacuum is given by

$$\langle N_i \rangle = \langle\text{in}|\, (a_i^{\text{out}})^{+}\, a_i^{\text{out}}\, |\text{in}\rangle = \sum_j |T_{ij}|^2 \tag{3.9}$$

where $T_{ij} \equiv (p_i^{\text{out}}, n_j^{\text{in}})$ is the (standard Klein-Gordon) scalar product, $i.e.$, physically the *transition amplitude*, between the incoming negative-frequency mode n_j^{in} and the outgoing positive-frequency one p_i^{out}. In the black-hole case the in-vacuum will be defined by focussing on high-frequency wave packets propagating in the vicinity of the horizon $\mathcal{H}$, say some $n_j^{\text{in}}(x)$, and frequency-analyzed in a local freely-falling frame. We will then be interested in the transition amplitude between such an incoming $n_j^{\text{in}}(x)$ and an outgoing mode which reaches (future) infinity as a positive-frequency wave-packet (in the usual sense of the decomposition (3.5) there).

Inserting the explicit form (1.1) of the Schwarzschild metric into Eq. (3.6), using the mode decomposition (3.5), and introducing the usual "tortoise" radial coordinate [4]

$$r_* = \int \frac{dr}{1 - \frac{2M}{r}} = r + 2M \ln\left(\frac{r - 2M}{2M}\right), \tag{3.10}$$

leads to the following radial equation for $u_{\omega\ell m}(r)$:

$$\frac{\partial^2 u}{\partial r_*^2} + \left(\omega^2 - V_\ell\,[r(r_*)]\right) u = 0, \tag{3.11}$$

where the effective radial potential V_ℓ reads

$$V_\ell(r) = \left(1 - \frac{2M}{r}\right)\left(\frac{\ell(\ell + 1)}{r^2} + \frac{2M}{r^3}\right). \tag{3.12}$$

In the coordinate r_*, the horizon $(r = 2M)$ is located at $r_* \to -\infty$, $i.e.$, in the left asymptotic region of the potential $V_\ell\,[r(r_*)]$, where V_ℓ tends exponentially towards zero. [Note that $V_\ell\,[r(r_*)]$ tends to zero both as $r_* \to -\infty$ $(r \to 2M)$ and as $r_* \to +\infty$ $(r \to +\infty)$, and creates a positive potential barrier around $r \sim 3M$.] Neglecting the potential V_ℓ in Eq. (3.11), we see that, near the horizon, the radial function $u(r_*)$ behaves as $e^{\pm i\omega r_*}$, so that (after factoring $1/r$ and $Y_{\ell m}(\theta, \varphi)$) the mode function (3.5) behaves there as

$$\varphi_\omega(t, r) \propto e^{-i\omega(t \pm r_*)} . \tag{3.13}$$

At this point we need to recall that the "singularity" of the Schwarzschild metric (1.1) on the horizon $r = 2M$ can be eliminated by introducing new coordinates, better adapted to the intrinsically regular geometry of the (future) horizon $\mathcal{H}$. As above a simple choice are the (ingoing) Eddington-Finkelstein coordinates (v, r) with the transformation $(t, r) \to (v, r)$ given by $v \equiv t + r_*$, $r \equiv r$ [see, e.g., [4]]. Using these coordinates, let us write down the expression, near $\mathcal{H}$ ($r = 2M$, v finite), of the mode (3.13) describing a wave which is *outgoing* from $\mathcal{H}$, and which will have (when it reaches infinity) a *positive* frequency.

It is given by (3.13) with a *minus* sign in front of r_*, and with $\omega > 0$. Using

$$t - r_* = t + r_* - 2r_* = v - 2r_* = v - 2r - 4M \ln \left(\frac{r - 2M}{2M} \right) \tag{3.14}$$

this outgoing mode $\varphi_\omega^{\text{out}} \propto e^{-i\omega(t - r_*)}$ reads

$$[\varphi_\omega^{\text{out}}(v, r)]_{\mathcal{H}}^{\text{near}} \propto e^{-i\omega v} e^{+2i\omega r} \left(\frac{r - 2M}{2M} \right)^{i4M\omega} . \tag{3.15}$$

Because of the last factor on the R.H.S. of Eq. (3.15), this outgoing mode seems to be rather singular on $\mathcal{H}$: (i) it piles up an infinite number of oscillations of shorter and shorter wavelengths as $r \to 2M + 0$, and (ii) it is undefined inside the black hole, *i.e.*, for $r < 2M$. Let us now *extend* the definition of the mode $\varphi_\omega^{\text{out}}$ so that it defines, in a local (regular) Fermi-frame in the vicinity of $\mathcal{H}$, a high-frequency wave packet made only of (locally) *negative frequencies*. (This will define a mode $n_j^{\text{in}}(x)$ in the sense of Eq. (3.7).) Consider, in Minkowski space, a wave packet

$$\varphi_-(x) = \int_{\mathcal{C}^-} d^4k \, \tilde{\varphi}(k) \, e^{ik_\mu x^\mu} \tag{3.16}$$

made only of *negative frequencies*, *i.e.*, such that the 4-momenta k^μ entering (3.16) are all contained in the *past light cone* $\mathcal{C}^-$ of k^μ. A convenient technical criterion for characterizing, in x-space, such a negative-frequency wave packet is the well-known condition that $\varphi_-(x)$ be analytically continuable to *complexified* spacetime points $x^\mu + i y^\mu$ with y^μ lying in the *future* light cone, $y^\mu \in \mathcal{C}^+$. This local, x-space criterion can be applied (in a local frame near $\mathcal{H}$) to the wave packet (3.15). As the infinitesimal displacement $r \to r - \varepsilon$, $v \to v$ is seen to be future directed (and null), this criterion finally tells us that the following new, extended wavepacket (defined by applying the analytic continuation $r \to r - i\varepsilon$ to (3.15))

$$n_\omega(v, r) \equiv N_\omega \, \varphi_\omega^{\text{out}}(v, r - i\varepsilon) \propto e^{-i\omega v} e^{2i\omega r} \left(\frac{r - 2M - i\varepsilon}{2M} \right)^{i4M\omega} \tag{3.17}$$

is (when Fourier analyzed in the vicinity of $\mathcal{H}$) a *negative frequency* wave packet. Note that the new wavepacket $n_\omega(v,r)$ is defined on both sides of $\mathcal{H}$ and that we have included a new normalizing factor N_ω in its definition in terms of the analytic continuation of the "old" mode $\varphi_\omega^{\mathrm{out}}$ (which had its own normalization).

The normalization factor N_ω will give the main effect of the Hawking radiance and is simply determined as follows. The original modes (3.5) (which vanished inside $\mathcal{H}$) were normalized so that the invariant scalar product defined by the (massless) Klein-Gordon equation took the form $(\varphi_{\omega_1\ell_1 m_1}, \varphi_{\omega_2\ell_2 m_2}) = +\delta(\omega_1 - \omega_2)\,\delta_{\ell_1\ell_2}\,\delta_{m_1 m_2}$. The analytic continuation $r \to r - i\varepsilon$ introduces, via the rotation by $e^{-i\pi}$ of $r - 2M$ in $(r - 2M)^{i4M\omega}$, a factor $(e^{-i\pi})^{i4M\omega} = e^{+4\pi M\omega}$ in the left part $(r < 2M)$ of n_ω. In other words

$$n_\omega(r) = N_\omega[\theta(r - 2M)\,\varphi_\omega^{\mathrm{out}}(r - 2M) + e^{4\pi M\omega}\,\theta(2M - r)\,\varphi_\omega^{\mathrm{out}}(2M - r)], \quad (3.18)$$

where $\theta(x)$ is the step function. Computing the scalar product $(n_{\omega_1\ell_1 m_1}, n_{\omega_2\ell_2 m_2})$ one gets, besides the same delta functions as above, a factor $|N_\omega|^2\,[1 - (e^{4\pi M\omega})^2]$, where the minus sign is due to the (essentially) negative-frequency aspect of $\varphi_\omega^{\mathrm{out}}(2M - r)\,\theta(2M - r)$.

The correct normalization (for a negative-frequency mode),

$$(n_{\omega_1\ell_1 m_1}, n_{\omega_2\ell_2 m_2}) = -\delta(\omega_1 - \omega_2)\,\delta_{\ell_1\ell_2}\,\delta_{m_1 m_2}$$

is then obtained for

$$|N_\omega|^2 = \frac{1}{e^{8\pi M\omega} - 1}. \quad (3.19)$$

Finally, the result (3.18) says that the initial negative-frequency mode n_ω straddling the horizon (and looking as a high-frequency wavepacket there) splits, as time evolves, into an outgoing mode $\varphi_\omega^{\mathrm{out}}(r - 2M)$ (which will appear as having the positive-frequency ω at infinity), and another mode $\varphi_\omega^{\mathrm{out}}(2M - r)$ which falls towards the singularity. By the general formalism of particle production in curved spacetime recalled above, the number of outgoing particles in the mode ω, ℓ, m is given by the sum (3.9) where T_{ij} is the scalar product between $n_j^{\mathrm{in}} = n_{\omega_1\ell_1 m_1}$ and $p_i^{\mathrm{out}} \propto \varphi_{\omega\ell m}^{\mathrm{out}}$. Actually, the mode p_ω^{out}, normalized at infinity, differs from the mode $\varphi_\omega^{\mathrm{out}}$ used above, and normalized on the horizon, by the effect of the potential barrier $V_\ell(r)$ in Eq. (3.11) (which was negligible near $r = 2M$). If $V_\ell(r)$ were absent the looked-for transition amplitude would be

$$(\varphi_{\omega\ell m}^{\mathrm{out}}, n_{\omega_1\ell_1 m_1}) = N_\omega\,\delta(\omega - \omega_1)\,\delta_{\ell\ell_1}\,\delta_{mm_1}, \quad (3.20)$$

and the number of created particles (3.9) would contain $|N_\omega|^2$ times the square of $\delta(\omega - \omega_1)$, which, by Fermi's Golden Rule, is simply $\delta(\omega - \omega_1) \times \int dt/2\pi$. Finally, adding the *grey body* factor $\Gamma_\ell(\omega)$, giving the fraction of the flux of $\varphi_\omega^{\mathrm{out}}$ which ends up at infinity because of the effect of $V_\ell(r)$ in Eq. (3.11) (see Fig. 2), we get a *rate* of particle creation by the black hole of the form

$$\frac{d\langle N\rangle}{dt} = \sum_{\ell,m}\int \frac{d\omega}{2\pi}\,|N_\omega|^2\,\Gamma_\ell(\omega) = \sum_{\ell,m}\int \frac{d\omega}{2\pi}\,\frac{\Gamma_\ell(\omega)}{e^{8\pi M\omega} - 1}. \quad (3.21)$$

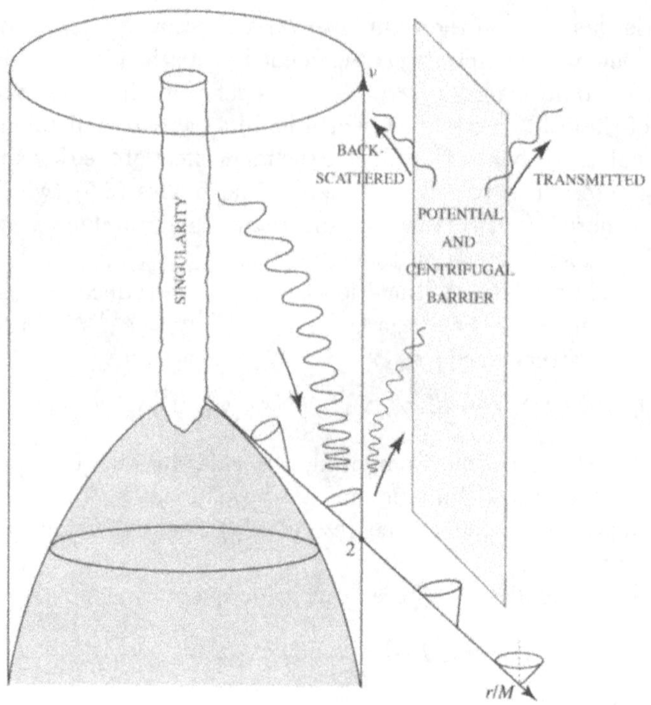

Figure 2: Splitting of an initial negative-frequency mode straddling the horizon into a mode falling into the black hole, and an outgoing mode which, after being partially reflected back into the black hole by the potential barrier representing gravitational and centrifugal effects, ends up as positive-frequency Hawking radiation at infinity. The antiparticle mode falling into the black hole can be interpreted as a particle traveling backwards in time, from the singularity down to the horizon [25] (hence the downwards orientation of the arrow).

The formula (3.21) exhibits, in the simple case of a Schwarzschild black hole, the Hawking phenomenon: a black hole radiates as if it were a black body of temperature $T_{\mathrm{BH}} = 1/(8\pi M)$, covered up by a "blanket" with transmission factor $\Gamma_\ell(\omega)$. [The blanket being due to the effect of the potential $V_\ell(r)$ in Eq. (3.11), i.e., physically to the combined effect of gravitational attraction and centrifugal forces on the waves outgoing from the horizon.] For a Schwarzschild black hole the surface gravity is $\kappa = GM/r_S^2 = 1/(4M)$ (with $c = G = 1$), so that the Planck spectrum present in (3.21) agrees with the temperature (3.3). Note that the technical origin of a Planck spectrum with temperature $1/8\pi M$ is the logarithm $-4M \ln((r - 2M)/2M)$ in Eq. (3.14) which is itself linked to the logarithm $+2M \ln((r - 2M)/2M)$ in Eq. (3.10). In other words, it is the logarithmically divergent link between the coordinate time of a local freely falling frame near $\mathcal{H}$

and the time at infinity which is the source of a Planck spectrum. Note that the calculation above technically relies on using such a logarithmically divergent link between the negative-frequency wave-packet near the horizon, and the positive-frequency wave-packet at infinity. When considering finite-frequency packets at infinity, this formally means that one has worked with packets containing *arbitrarily high* frequencies near the horizon.

When generalizing the analysis to a general black hole, one finds that the factor $4M$ gets replaced by the inverse of the surface gravity of the hole. One then finds a Planck-like spectrum containing (for bosons), besides a grey-body factor Γ, a factor $(e^{2\pi(\omega - p_\varphi \Omega - e\Phi)/\kappa} - 1)^{-1}$ exhibiting the combined effect of the general temperature (3.3) and of the couplings of the conserved angular momentum p_φ and electric charge e to, respectively, the angular velocity Ω and electric potential Φ of the hole, see Eq. (2.10).

Numerically, $T_{\mathrm{BH}} \sim 10^{-6} \, K \, M_\odot/M$ so that Hawking's radiation is not astrophysically relevant for stellar-mass or larger black holes. Conceptually, it is, however, a beautiful discovery which connects relativistic gravity and quantum mechanics.

4   Black holes, quantum mechanics and string theory

4.1   Conceptual puzzles

The discovery of Hawking radiation led to new conceptual challenges. First, the obtention of a Planck-like spectrum with the temperature (3.3) fixed the dimensionless coefficient $\hat\alpha$ in the general form (3.2) to the value $\hat\alpha = \frac{1}{4}$. By the general "first law of black-hole thermodynamics" (2.10) this would then correspond to an entropy (3.1) with the same value of $\hat\alpha$, *i.e.*,

$$S_{\mathrm{BH}} = \frac{1}{4} \frac{c^3}{\hbar G} A \equiv \frac{1}{4} \frac{A}{\ell_P^2} . \tag{4.1}$$

However, Hawking's derivation of a temperature does not help in understanding the physical meaning of the Bekenstein-Hawking "entropy" (4.1). Beyond the original hints of Bekenstein that S_{BH} is a measure of the information loss about what went into the black hole, it remains the challenge of interpreting in a precise way S_{BH}, à la Boltzmann, as the logarithm of the number of quantum micro-states of a macroscopic black hole.

Another challenge comes from considering the end-point of the Hawking quantum evaporation process. In order of magnitude (and in units $c = \hbar = G = 1$), Hawking's radiation corresponds to an outgoing flux of energy of order (by Stephan's law) $\sim A T_{\mathrm{BH}}^4 \sim R_{\mathrm{BH}}^2 T_{\mathrm{BH}}^4 \sim M^2 M^{-4} \sim M^{-2}$. This loss should correspond to a secular decrease of the mass M of the black hole so that $dM/dt \sim -M^{-2}$, *i.e.*, $M^3 \sim M_0^3 - (t - t_0)$. This leads to a finite evaporation lifetime for a black hole of order

$$t_{\text{evap}} \sim t_P \left(\frac{M}{M_P}\right)^3 \sim 10^{-44}s \left(\frac{M}{10^{-5}g}\right)^3 \sim 10^{10}\text{yr} \left(\frac{M}{10^{14}g}\right)^3 \qquad (4.2)$$

where $t_P \equiv (\hbar G/c^5)^{1/2} \simeq 5.4 \times 10^{-44}s$ is the Planck time and $M_P \equiv (\hbar c/G)^{1/2} \simeq 2.2 \times 10^{-5}g$ the Planck mass. Besides the (remote) astrophysical possibility [25] of seeing today in the sky the explosions of black holes of initial mass $M \sim 10^{14}g$ formed in the big bang, the main consequence of the finiteness of the evaporation time (4.2) is to oblige us to face the following questions: (i) what is the end point of the evaporation process?, and (ii) is there an "information loss" paradox due to the eventual final disappearance of the part of the wave packets in Fig. 2 which fell into the hole, and the concomitant emission of an outgoing flux of uncorrelated thermal radiation. Hawking suggested [37] that if the quantum radiance phenomenon ultimately leads to the disappearance of the black hole, then the information about what has fallen in is completely lost, and one must describe the evaporation process by a density matrix rather than by a unitary quantum evolution. In addition, we saw above that the calculation of the Hawking process formally involved *arbitrarily high* frequencies near the horizon. Many authors felt uncomfortable about this intermediate use of arbitrarily high (so-called "trans-Planckian") frequencies. A lot of work was devoted to these issues, as well as to the issue of understanding the statistical meaning of the Bekenstein-Hawking entropy.

We will not try here to summarize all the work done on these issues. Let us mention the existence of various lines of work, and give entries to the literature. Several authors explored the possibility that quantum black holes have a discrete mass spectrum, *e.g.*, with the area $A = 16\pi M_{\text{irr}}^2$ having an uniformly spaced spectrum of eigenvalues $\frac{1}{4} A = \ell_P^2(n + \text{cst.})$. Then, to give a statistical meaning to the entropy S_{BH} one needs to assume that each discrete energy level A_n has a huge degeneracy $g_n = \exp\left(\frac{1}{4} A_n/\ell_P^2\right)$. See Refs. [28, 29, 30, 31, 32]. Some authors have tried to identify the microstates responsible for black-hole entropy as near-horizon excitations [33, 34, 35, 36]. For entries in the literature on the "black-hole information puzzle", see [37, 38, 39, 40, 41, 42]. For discussions of the insensitivity of the Hawking process to the arbitrarily high frequencies that seem to play a crucial role in all its derivations, see [43, 44, 45, 46]. For suggestions that the so-called Loop Quantum Gravity may account for black-hole entropy see [47, 48]. For a possible link between black-hole entropy and the quasinormal modes of black holes see [49, 50, 51]. Finally, let us mention the suggestion of a (black-hole motivated) *upper bound*, on the entropy of any system, of the form $S < \beta ER/\hbar c$, where β is a constant of order unity, E the energy of the system, and R its (effective) radius [52]. See, however, [53] and references therein, for a discussion of possible loopholes in this, and other ("holographic"-type) entropy bounds.

4.2 Extremal black holes and supersymmetric solitonic states in string theory

4.2.1 BPS extremal black holes and D branes

Here we shall focus on the most striking "explanation" of black-hole entropy, which was obtained within the framework of string theory. For early suggestions of a link between black-hole entropy and "string entropy" (*i.e.*, the huge degeneracy of very massive string states) see [54, 55, 56]. After a seminal paper of Sen [57], a breakthrough was brought by the works of Strominger and Vafa [58] and Callan and Maldacena [59]. These works on the microscopic origin of the Bekenstein-Hawking entropy of certain (Bogomolnyi-Prasad-Sommerfield; BPS) *extremal* black holes initiated a huge activity, which is reviewed in [60, 61]. For textbook reviews see section 14.8 in volume 2 of [62] and chapter 17 of [63] (see also [64] for an entry into the literature on stringy black holes).

The line of work just mentioned on the micro-structure of certain BPS extremal black holes has led to very beautiful results. The factor $\frac{1}{4}$ [4] in the entropy (4.1) has been verified in the explicit counting of the degeneracy of some *extremal* black holes, and one could also verify some aspects of the physics of *near-extremal* holes, such as explaining Hawking radiation as a process of quantum decay of some unstable states [59, 66], and even verifying the presence of the correct grey-body factor $\Gamma_\ell(\omega)$ in Eq. (3.21) (at least in the long-wavelength limit) [67, 68]. Note that the Hawking temperature (3.3), being proportional to the surface gravity κ, Eq. (2.12), *vanishes* in the limit of *extremal* black holes, *i.e.*, black holes saturating the inequality (1.7). [The BPS black holes automatically saturate this inequality.] One therefore expects any quantum representation of an extremal black hole to be a stable (non radiating) object. This is the case of the supersymmetric BPS states studied in the above references. Then, allowing for a small deviation from the BPS condition, and therefore a small deviation from extremality, is expected to lead to a weakly unstable object, which can be treated by perturbation theory. [Even so, the problem is more tricky than was first thought [61].] Let us also mention that the study of extremal BPS black holes in certain limits led Maldacena [69] to conjecture a correspondence between string theory in anti-de-Sitter (AdS) spacetimes (times a compact manifold) and certain Conformal Field Theories (CFT). In turn, the study of this correspondence led to new insights into black-hole physics and to a confirmation that the black-hole evaporation process should be describable as a unitary process in quantum mechanics (see [70] for a review).

However, in spite of its striking successes the work on the string-theory modeling of BPS black holes has some drawbacks and limitations. The main point is that this approach describes the quantum micro-structure of BPS black holes in terms of a (nearly non-interacting) gas of massive, charged string solitons, called

---

[4]Though Section II above discussed only 4-dimensional black holes, it has been shown that the factor $\hat{\alpha} = \frac{1}{4}$ is independent of the dimension. Note that many (but not all [65]) string calculations deal with higher-dimensional black holes, *e.g.*, 5-dimensional ones, after compactification of 5 spatial dimensions.

"Dirichlet branes", or simply D-branes [71, 62]. For a recent extensive review of D-branes see [63]. These solitons come in various dimensionalities: a D-p brane denotes a D-brane which is extended in p spatial dimensions, e.g., a D-0 brane denotes a Dirichlet point particle, while a D-1 brane denotes a Dirichlet *string*. These solitons carry "Ramond-Ramond" charges, *i.e.*, they create electric-like or magnetic-like p'-form fields called Ramond-Ramond fields. [p'-form fields are generalizations of Maxwell fields: instead of a (one-index) vector potential A_μ, now called a 1-form field, one considers p'-index antisymmetric potentials $A_{\mu_1\cdots\mu_{p'}}$. In a $\mathcal{D}$-dimensional spacetime, a D-p brane naturally couples to an electric-like $(p+1)$-form field, or to a magnetic-like $(\mathcal{D}-p-3)$-form field.]

Roughly speaking the interaction between these solitons vanishes because of a cancellation between gravitational attraction and electrostatic repulsion. [Recall that the non-spinning limit of the extremality constraint (1.7) yields $|Q| = G^{\frac{1}{2}}M$ which corresponds exactly to the pairwise cancellation between $-GM_1M_2/r^{d-2}$ and $+Q_1Q_2/r^{d-2}$ in space dimension d.] Because of the BPS condition (which requires the preservation of part of the original maximal supersymmetry), the classical cancellation just mentioned extends also to quantum cancellations between couplings involving bosonic fields and couplings involving fermionic ones. These quantum cancellations then entail the existence of certain *non-renormalization theorems*, stating that certain, specific quantities do not depend on the value of the string coupling constant, say g. In turn, these non-renormalization theorems are crucial to the whole enterprise because of the following fact, that I have not yet spelled out. All the string calculations of supposedly "black hole" properties (such as the degeneracy of micro-states or the decay rate of nearly-BPS states) actually refer to a system of D-brane solitons in the limit $g \to 0$ where these solitons admit a string-theory description. Roughly speaking, the mass of those solitons scale like g^{-1}, while the gravitational constant scales like g^2. Therefore the $g \to 0$ limit corresponds to a limit where the product $GM \propto g$ tends to zero, which means physically that the "Schwarzschild radius" of the considered assembly of (nearly coincident) D-branes is much smaller than the string length $\ell_s \sim \sqrt{\alpha'}$. This is the limit where the assembly of D-branes is physically a "point particle", rather than a classically describable "black hole", having an horizon much larger than ℓ_s.

4.2.2 The example of D-1–D-5 branes in type IIB string theory

One of the most studied system is a configuration of N_1 D-1 branes and N_5 D-5 branes in type IIB string theory, carrying some momentum along the direction of the D-1 branes (which are all parallel to each other and embedded within the D-5 branes). One compactifies, à la Kaluza-Klein, the 5 spatial dimensions along which lie the D-5 and the D-1 branes. More precisely, the compactified manifold is $T^4 \times S^1$ where the 4-torus T^4 has volume $(2\pi)^4 V$ and the circle S^1 has length $2\pi R$. This has the effect of quantizing the (internal) momentum moving along the branes, say $P = N\hbar/R$, where R is the compactified radius along the D-1 direction. Finally, the configuration appears as being "point-like" in the 4 remaining uncompactified spatial dimensions, and this "point" carries three types of charges (quantized in integers): N_1, N_5 and N. The total mass-energy of the configuration reads (in string units $\alpha' = 1$)

$$M = \frac{N_1 R}{g} + \frac{N_5 R V}{g} + \frac{N}{R}. \tag{4.3}$$

D-brane techniques (together with the general Cardy formula [72] giving the density of states in two-dimensional conformal field theory) allow one to evaluate the quantum degeneracy $\mathcal{D}$ of the (supersymmetric) ground state of this system of D-1 and D-5 branes [58, 59, 61]. Indeed, though the full structure of those solitons cannot be described by perturbative string theory, one *can* describe, in the limit $g \to 0$ where the D-branes become infinitely massive, the excitations of collections of D-branes in terms of (perturbative) *open strings* whose end points are constrained to move on the (nearly fixed) D-brane submanifolds [71, 62, 63]. In the considered case, one finds that the excitations of the D-1, D-5 brane configuration are described by a two-dimensional superconformal field theory with $4N_1 N_5$ real scalars and, by supersymmetry, $4N_1 N_5$ Majorana fermions. This yields a central charge $c = 6N_1 N_5$ for a system of left-movers[5] with L_0 level given by N. The general Cardy formula $\mathcal{D}(N) \sim \exp(2\pi\sqrt{cN/6})$ (asymptotically valid for large values of N; see [73] for power-law corrections to the general exponential Cardy formula) then gives the following result for the quantum degeneracy of the D-brane configuration

$$\mathcal{D} \sim e^{2\pi\sqrt{N_1 N_5 N}}, \tag{4.4}$$

which will be compared below to the degeneracy of the corresponding black hole expected from the Bekenstein-Hawking entropy.

At some larger value of g one expects the collection of D-branes to smoothly turn into a quasi-classical black hole. However, the present state of development of string theory does not allow one to explicitly study this transition, and to set up a well-defined quantum Hilbert-space description of quasi-classical black holes. One has to crucially rely on non-renormalization theorems to "blindly" continue the result of calculations well-defined as $g \to 0$ into physical properties of quasi-classical black holes (such as the degeneracy of a certain quantum BPS state which is explicitly describable when $g \to 0$, but whose "physical form" for larger g cannot be concretely described as a quantum state). For instance, the D-1 D-5 system mentioned above is expected to appear, when g (or rather some effective value of g, say g_{eff}, which is a combination of g and of the integers N_1, N_5, N) gets larger than one, as a (BPS) black hole, *i.e.*, a generalization of the extremal Reissner-Nordstrøm black hole ($G^{1/2}M = |Q|$), carrying the (conserved) quantized charges N_1, N_5, N. In other words, when g_{eff} gets large one can no longer describe this system of massive solitons as a "point charge", because the combined gravitational

---

[5] Only left movers are excited in supersymmetric states, corresponding to extremal BPS black holes. By adding a little bit of excitation energy in right movers one can then describe near supersymmetric D-brane configurations corresponding to near extremal black holes [59].

effect of the total mass of the soliton configuration, and of the mass-energy associated to the various Ramond-Ramond (and internal-momentum related) fields it creates, generates a strong curvature of space around it, and ultimately gives birth to a finite area horizon, instead of a "point with nothing inside it". For instance, the $(4+1)$-dimensional (Einstein-frame) metric generated, in the uncompactified dimensions, by a D-1 D-5 system carrying the quantized charges N_1, N_5, N, reads

$$ds_E^2 = -(f_1\,f_5\,f)^{-\frac{2}{3}}\,dt^2 + (f_1\,f_5\,f)^{\frac{1}{3}}\,(dx_1^2 + \ldots + dx_4^2)\,, \tag{4.5}$$

with

$$f_1 = 1 + \frac{4GR}{g\pi\alpha'}\,\frac{N_1}{r^2}\,,\quad f_5 = 1 + \alpha'g\,\frac{N_5}{r^2}\,,\quad f = 1 + \frac{4G}{\pi R}\,\frac{N}{r^2}\,, \tag{4.6}$$

where $r^2 \equiv x_1^2 + \ldots + x_4^2$ denotes a (squared) radial coordinate in the 4 uncompactified spatial dimensions. See references [59, 60, 61, 62, 63] for the explicit form of the other fields generated by a system of coincident D-branes. We shall content ourselves here with illustrating the evolution, with g, of the physical structure of a system of coincident D-branes in Fig. 3.

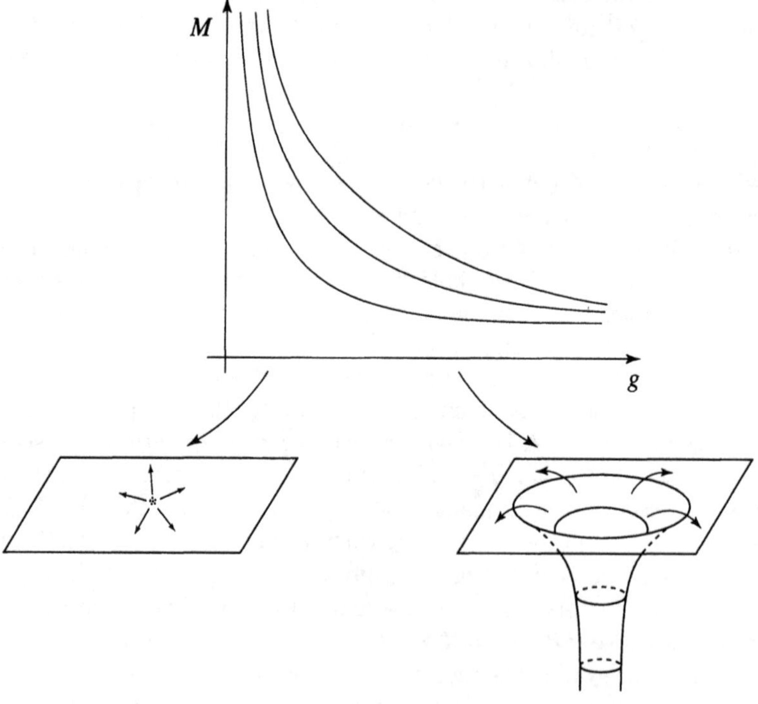

Figure 3: Evolution of the mass-energy, and of the physical structure, of an initially point-like configuration of D-branes as the string coupling g increases.

The resulting extremal black hole is uniquely characterized by the knowledge of the conserved quantized charges N, N_1, N_5. In particular, the horizon of the black-hole metric (4.5) is located at the radial coordinate $r = 0$. [Note that, as $r \to 0$, the time-time component of the metric (4.5) is proportional to $((r^{-2})^3)^{-\frac{2}{3}} = r^{+4}$.] The "area" (which is that of a 3-dimensional sphere) of the black-hole horizon is finite, because $(f_1 f_5 f)^{\frac{1}{3}} \propto r^{-2}$ compensates the factor r^2 in the non-radial part of the spatial metric $r^2 d\Omega_3^2$. It is an easy matter to evaluate this "area", and thereby, using formula Eq.(4.1) above, the corresponding Bekenstein-Hawking entropy. One finds that the dependences on G, R, g and α' cancell out so that the latter entropy depends only on the quantized charges carried by the BPS black hole, and equals

$$S_{BH} = 2\pi \sqrt{N_1 N_5 N}\,, \tag{4.7}$$

which *exactly coincides with the logarithm of the D-brane degeneracy* Eq.(4.4).

4.3 Schwarzschild black holes and string states

4.3.1 String-black-hole transition and conjectured correspondence between string and black-hole states

Because of the drawback of not being able to follow in detail the "transformation" between a point-like system of D-branes and a black hole, we shall focus henceforth on another type of string-theory calculation of the properties of massive string states and on their transition to black-hole states [74, 75, 76]. As just said, most of the stringy literature has concentrated on some special, supersymmetric extremal black holes (BPS black holes). These black holes carry special (Ramond-Ramond) charges and their microscopic structure seem to be describable (when $g^2 < g_c^2$) in terms of Dirichlet-branes. By contrast, we consider here the simplest, Schwarzschild black holes (in any space dimension d). It will be argued that their "microscopic structure" at low g^2 involves only *fundamental string states* (instead of solitonic string states, such as D-branes). However, the lack of supersymmetry means that it becomes essential to deal with *self-gravity effects*.

To start with, let us recall that thirty years ago the study of the spectrum of string theory revealed [77] a huge degeneracy of states growing as an exponential of the mass. This string result was obtained a few years before Bekenstein [21] proposed that the entropy of a black hole should be proportional to the area of its horizon in Planck units, and Hawking [26] fixed the constant of proportionality after discovering that black holes do emit thermal radiation at a temperature $T_{\mathrm{Haw}} \sim R_{\mathrm{BH}}^{-1}$.

When string and black-hole entropies are compared one immediately notices a striking difference: string entropy[6] is proportional to the first power of mass in

---

[6]As we shall discuss, the self-interaction of a string lifts the huge degeneracy of free string states. One then defines the entropy of a narrow band of string states, defined with some energy resolution $M_s \lesssim \Delta E \ll M$, as the logarithm of the number of states within the band ΔE.

any number of spatial dimensions d, while black-hole entropy is proportional to a d-dependent power of the mass, always larger than 1. In formulae:

$$S_s \sim \frac{\alpha' M}{\ell_s} \sim M/M_s \quad , \quad S_{\mathrm{BH}} \sim \frac{\mathrm{Area}}{G} \sim \frac{R_{\mathrm{BH}}^{d-1}}{G} \sim \frac{(g^2 \, M/M_s)^{\frac{d-1}{d-2}}}{g^2} \quad , \tag{4.8}$$

where, as usual, α' is the inverse of the classical string tension, $\ell_s \sim \sqrt{\alpha' \hbar}$ is the quantum length associated with it[7], $M_s \sim \sqrt{\hbar/\alpha'}$ is the corresponding string mass scale, R_{BH} is the Schwarzschild radius associated with M:

$$R_{\mathrm{BH}} \sim (G \, M)^{1/(d-2)} \, , \tag{4.9}$$

and we have used that, at least at sufficiently small coupling, the Newton constant and α' are related via the string coupling by

$$G \sim g^2 (\alpha')^{(d-1)/2} \tag{4.10}$$

(or more geometrically, $\ell_P^{d-1} \sim g^2 \ell_s^{d-1}$).

Given their different mass dependence, it is obvious that, for a given set of the fundamental constants G, α', g^2, $S_s > S_{\mathrm{BH}}$ at sufficiently small M, while the opposite is true at sufficiently large M. Obviously, there has to be a critical value of M, M_c, at which $S_s = S_{\mathrm{BH}}$. This observation had led Bowick et al. [54] to conjecture that large black holes end up their Hawking-evaporation process when $M = M_c$, and then transform into a higher-entropy string state without ever reaching the singular zero-mass limit. This reasoning is confirmed [55] by the observation that, in string theory, the fundamental string length ℓ_s should set a minimal value for the Schwarzschild radius of any black hole (and thus a maximal value for its Hawking temperature). It was also noticed [54, 56, 78] that, precisely at $M = M_c$, $R_{\mathrm{BH}} = \ell_s$ and the Hawking temperature equals the Hagedorn temperature of string theory. For any d, the value of M_c is given by:

$$M_c \sim M_s g^{-2} \, . \tag{4.11}$$

Susskind and collaborators [56, 79] went a step further and proposed that the spectrum of black holes and the spectrum of single string states be "identical", in the sense that there be a one to one *correspondence* between (uncharged) fundamental string states and (uncharged) black-hole states. Such a "correspondence principle" has been generalized by Horowitz and Polchinski [74] to a wide range of charged black-hole states (in any dimension). Instead of keeping fixed the fundamental constants and letting M evolve by evaporation, as considered above, one can (equivalently) describe the physics of this conjectured correspondence by following a narrow band of states, on both sides of and through, the string $\rightleftharpoons$

---

[7]Below, we shall use the precise definition $\ell_s \equiv \sqrt{2\alpha' \hbar}$, but, in this paragraph, we neglect factors of order unity.

black hole transition, by keeping fixed the entropy[8] $S = S_s = S_{\mathrm{BH}}$, while adiabatically[9] varying the string coupling g, *i.e.*, the ratio between ℓ_P and ℓ_s. The correspondence principle then means that if one increases g each (quantum) string state should turn into a (quantum) black-hole state at sufficiently strong coupling, while, conversely, if g is decreased, each black-hole state should "decollapse" and transform into a string state at sufficiently weak coupling. For all the reasons mentioned above, it is very natural to expect that, when starting from a black hole state, the critical value of g at which a black hole should turn into a string is given, in clear relation to (4.11), by

$$g_c^2 \, M \sim M_s \,, \tag{4.12}$$

and is related to the common value of string and black-hole entropy via

$$g_c^2 \sim \frac{1}{S_{\mathrm{BH}}} = \frac{1}{S_s} \,. \tag{4.13}$$

Note that $g_c^2 \ll 1$ for the very massive states $(M \gg M_s)$ that we consider. This justifies our use of the perturbative relation (4.10) between G and α'.

As we said above, in the case of extremal BPS, and nearly extremal, black holes the conjectured correspondence was dramatically confirmed through the work of Strominger and Vafa [58] and others [57, 59, 61] leading to a statistical mechanics interpretation of black-hole entropy in terms of the number of microscopic states sharing the same macroscopic quantum numbers. However, little is known about whether and how the correspondence works for non-extremal, non BPS black holes, such as the simplest Schwarzschild black hole[10]. By contrast to BPS states whose mass is protected by supersymmetry, we shall consider here the effect of varying g on the mass and size of non-BPS string states.

Although it is remarkable that black-hole and string entropy coincide when $R_{\mathrm{BH}} = \ell_s$, this is still not quite sufficient to claim that, when starting from a string state, a string becomes a black hole at $g = g_c$. In fact, the process in which one starts from a string state in flat space and increases g poses a serious puzzle [56]. Indeed, the radius of a typical excited string state of mass M is generally thought of being of order

$$R_s^{\mathrm{rw}} \sim \ell_s (M/M_s)^{1/2} \,, \tag{4.14}$$

as if a highly excited string state were a random walk made of $M/M_s = \alpha' M/\ell_s$ segments of length ℓ_s [80]. [The number of steps in this random walk is, as is natural, the string entropy (4.8).] The "random walk" radius (4.14) is much larger

---

[8]One uses here the fact that, during an adiabatic variation of g, the entropy of the black hole $S_{\mathrm{BH}} \sim (\mathrm{Area})/G \sim R_{\mathrm{BH}}^{d-1}/G$ stays constant. This result (known to hold in the Einstein conformal frame) applies also in string units because S_{BH} is dimensionless.

[9]The variation of g can be seen, depending on one's taste, either as a real, adiabatic change of g due to a varying dilaton background, or as a mathematical way of following energy states.

[10]For simplicity, we shall consider in this work only Schwarzschild black holes, in any number $d \equiv D - 1$ of non-compact spatial dimensions.

than the Schwarzschild radius for all couplings $g \leq g_c$, or, equivalently, the ratio of self-gravitational binding energy to mass (in d spatial dimensions)

$$\frac{G\,M}{(R_s^{\mathrm{rw}})^{d-2}} \sim \left(\frac{R_{\mathrm{BH}}(M)}{R_s^{\mathrm{rw}}}\right)^{d-2} \sim g^2 \left(\frac{M}{M_s}\right)^{\frac{4-d}{2}} \tag{4.15}$$

remains much smaller than one (when $d > 2$, to which we restrict ourselves) up to, and including, the transition point. In view of (4.15) it does not seem natural to expect that a string state will "collapse" to a black hole when g reaches the value (4.12). One would expect a string state of mass M to turn into a black hole only when its typical size is of order of $R_{\mathrm{BH}}(M)$ (which is of order ℓ_s at the expected transition point (4.12)). According to Eq. (4.15), this seems to happen for a value of g much larger than g_c.

Horowitz and Polchinski [75] have addressed this puzzle by means of a "thermal scalar" formalism [81]. Their results suggest a resolution of the puzzle when $d = 3$ (four-dimensional spacetime), but lead to a rather complicated behaviour when $d \geq 4$. Moreover, even in the simple $d = 3$ case, the formal nature of the auxiliary "thermal scalar" renders unclear (at least to me) the physical interpretation of their analysis. In the next section, I will review the results of Ref. [76] whose aim was to clarify the string $\rightleftharpoons$ black-hole transition by a direct study, in real spacetime, of the size and mass of a *typical* excited string, within the microcanonical ensemble of *self-gravitating* strings.

4.3.2 A microcanonical approach to self-gravitating strings

The results of [76] lead to a rather simple picture of the transition, in any dimension. Let us summarize them before entering into the technical details of the analysis.

The critical value for the transition is (4.12), or (4.13) in terms of the entropy S, for both directions of the string $\rightleftharpoons$ black hole transition. In *three spatial dimensions*, one finds that the size (computed in real spacetime) of a *typical self-gravitating* string is given by the random walk value (4.14) when $g^2 \leq g_0^2$, with $g_0^2 \sim (M/M_s)^{-3/2} \sim S^{-3/2}$, and by

$$R_{\mathrm{typ}} \sim \frac{1}{g^2\,M}, \tag{4.16}$$

when $g_0^2 \leq g^2 \leq g_c^2$. Note that R_{typ} smoothly interpolates between R_s^{rw} and ℓ_s. This result confirms the picture proposed by Ref. [75] when $d = 3$, but with the bonus that Eq. (4.16) refers to a radius which is estimated directly in physical space (see below), and which is the size of a typical member of the microcanonical ensemble of self-gravitating strings. In all higher dimensions[11], we find that the size of a typical self-gravitating string remains fixed at the random walk value (4.14)

---

[11]With the proviso that the consistency of the analysis is open to doubt when $d \geq 8$.

when $g \leq g_c$. However, when g gets close to a value of order g_c, the ensemble of self-gravitating strings becomes (smoothly in $d = 4$, but suddenly in $d \geq 5$) dominated by very compact strings of size $\sim \ell_s$ (which are then expected to collapse with a slight further increase of g because the dominant size is only slightly larger than the Schwarzschild radius at g_c). The results of [76] confirm and clarify the main idea of a correspondence between string states and black hole states [56, 79, 74, 75], and suggest that the transition between these states is rather smooth, symmetrical with no apparent "hysteresis effect", and with continuity in entropy, mass, typical size, and luminosity. It is, however, beyond the technical grasp of this analysis to compute any precise number at the transition (such as the famous factor 1/4 in the Bekenstein-Hawking entropy formula). Let us now enter into some of the technical details.

For simplicity, we deal with open bosonic strings ($\ell_s \equiv \sqrt{2\,\alpha'}$, $0 \leq \sigma \leq \pi$)

$$X^\mu(\tau,\sigma) = X^\mu_{\mathrm{cm}}(\tau,\sigma) + \widetilde{X}^\mu(\tau,\sigma)\,, \tag{4.17}$$

$$X^\mu_{\mathrm{cm}}(\tau,\sigma) = x^\mu + 2\,\alpha'\,p^\mu\,\tau\,, \tag{4.18}$$

$$\widetilde{X}^\mu(\tau,\sigma) = i\,\ell_s \sum_{n \neq 0} \frac{\alpha^\mu_n}{n}\, e^{-in\tau}\,\cos n\sigma\,. \tag{4.19}$$

Here, we have explicitly separated the center of mass motion X^μ_{cm} (with $[x^\mu, p^\nu] = i\,\eta^{\mu\nu}$) from the oscillatory one $\widetilde{X}^\mu$, which is expressed in terms of two parameters: the worldsheet time parameter τ and the worldsheet spatial parameter σ ($[\alpha^\mu_m, \alpha^\nu_n] = m\,\delta^0_{m+n}\,\eta^{\mu\nu}$). The free spectrum is given by $\alpha'\,M^2 = N - 1$ where ($\alpha \cdot \beta \equiv \eta_{\mu\nu}\,\alpha^\mu\,\beta^\nu \equiv -\alpha^0\,\beta^0 + \alpha^i\,\beta^i$)

$$N = \sum_{n=1}^{\infty} \alpha_{-n} \cdot \alpha_n = \sum_{n=1}^{\infty} n\,N_n\,. \tag{4.20}$$

Here $N_n \equiv a^\dagger_n \cdot a_n$ is the occupation number of the nth oscillator ($\alpha^\mu_n = \sqrt{n}\,a^\mu_n$, $[a^\mu_n, a^{\nu\dagger}_m] = \eta^{\mu\nu}\,\delta_{nm}$, with n, m positive).

The decomposition (4.17)–(4.19) holds in any conformal gauge (($\partial_\tau X^\mu \pm \partial_\sigma X^\mu)^2 = 0$). One can further specify the choice of worldsheet coordinates by imposing

$$n_\mu X^\mu(\tau,\sigma) = 2\alpha'(n_\mu p^\mu)\,\tau\,, \tag{4.21}$$

where n^μ is an arbitrary timelike or null vector ($n \cdot n \leq 0$) [82]. Eq. (4.21) means that the n-projected oscillators $n_\mu \alpha^\mu_m$ are set equal to zero. As we shall be interested in quasi-classical, very massive string states ($N \gg 1$) it should be possible to work in the "center of mass" gauge, where the vector n^μ used in Eq. (4.21) to define the τ-slices of the world-sheet is taken to be the total momentum p^μ of the string. This gauge is the most intrinsic way to describe a string in the classical limit. Using this intrinsic gauge, one can covariantly *define the proper rms size of a massive string state* as

$$R^2 \equiv \frac{1}{d}\,\langle(\widetilde{X}^\mu_\perp(\tau,\sigma))^2\rangle_{\sigma,\tau}\,, \tag{4.22}$$

where $\widetilde{X}_{\perp}^{\mu} \equiv \widetilde{X}^{\mu} - p^{\mu}(p \cdot \widetilde{X})/(p \cdot p)$ denotes the projection of $\widetilde{X}^{\mu} \equiv X^{\mu} - X_{\mathrm{cm}}^{\mu}(\tau)$ orthogonally to p^{μ}, and where the angular brackets denote the (simple) average with respect to σ and τ.

In the center of mass gauge, $p_{\mu} \widetilde{X}^{\mu}$ vanishes by definition, and Eq. (4.22) yields simply

$$R^2 = \frac{1}{d}\, \ell_s^2\, \mathcal{R}\,, \tag{4.23}$$

with (after discarding a logarithmically infinite, but state independent, contribution)

$$\mathcal{R} \equiv \sum_{n=1}^{\infty} \frac{\alpha_{-n} \cdot \alpha_n}{n^2} = \sum_{n=1}^{\infty} \frac{a_n^{\dagger} \cdot a_n}{n} = \sum_{n=1}^{\infty} \frac{N_n}{n}\,. \tag{4.24}$$

We wish to estimate the *distribution function in size of the ensemble of free string states* of mass M, *i.e.*, to count the number of string states, having some fixed values of M and R (or, equivalently, N and $\mathcal{R}$). An approximate estimate of this number ("degeneracy") is [76]

$$\mathcal{D}(M, R) \sim \exp\left[c(R)\, a_0\, M\right], \tag{4.25}$$

where $a_0 = 2\,\pi\,((d-1)\,\alpha'/6)^{1/2}$ and

$$c(R) = \left(1 - \frac{c_1}{R^2}\right)\left(1 - c_2\,\frac{R^2}{M^2}\right), \tag{4.26}$$

with the coefficients c_1 and c_2 being of order unity in string units. The coefficient $c(R)$ gives the fractional reduction in entropy brought by imposing a size constraint. Note that (as expected) this reduction is minimized when $c_1\,R^{-2} \sim c_2\,R^2/M^2$, *i.e.*, for $R \sim R_{\mathrm{rw}} \sim \ell_s\,\sqrt{M/M_s}$.

We also need to estimate the *mass shift of string states* (of mass M and size R) due to the exchange of the various long-range fields which are universally coupled to the string: graviton, dilaton and axion. As we are interested in very massive string states, $M \gg M_s$, in extended configurations, $R \gg \ell_s$, we expect that massless exchange dominates the (state-dependent contribution to the) mass shift. [The exchange of spin 1 fields (for open strings) becomes negligible when $M \gg M_s$ because it does not increase with M.]

The evaluation, in string theory, of (one loop) mass shifts for massive states is technically quite involved, and can only be tackled for the states which are near the leading Regge trajectory [83]. [Indeed, the vertex operators creating these states are the only ones to admit a manageable explicit oscillator representation.] As we consider states which are very far from the leading Regge trajectory, there is no hope of computing exactly (at one loop) their mass shifts.

In Ref. [76] we could estimate the one-loop mass-shift by resorting to a semi-classical approximation. The starting point of this semi-classical approximation is the effective action of self-gravitating fundamental strings derived in Ref. [84].

Using coherent-state methods [85, 82, 86] and a generalization of Bloch's theorem (see Eq. (3.13) of [76]) one finds

$$\delta M \simeq -c_d \, G \, \frac{M^2}{R^{d-2}} \, , \tag{4.27}$$

with the (positive) numerical constant

$$c_d = \left[\frac{d-2}{2} \, (4\pi)^{\frac{d-2}{2}} \right]^{-1} , \tag{4.28}$$

equal to $1/\sqrt{\pi}$ in $d = 3$.

The result (4.27) was expected in order of magnitude, but it is important to check that it approximately comes out of a detailed calculation of the mass shift which incorporates both relativistic and quantum effects and which uses the precise definition (4.24) of the squared size.

Finally, let us mention that, by using the same tools Ref. [76] has computed the imaginary part of the mass shift $\delta M = \delta M_{\rm real} - i\,\Gamma/2$, $i.e.$, the total decay rate Γ in massless quanta, as well as the total power radiated P. In order of magnitude these quantities are

$$\Gamma \sim g^2 \, M \, , \quad P \sim g^2 \, M \, M_s \, . \tag{4.29}$$

Finally one combines the results above, Eqs. (4.25) and (4.27), and heuristically extend them at the limit of their domain of validity. We consider a narrow band of string states that we follow when increasing adiabatically the string coupling g, starting from $g = 0$. Let M_0, R_0 denote the "bare" values ($i.e.$, for $g \to 0$) of the mass and size of this band of states. Under the adiabatic variation of g, the mass and size, M, R, of this band of states will vary. However, the entropy $S(M, R)$ remains constant under this adiabatic process: $S(M, R) = S(M_0, R_0)$. We consider states with sizes $\ell_s \ll R_0 \ll M_0$ for which the correction factor,

$$c\,(R_0) \simeq (1 - c_1 \, R_0^{-2}) \, (1 - c_2 \, R_0^2/M_0^2) \, , \tag{4.30}$$

in the entropy

$$S(M_0, R_0) = c\,(R_0) \, a_0 \, M_0 \, , \tag{4.31}$$

is near unity. [We use Eq. (4.25) in the limit $g \to 0$, for which it was derived.] Because of this reduced sensitivity of $c\,(R_0)$ on a possible direct effect of g on R ($i.e.$, $R(g) = R_0 + \delta_g R$), the main effect of self-gravity on the entropy (considered as a function of the actual values M, R when $g \neq 0$) will come from replacing M_0 as a function of M and R. The mass-shift result (4.27) gives $\delta M = M - M_0$ to first order in g^2. To the same accuracy[12], (4.27) gives M_0 as a function of M and R:

$$M_0 \simeq M + c_3 \, g^2 \, \frac{M^2}{R^{d-2}} = M \left(1 + c_3 \, \frac{g^2 \, M}{R^{d-2}} \right) , \tag{4.32}$$

where c_3 is a positive numerical constant.

---

[12] Actually, Eq. (4.32) is probably a more accurate version of the mass-shift formula because it exhibits the real mass M (rather than the bare mass M_0) as the source of self-gravity.

Finally, combining Eqs. (4.30)–(4.32) (and neglecting, as just said, a small effect linked to $\delta_g R \neq 0$) leads to the following *relation between the entropy, the mass and the size (all considered for self-gravitating states, with $g \neq 0$)*

$$S(M,R) \simeq a_0 M \left(1 - \frac{1}{R^2}\right)\left(1 - \frac{R^2}{M^2}\right)\left(1 + \frac{g^2 M}{R^{d-2}}\right). \qquad (4.33)$$

For notational simplicity, we henceforth set to unity (by using some redefinitions and working in some suitably defined "string units")) the coefficients c_1, c_2 and c_3. The possibility of smoothly transforming self-gravitating string states into black-hole states come from the peculiar radius dependence of the entropy $S(M,R)$. Eq. (4.32) exhibits two effects varying in opposite directions: (i) self-gravity favors small values of R (because they correspond to larger values of M_0, *i.e.*, of the "bare" entropy), and (ii) the constraint of being of some fixed size R disfavors both small ($R \ll \sqrt{M}$) and large ($R \gg \sqrt{M}$) values of R. For given values of M and g, the most numerous (and therefore most probable) string states will have a size $R_*(M;g)$ which maximizes the entropy $S(M,R)$. Said differently, the total degeneracy of the complete ensemble of self-gravitating string states with total energy M (and no *a priori* size restriction) will be given by an integral (where ΔR is the rms fluctuation of R)

$$\mathcal{D}(M) \sim \int \frac{dR}{\Delta R}\, e^{S(M,R)} \sim e^{S(M,R_*)} \qquad (4.34)$$

which will be dominated by the saddle point R_* which maximizes the exponent.

The value of the most probable size R_* is a function of M, g and the space dimension d. We refer to Ref. [76] for a full treatment. Let us only indicate the results in the (actual) *three-dimensional* case, $d = 3$. When maximizing the entropy $S(M,R)$ with respect to R one finds that: (i) when $g^2 \ll M^{-3/2}$, the most probable size $R_*(M,g) \sim \sqrt{M}$, (ii) when $g^2 \gg M^{-3/2}$,

$$R_*(M,g) \simeq \frac{1 + \sqrt{1 + 3\lambda^2}}{\lambda} \qquad (4.35)$$

where $\lambda \equiv g^2 M$.

Eq. (4.35) says that, when g^2 increases, and therefore when λ increases (beyond $M^{-1/2}$) the typical size of a self-gravitating string decreases, and (formally) tends to a limiting size of order unity, $R_\infty = \sqrt{3}$ (*i.e.*, of order the string length scale $\ell_s = \sqrt{2\alpha'}$) when $\lambda \gg 1$. However, the fractional self-gravity $GM/R_* \simeq \lambda/R_*$ (which measures the gravitational deformation away from flat space) becomes unity for $\lambda = \sqrt{5}$ and formally increases without limit when λ further increases. Therefore, we expect that for some value of λ of order unity, the self-gravity of the compact string state already reached when $\lambda \sim 1$ (indeed, Eq. (4.35) predicts $R_* \sim 1$ when $\lambda \sim 1$) will become so strong that it will (continuously) turn into a black-hole state. Having argued that the dynamical threshold for the transition string $\rightarrow$ black hole is $\lambda \sim 1$, we now notice that, for such a value of λ the entropy $S(M) = S(M, R_*(M)) \simeq a_0 M \left[1 + \frac{1}{4} (g^2 M)^2\right]$ of the string state (of mass M) *matches the Bekenstein-Hawking entropy* $S_{\mathrm{BH}}(M) \sim g^2 M^2 = \lambda M$ of the formed black hole. One further checks that the other global physical characteristics of the string state (radius R_*, luminosity P, Eq. (4.29)) match those of a Schwarzschild black hole of the same mass ($R_{\mathrm{BH}} \sim GM \sim g^2 M$, $P_{\mathrm{Hawking}} \sim R_{\mathrm{BH}}^{-2}$) when $\lambda \sim 1$.

Conceptually, the main new result of the analysis summarized above concerns the most probable state of a very massive single[13] self-gravitating string. By combining our estimates of the entropy reduction due to the size constraint, and of the mass shift we come up with the expression (4.33) for the logarithm of the number of self-gravitating string states of size R. Our analysis of the function $S(M, R)$ clarifies the correspondence [56, 79, 74, 75] between string states and black holes. In particular, our results confirm many of the results of [75], but make them (in our opinion) physically clearer by dealing directly with the size distribution, in real space, of an ensemble of string states. When our results differ from those of [75], they do so in a way which simplifies the physical picture and make even more compelling the existence of a correspondence between strings and black holes. The simple physical picture suggested[14] by our results is the following: In any dimension, if we start with a massive string state and increase the string coupling g, a *typical* string state will, eventually, become more compact and will end up, when $\lambda_c = g_c^2 M \sim 1$, in a "condensed state" of size $R \sim 1$, and mass density $\rho \sim g_c^{-2}$. Note that the basic reason why small strings, $R \sim 1$, dominate in any dimension the entropy when $\lambda \sim 1$ is that they descend from string states with bare mass $M_0 \simeq M(1 + \lambda/R^{d-2}) \sim 2M$ which are exponentially more numerous than less condensed string states corresponding to smaller bare masses (see Fig. 4).

The nature of the transition between the initial "dilute" state and the final "condensed" one depends on the value of the space dimension d. In $d = 3$, the transition is gradual: when $\lambda < M^{-1/2}$ the size of a typical state is $R_*^{(d=3)} \simeq M^{1/2}(1 - M^{1/2}\lambda/8)$, when $\lambda > M^{1/2}$ the typical size is $R_*^{(d=3)} \simeq (1 + (1 + 3\lambda^2)^{1/2})/\lambda$. In $d = 4$, the transition toward a condensed state is still continuous, but most of the size evolution takes place very near $\lambda = 1$: when $\lambda < 1$, $R_*^{(d=4)} \simeq M^{1/2}(1 - \lambda)^{1/4}$, and when $\lambda > 1$, $R_*^{(d=4)} \simeq (2\lambda/(\lambda - 1))^{1/2}$, with some smooth blending between the two evolutions around $|\lambda - 1| \sim M^{-2/3}$. In $d \geq 5$, the transition is discontinuous (like a first order phase transition between, say, gas and liquid states). Barring the consideration of metastable (supercooled) states, on expects that when $\lambda = \lambda_2 \simeq \nu^\nu/(\nu - 1)^{\nu-1}$ (with $\nu = (d - 2)/2$), the most

---

[13]We consider states of a single string because, for large values of the mass, the single-string entropy approximates the total entropy up to subleading terms.

[14]Our conclusions are not rigorously established because they rely on assuming the validity of the result (4.33) beyond the domain ($R^{-2} \ll 1$, $g^2 M/R^{d-2} \ll 1$) where it was derived. However, we find heuristically convincing to believe in the presence of a reduction factor of the type $1 - R^{-2}$ down to sizes very near the string scale. Our heuristic dealing with self-gravity is less compelling because we do not have a clear signal of when strong gravitational field effects become essential.

probable size of a string state will jump from R_{rw} (when $\lambda < \lambda_2$) to a size of order unity (when $\lambda > \lambda_2$).

One can think of the "condensed" state of (single) string matter, reached (in any d) when $\lambda \sim 1$, as an analog of a neutron star with respect to an ordinary star (or a white dwarf). It is very compact (because of self gravity) but it is stable (in some range for g) under gravitational collapse. However, if one further increases g or M (in fact, $\lambda = g^2 M$), the condensed string state is expected (when λ reaches some $\lambda_3 > \lambda_2$, $\lambda_3 = \mathcal{O}(1)$) to collapse down to a black-hole state (analogously to a neutron star collapsing to a black hole when its mass exceeds the Landau-Oppenheimer-Volkoff critical mass). Still in analogy with neutron stars, one notes that general relativistic strong gravitational field effects are crucial for determining the onset of gravitational collapse; indeed, under the "Newtonian" approximation (4.33), the condensed string state could continue to exist for arbitrary large values of λ.

It is interesting to note that the value of the mass density at the formation of the condensed string state is $\rho \sim g^{-2}$. This is reminiscent of the prediction by Atick and Witten [87] of a first-order phase transition of a self-gravitating thermal gas of strings, near the Hagedorn temperature[15], towards a dense state with energy density $\rho \sim g^{-2}$ (typical of a genus-zero contribution to the free energy). Ref. [87] suggested that this transition is first-order because of the coupling to the dilaton. This suggestion agrees with our finding of a discontinuous transition to the single string condensed state in dimensions ≥ 5 (Ref. [87] works in higher dimensions, $d = 25$ for the bosonic case). It would be interesting to deepen these links between self-gravitating single string states and multi-string states.

Let us come back to the consequences of the picture summarized above for the problem of the end point of the evaporation of a Schwarzschild black hole and the interpretation of black-hole entropy. See Fig. 4.

In that case one fixes the value of g (assumed to be $\ll 1$) and considers a black hole which slowly looses its mass via Hawking radiation. When the mass gets as low as a value[16] $M \sim g^{-2}$, for which the radius of the black hole is of order one (in string units), one expects the black hole to transform (in all dimensions) into a typical string[17] state corresponding to $\lambda = g^2 M \sim 1$, which is a dense state (still of radius $R \sim 1$). This string state will further decay and loose mass, predominantly via the emission of massless quanta, with a quasi thermal spectrum with temperature $T \sim T_{\mathrm{Hagedorn}} = a_0^{-1}$, with a_0 defined after (4.25), and which smoothly matches the previous black hole Hawking temperature. This mass loss will further decrease the product $\lambda = g^2 M$, and this decrease will, either gradually

---

[15]Note that, by definition, in our *single* string system, the formal temperature $T = (\partial S/\partial M)^{-1}$ is always near the Hagedorn temperature.

[16]Note that the mass at the black hole $\rightarrow$ string transition is larger than the Planck mass $M_P \sim (G)^{-1/2} \sim g^{-1}$ by a factor $g^{-1} \gg 1$.

[17]A check on the single-string dominance of the transition black hole $\rightarrow$ string is to note that the single string entropy $\sim M/M_s$ is much larger than the entropy of a ball of radiation $S_{\mathrm{rad}} \sim (RM)^{d/(d+1)}$ with size $R \sim R_{\mathrm{BH}} \sim \ell_s$ at the transition.

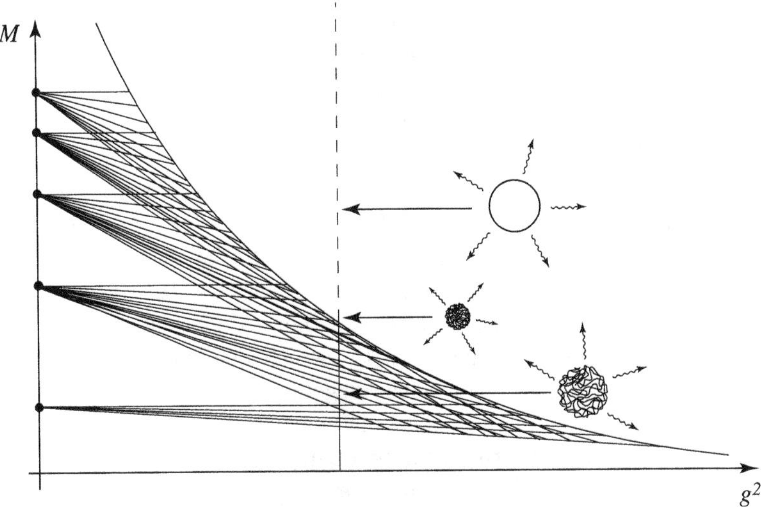

Figure 4: Evolution of the energy levels of massive string states as the string coupling g increases. String states are expected to transform into black-hole states on the correspondence line $g^2 M \sim M_s$. Vertical line: Corresponding transformation of a black hole, loosing mass at fixed g under Hawking radiation, into an initially compact string state, which inflates before decaying into massless radiation.

or suddenly, cause the initially compact string state to inflate to much larger sizes. For instance, if $d \geq 4$, the string state will quickly inflate to a size $R \sim \sqrt{M}$. Later, with continued mass loss, the string size will slowly shrink again toward $R \sim 1$ until a remaining string of mass $M \sim 1$ finally decays into stable massless quanta. In this picture, the black-hole entropy acquires a somewhat clear statistical significance (as the degeneracy of a corresponding typical string state) *only* when M and g are related by $g^2 M \sim 1$. If we allow ourselves to vary (in a Gedanken experiment) the value of g this gives a potential statistical significance to any black-hole entropy value S_{BH} (by choosing $g^2 \sim S_{\mathrm{BH}}^{-1}$). We do not claim, however, to have a clear idea of the direct statistical meaning of S_{BH} when $g^2 S_{\mathrm{BH}} \gg 1$. Neither do we clearly understand the fate of the very large space (which could be excited in many ways) which resides inside very large classical black holes of radius $R_{\mathrm{BH}} \sim (g^2 S_{\mathrm{BH}})^{1/(d-1)} \gg 1$. The fact that the interior of a black hole of given mass could be *arbitrarily* large[18], and therefore arbitrarily complex, suggests that black-hole physics is not exhausted by the idea (discussed here) of a reversible transition between string-length-size black holes and string states.

---

[18] E.g., in the Oppenheimer-Snyder model, one can join an arbitrarily large closed Friedmann dust universe, with hyperspherical opening angle $0 \leq \chi_0 \leq \pi$ arbitrarily near π, onto an exterior Schwarzschild spacetime of given mass M.

On the string side, we also do not clearly understand how one could follow in detail (in the present non BPS framework) the "transformation" of a strongly self-gravitating string state into a black-hole state.

Finally, let us note that we expect that self-gravity will lift nearly completely the degeneracy of string states. [The degeneracy linked to the rotational symmetry, *i.e.*, $2J + 1$ in $d = 3$, is probably the only one to remain, and it is negligible compared to the string entropy.] Therefore we expect (contrary to what is assumed in many works [28, 29, 30, 31, 32, 49, 50]) that the separation δE between subsequent (string and black-hole) energy levels will be exponentially small: $\delta E \sim \Delta M \exp(-S(M))$, where ΔM is the canonical-ensemble fluctuation in M. See Fig. 4 which sketches how self-gravity (which depends on the "size" of the considered massive string state) lifts the degeneracy of free string states, and is expected to lead to a very high density of (nearly non-degenerate) energy levels. [This picture of the level structure of non-supersymmetric string/black-hole states should be contrasted with the case of supersymmetric BPS D-brane/black-hole states, which features well-separated discrete levels having exponentially large degeneracies. See Fig. 3 and Eq. (4.3).] Such a δE is negligibly small compared to the radiative width $\Gamma \sim g^2 M$ of the levels. This seems to mean that the discreteness of the quantum levels of strongly self-gravitating strings and black holes is very much blurred, and difficult to see observationally. Such exponentially small δE is also expected to play an important role in "explaining" why sufficiently massive string states behave as a thermal state (see, in this respect, the recent discussion [88] of the "self-thermalization" of pure states).

5   Conclusions

We hope that this short review has allowed the reader to grasp the richness of the issue of black-hole entropy. Thirty years after the original proposal of Bekenstein and the discovery by Hawking of the quantum radiance of black holes, the issue remains somewhat cloudy. String theory has brought very significant progress by allowing one to interpret the microscopic degrees of freedom of certain *extremal* (supersymmetric) black holes in terms of a sort of "analytic continuation in the string coupling g" of systems of very massive, (Ramond-Ramond) charged string solitons (Dirichlet branes). However, it is frustrating that the D-brane description does not apply to the regime of large (effective) string coupling g where these systems are believed to "transform into" black holes. We have sketched the limited progress made in string theory for understanding the microscopic degrees of freedom of *non extremal* (non supersymmetric) black holes, and, in particular, the simplest Schwarzschild ones. A plausible picture emerged of a reversible transformation between Schwarzschild black-hole states (when $g^2 M \gtrsim M_s$) into self-gravitating massive string states (when $g^2 M \lesssim M_s$). However, the lack of supersymmetry for such states limits the secure applicability of present string technology to the regime $g^2 M \ll M_s$. The extension of the results to $g^2 M \sim M_s$ is physically plausible but not technically justified. In addition, all stringy calculations fail to give a precise Hilbert-space description of the microscopic structure of black holes in the regime $(g^2 M \gg M_s$, or its BPS equivalent) where the considered "internal structure" has become a large, quasi-classical black hole. In particular, the quantum description of the large spacetime region constituting the interior of a quasi-classical black hole (see Fig. 1), and of all the excitations it contains (such as the "lost" parts of the Hawking particle creation phenomenon), remain somewhat mysterious (see [89] for a recent attempt at describing the quantum state of a quasi-classical black hole in a unitary way). Clearly the issue of black-hole entropy will continue to exercise the sagacity of many researchers for many years to come, and will continue to provide one of our few handles on the physics of quantum relativistic gravity, i.e., the domain where $\hbar$, c and G all play important roles.

Acknowledgments

I wish to thank Bertrand Duplantier for useful comments, Vincent Pasquier for bringing Wheeler's recollection to my attention, and the Kavli Institute for Theoretical Physics for hospitality while this review was completed. This work was supported in part by the National Science Foundation under Grant No. PHY99-07949.

Appendix

Here is an excerpt from John Wheeler's book [22] about the genesis of the discovery of the concept of black-hole entropy by Jacob Bekenstein:

"One afternoon in 1970 Bekenstein – then a graduate student – and I were discussing black-hole physics in my office in Princeton's Jadwin Hall. I told him the concern I always feel when a hot cup of tea exchanges heat energy with a cold cup of tea. By allowing that transfer of heat I do not alter the energy of the universe, but I do increase its microscopic disorder, its information loss, its entropy. The entropy of the world always increases in an irreversible process like that.

"The consequence of my crime, Jacob, echo down to the end of time," I noted. *"But if a black hole swims by, and I drop the teacups into it, I conceal from all the world the evidence of my crime. How remarkable!"*

Bekenstein, a man of deep integrity, takes the lawfulness of creation as a matter of utmost seriousness. Several months later, he came back with a remarkable idea. *"You don't destroy entropy when you drop those teacups into the black hole. The black hole already has entropy, and you only increase it!"*

Bekenstein went on to explain that the surface area of a black hole is not only analogous to entropy, it is entropy, and the surface gravity of a black hole (measured, for example, by the downward acceleration of a rock as it crosses the horizon) is not only analogous to temperature, it is temperature. A black hole is not totally cold!..."

References

[1] K. Schwarzschild, "On the gravitational field of a masspoint according to Einstein's theory" (in German), *Sitzungsber. Preuss. Akad. Wiss. Berlin (Math. Phys.)* **1916**, 189 (1916); for an English translation see arXiv:physics/9905030.

[2] C. DeWitt and B.S. DeWitt, eds., *Black Holes* (Les Houches 1972) Gordon and Breach, New York, 1973.

[3] S.W. Hawking and G.F.R. Ellis, *The large scale structure of space time*, Cambridge U. Press, Cambridge, 1973.

[4] C.W. Misner, K.S. Thorne and J.A. Wheeler, *Gravitation*, Freeman, San Francisco, 1972.

[5] R.M. Wald, *General Relativity*, Chicago U. Press, Chicago, 1984.

[6] M. Heusler, *Black Hole Uniqueness Theorems*, Cambridge U. Press, Cambridge, 1996.

[7] R. Penrose, *Rivista Nuovo Cimento* **1**, 252 (1969);
R. Penrose and R.M. Floyd, *Nature* **229**, 177 (1971).

[8] D. Christodoulou, *Phys. Rev. Lett.* **25**, 1596 (1970).

[9] D. Christodoulou and R. Ruffini, *Phys. Rev.* **D4**, 3552 (1971).

[10] S.W. Hawking, in Ref. [2].

[11] B. Carter, in Ref. [2].

[12] J.M. Bardeen, B. Carter and S.W. Hawking, *Comm. Math. Phys.* **31**, 161 (1973).

[13] R.M. Wald, "The thermodynamics of black holes", Living Rev. Relativity **4**, 6 (2001). [Online article]: cited on 15 Aug 2001;
http://www.livingreviews.org/lrr-2001-6.

[14] T. Damour, *Phys. Rev.* **D18**, 3598 (1978).

[15] R.L. Znajek, *Mon. Not. R. Astr. Soc.* **185**, 833 (1978).

[16] T. Damour, Thèse de Doctorat d'Etat, Université de Paris VI, January 1979; and in *Proceedings of the Second Marcel Grossmann Meeting on General Relativity*, ed. by R. Ruffini (North Holland, Amsterdam, 1982), pp. 587-608.

[17] S.W. Hawking and J.B. Hartle, *Comm. Math. Phys.* **27**, 283 (1972).

[18] J.B. Hartle, *Phys. Rev.* **D9**, 2749 (1974).

[19] J.D. Bekenstein, Ph. D. thesis, Princeton University, 1972.

[20] J.D. Bekenstein, *Lett. Nuovo Cimento* **4**, 737 (1972).

[21] J.D. Bekenstein, *Phys. Rev.* **D7**, 2333 (1973).

[22] J.A. Wheeler, *A Journey into Gravity and Spacetime*, Scientific American Library, (W.H. Freeman & Co., New York, 1990).

[23] J.D. Bekenstein, *Physics Today* January 1980, pp. 24-31.

[24] L. Brillouin, *Science and Information Theory*, Academic Press, New York, 1956.

[25] S.W. Hawking, *Nature* **248**, 30 (1974).

[26] S.W. Hawking, *Comm. Math. Phys.* **43**, 199 (1975).

[27] T. Damour and R. Ruffini, *Phys. Rev.* **D14**, 332 (1976).

[28] J.D. Bekenstein, *Lett. Nuovo Cimento* **11**, 467 (1974).

[29] V. Mukhanov, *JETP Letters* **44**, 63 (1986).

[30] I.I. Kogan, *JETP Letters* **44**, 267 (1986); and hep-th/9412232.

[31] M. Maggiore, *Nucl. Phys.* **B429**, 205 (1994).

[32] J.D. Bekenstein, gr-qc/9710076, in *Proceedings of the 8th Marcel Grossmann Meeting*, ed. T. Piran,(World Scientific, 1988), p. 92.

[33] W.H. Zurek and K.S. Thorne, *Phys. Rev. Lett.* **54**, 2171 (1985).

[34] S. Carlip, *Phys. Rev. D* **51**, 632 (1995).

[35] A. Strominger, *J. High Energy Phys.* **2**, 009 (1998).

[36] S. Carlip, *Phys. Rev. Lett.* **82**, 2828 (1999).

[37] S.W. Hawking, *Phys. Rev. D* **14**, 2460 (1976).

[38] T. Banks, M. O'Laughlin and A. Strominger, *Phys. Rev. D* **47**, 4476 (1993).

[39] D.N. Page, *Phys. Rev. Lett.* **71**, 3743 (1993).

[40] J.R. Anglin, R. Laflamme, W.H. Zurek and J.P. Paz, *Phys. Rev. D* **52**, 2221 (1995).

[41] S.B. Giddings, "Quantum mechanics of black holes," arXiv:hep-th/9412138.

[42] D.A. Lowe and L. Thorlacius, *Phys. Rev. D* **60**, 104012 (1999).

[43] K. Fredenhagen and R. Haag, *Comm. Math. Phys.* **127**, 273 (1990).

[44] W.G. Unruh, *Phys. Rev. D* **51**, 2827 (1995).

[45] S. Corley and T. Jacobson, *Phys. Rev. D* **54**, 1568 (1996).

[46] N. Hambli and C.P. Burgess, *Phys. Rev. D* **53**, 5717 (1996).

[47] A. Ashtekar, J. Baez, A. Corichi and K. Krasnov, *Phys. Rev. Lett.* **80**, 904 (1998).

[48] A. Ashtekar, J.C. Baez and K. Krasnov, *Adv. Theor. Math. Phys.* **4**, 1 (2000).

[49] S. Hod, *Phys. Rev. Lett.* **81**, 4293 (1998).

[50] O. Dreyer, *Phys. Rev. Lett.* **90**, 081301 (2003).

[51] L. Motl, *Adv. Theor. Math. Phys.* **6**, 1135 (2003).

[52] J.D. Bekenstein, *Phys. Rev. D* **23**, 287 (1981); **49**, 1912 (1994).

[53] D. Marolf, and R. D. Sorkin, hep-th/0309218.

[54] M. Bowick, L. Smolin and L.C.R. Wijewardhana, *Gen. Rel. Grav.* **19**, 113 (1987).

[55] G. Veneziano, *Europhys. Lett.* **2**, 199 (1986).

[56] L. Susskind, hep-th/9309145.

[57] A. Sen, *Mod. Phys. Lett. A* **10**, 2081 (1995).

[58] A. Strominger and C. Vafa, *Phys. Lett. B* **379**, 99 (1996).

[59] C.G. Callan and J.M. Maldacena, *Nucl. Phys. B* **472**, 591 (1996); see also J.M. Maldacena, *Black Holes in String Theory*, Ph.D. Thesis, Princeton University, June 1996 [hep-th/9607235].

[60] A.W. Peet, "TASI Lectures on Black Holes in String Theory", in TASI 1999: *Strings, Branes and Gravity* (World Scientific, 2001), hep-th/0008241.

[61] J.R. David, G. Mandal and S.R. Wadia, *Phys. Rep.* **369**, 549–686 (2002).

[62] J. Polchinski, *String theory*, Vols. 1 and 2 (Cambridge University Press, 1998).

[63] C.V. Johnson, *D-Branes* (Cambride University Press, 2003).

[64] R.C. Myers, "Black holes and string theory," *Rev. Mex. Fis.* **49S1**, 14 (2003) [arXiv:gr-qc/0107034].

[65] G.T. Horowitz, D.A Lowe and J.M. Maldacena, *Phys. Rev. Lett.* **77**, 430 (1996).

[66] S.R. Das and S.D. Mathur, *Nucl. Phys. B* **478**, 561 (1996).

[67] J.M. Maldacena and A. Strominger, *Phys. Rev. D* **55**, 861 (1997).

[68] J.R. David, G. Mandal and S.R. Wadia, *Nucl. Phys.* **B 544**, 590 (1999).

[69] J.M. Maldacena, *Adv. Theor. Math. Phys.* **2**, 231 (1998), [*Int. J. Theor. Phys.* **38**, 1113 (1999)].

[70] O. Aharony, S.S. Gubser, J.M. Maldacena, H. Ooguri and Y. Oz, *Phys. Rep.* **323**, 183–386 (2000).

[71] J. Polchinski, *Phys. Rev. Lett.* **75**, 4724 (1995).

[72] J.L. Cardy, *Nucl. Phys. B* **270**, 186 (1986); H.W.J. Blöte, J.L. Cardy and M.P. Nightingale, *Phys. Rev. Lett* **56**, 742 (1986).

[73] S. Carlip, *Class. Quant. Grav.* **17**, 4175 (2000).

[74] G.T. Horowitz and J. Polchinski, *Phys. Rev. D* **55**, 6189 (1997).

[75] G.T. Horowitz and J. Polchinski, *Phys. Rev. D* **57**, 2557 (1998).

[76] T. Damour and G. Veneziano, *Nucl. Phys. B* **568**, 93 (2000).

[77] S. Fubini and G. Veneziano, *Nuovo Cim.* **64 A**, 811 (1969);
K. Huang and S. Weinberg, *Phys. Rev. Lett* **25**, 895 (1970).

[78] G. Veneziano, in *Hot Hadronic Matter: Theory and Experiments*, Divonne, June 1994, eds. J. Letessier, H. Gutbrod and J. Rafelsky, NATO-ASI Series B: Physics, **346** (Plenum Press, New York 1995), p. 63.

[79] E. Halyo, A. Rajaraman and L. Susskind, *Phys. Lett. B* **392**, 319 (1997);
E. Halyo, B. Kol, A. Rajaraman and L. Susskind, *Phys. Lett. B* **401**, 15 (1997).

[80] P. Salomonson and B.S. Skagerstam, *Nucl. Phys. B* **268**, 349 (1986); *Physica A* **158**, 499 (1989);
D. Mitchell and N. Turok, *Phys. Rev. Lett.* **58**, 1577 (1987); *Nucl.Phys. B* **294**, 1138 (1987).

[81] B. Sathiapalan, *Phys. Rev. D* **35**, 3277 (1987);
I.A. Kogan, *JETP Lett.* **45**, 709 (1987);
J.J. Atick and E. Witten, *Nucl. Phys. B* **310**, 291 (1988).

[82] J. Scherk, *Rev. Mod. Phys.* **47**, 123 (1975).

[83] H. Yamamoto, *Prog. Theor. Phys.* **79**, 189 (1988);
K. Amano and A. Touchiya, *Phys. Rev. D* **30**, 565 (1989);
B. Sundborg, *Nucl. Phys. B* **319**, 415 (1989); and **338**, 101 (1990).

[84] A. Buonanno and T. Damour, *Phys. Lett. B* **432**, 51(1998).

[85] V. Alessandrini, D. Amati, M. Le Bellac and D. Olive, *Phys. Rep.* **C1**, 269 (1971).

[86] M.B. Green, J.H. Schwarz and E. Witten, *Superstring Theory*, Volume 1, (Cambridge University Press, Cambridge, 1987).

[87] J.J. Atick and E. Witten, *Nucl. Phys. B* **310**, 291 (1988).

[88] N.D. Hari Dass, S. Kalyana Rama and B. Sathiapalan, *Int. J. Mod. Phys A* **18**, 2947 (2003).

[89] G.T. Horowitz and J.M. Maldacena, hep-th/0310281.

Thibault Damour
Institut des Hautes Etudes Scientifiques
35, route de Chartres
F-91440 Bures-sur-Yvette, France
email: damour@ihes.fr

Printed in the USA
CPSIA information can be obtained
at www.ICGtesting.com
LVHW020742231223
766698LV00038B/1610